Sp

CISM COURSES AND LECTURES

The series presents lecture notes, monographs, edited works and proceedings in the field of Mechanics, Engineering, Computer Science and Applied Mathematics.
Purpose of the series is to make known in the international scientific and technical community results obtained in some of the activities organized by CISM, the International Centre for Mechanical Sciences.

INTERNATIONAL CENTRE FOR MECHANICAL SCIENCES

COURSES AND LECTURES - No. 493

WIND EFFECTS ON BUILDINGS AND DESIGN OF WIND-SENSITIVE STRUCTURES

EDITED BY

TED STATHOPOULOS
CONCORDIA UNIVERSITY, MONTREAL, QC, CANADA

CHARALAMBOS C. BANIOTOPOULOS
ARISTOTLE UNIVERSITY OF THESSALONIKI, GREECE

SpringerWien NewYork

This volume contains 131 illustrations

SPIN 12076963

All contributions have been typeset by the authors.

ISBN 978-3-211-73075-1 SpringerWienNewYork

PREFACE

Aeolos was the Treasurer of the winds. Although he was not a God and had no temples in Ancient Greece, Aeolos is very often met in the Greek mythology and in the early Greek texts. In Odyssey, Odysseus (Ulysses) during his long trip back home to Ithaca, arrived to the island of Aeolos, a floating island close to Sicily. In the caves of this island the winds were imprisoned and Aeolos, following what Gods were dictating to him, used to let them out as soft breezes, gales, storms or whatever Gods wished... Aeolos entertained Ulysses for a whole month, and on Odysseus' departure, gave him as present, a bag containing all adverse winds to hold tight until he went home, so that Odysseus might reach Ithaca with a fair wind. Odysseus did as Aeolos advised him. However, approaching his homeland and while he was sleeping, his men wickedly opened the bag (thinking it was full of gold and silver), allowing all adverse winds to escape. The ships were then blown back to Aeolos' isle, and thence to the land of the savage Laestrygonians, far away from the home island...

Clearly the struggle of Ulysses and his men against the wind resembles the efforts of the contemporary structural engineers to dominate the power of wind and design buildings and wind-sensitive structures to be safe but also economical. However, although the effects of wind are extremely important for contemporary structural design, courses in this area are not generally available within the engineering curriculum of most universities around the world at present. Therefore, this book intends to cover the lack of relevant advanced professional training.

The book contains seven chapters written by wind engineering experts. **Chapter 1**, by Theodore Stathopoulos, addresses the fundamentals of wind engineering, wind velocities, turbulence and structure; it emphasizes the interaction of wind with buildings and refers to the difficulties associated with the evaluation of internal pressures, an issue still problematic in the development of contemporary wind standards and codes of practice. **Chapter 2**, written by Chris Geurts and Carine van Bentum, deals with the evaluation of wind loads on buildings experimentally, either in boundary layer wind tunnels or in full scale and makes particular reference to the application of the Eurocode for the evaluation of design wind loads on buildings. **Chapter 3**, by Jeorg Franke, introduces the concepts of Computational Wind Engineering, i.e. the application of Computational Fluid Dynamics (CFD) in wind engineering, for the prediction of wind loads on buildings. This is an emerging field with still several outstanding issues in the application of various simulation techniques and yields results, which may be unrealistic, therefore untrustworthy, in some cases. **Chapter 4**, authored by Ruediger Hoeffer, Mozes Galffy and Hans-Juergen Niemann, refers to the fundamentals of random vibrations of structures when excited by wind, whereas **Chapter 5**, by Giovanni Solari and Federica Tubino, provides a general framework of the dynamic approach to the wind loading of structures and addresses alongwind, crosswind and torsional structural response. Different methodologies are described with discussion on merits and defects, limits and implications, costs and benefits for each. **Chapter 6**, by Claudio Borri and Carlotta Costa, deals with aeroelastic phenomena with applications in bridge aerodynamics and large roof structures with special reference to galloping, torsional divergence and flutter. Finally, **Chapter 7**, written by Charalambos C. Baniotopoulos, addresses structural design questions for specific wind-sensitive structures, namely steel lattice masts, wind turbine towers and aluminum glass façades. This chapter includes a discussion of various design code recommendations with respect to serviceability, strength and safety criteria.

The editors would like to thank all contributors to this book for the excellence of their work and also extend their sincere appreciation to the CISM General Secretary, Professor Bernhard Schrefler, the CISM Rector, Professor Giulio Maier, the Editor of the Series Professor P. Serafini, as well as to all CISM staff in Udine for their excellent cooperation.

The book will be of interest for engineers, researchers and academicians who work on relevant scientific research or design topics in research centers, universities, industry and government agencies. The book is written to address the interests of practicing engineers and professionals as well.

The Editors

STATHOPOULOS T.

Concordia University
Montreal, Canada

BANIOTOPOULOS C.C.

Aristotle University of Thessaloniki
Thessaloniki, Greece

CONTENTS

Introduction to Wind Engineering, Wind Structure, Wind-Building Interaction

Theodore Stathopoulos

Department of Building, Civil and Environmental Engineering,
Concordia University, Montreal, Canada

Abstract. A brief introduction to wind engineering with the main characteristics of the wind structure are provided. Velocity profiles, turbulence and spectra are described along with the effects of upstream exposure. Wind speed and turbulence models for inhomogeneous upstream exposures are presented along with data comparisons from other sources. The basic elements of wind-building interaction in the time-averaged mode for uniform and boundary layer flows are described, external and internal pressures and forces on buildings with emphasis on design significance are discussed.

1 Introduction

Wind Engineering is best described as the rational treatment of interaction between wind in the atmospheric boundary layer and man and his works on the surface of earth (Cermak, 1975). It comprises a synthesis of knowledge from fluid mechanics, meteorology, structural mechanics, physiology and the like. Although aerodynamics is of central importance, most applications are non-aeronautical in nature. As far as structural engineering is concerned, the evaluation of wind-induced pressure loads on building surfaces, primary and secondary structural systems, and the consequent along wind, across wind and torsional response are clearly the most important applications. Good knowledge of fluid and structural mechanics is the fundamental background necessary for the understanding of details of interaction between wind flow and civil engineering structures or buildings.

The unsteady character of the wind regime, particularly in urban areas, combined with the additional unsteadiness generated by the separated flow after the wind impacts on a building generates highly fluctuating pressures depending on the flow characteristics and the building configuration. Naturally, the wind-induced pressure regime is more complex than the wind flow regime, so its evaluation becomes more cumbersome and analytical techniques fail in most cases. Consequently, boundary layer wind tunnels simulating atmospheric flows have been used and continue to use extensively for the evaluation of wind loads on buildings. Computational approaches have progressed through the last decade but they are still at a level that hesitation prevails when their results are suggested for use in practical applications.

To start with, wind derives its energy from the sun. Solar radiation is strongest at the equator and this produces temperature differences, which in turn produces pressure differences, which create the so-called atmospheric circulations in an effort for nature to show its fairness by opposing to inequities! Additional variations to the atmospheric circulations are caused by seasonal effects (annual march of sun north and south of the equator), geographical effects (uneven distribution of water and land) and rotation of earth (greater speed at equator than near the poles). There is a great confusion of names characterizing the various wind types. In the category of secondary circulations, the strongest wind is called hurricane and belongs in the tropical cyclones, which derive their energy by the latent heat released my the condensation of water vapor at low latitudes (near the equator). Hurricanes are known as typhoons in the Far East and cyclones in Australia and Indian Ocean. Extratropical cyclones (30-45 km/h) are produced by mountain barriers or by interaction of air masses along fronts.

Local winds have minimal influence on primary and secondary circulations but, regardless, they may have high intensity. Thunderstorms, caused by heavy precipitation (like wall jets) and tornadoes, which are the most powerful winds causing maximum damage, belong in this category.

2 Wind velocity and turbulence

For winds near the ground surface, frictional effects play a significant role. Ground obstructions retard the movement of air close to the ground surface, causing a reduction in wind speed.

At some height above ground, the movement of air is no longer affected by ground obstruction. This height is called gradient height, Z_G, which is a function of ground roughness. The unobstructed wind speed is called gradient wind speed, V_{Z_G} and it is considered to be constant above gradient height.

The power law, which is used by some engineers to represent variation of wind speed with height, is an empirical equation, which for the case of mean speeds takes the form of

$$\frac{V_z}{V_{z_G}} = (\frac{Z}{Z_G})^{\alpha} \tag{1}$$

where Z_G and α are functions of ground roughness. Typical values of Z_G and α are given in Table 1, which also includes the gust speed exponent β replacing α in Eq. (1) when gust speeds are used instead of mean speeds roughness length. It may be noticed that exponent β is approximately equal to 60% of the value of α.

The logarithmic law is used by both engineers and meteorologists. It is based on physics of the boundary layer and it is valid in the bottom 20 to 30% of the boundary layer.

$$\bar{V}_Z = 2.5 \cdot u^* \, ln \frac{Z}{Z_o} \tag{2}$$

where

Z_o = roughness length – see typical values in Table 1

u^* = friction velocity $= \sqrt{\frac{\tau_o}{\rho}} \cong 1\text{–}2 m/s$ (for extreme winds)

τ_o = shear stress at the ground surface
ρ = air density

Table 1. Typical values of parameters in wind profiles

Terrain category	Terrain description	Gradient height, Z_G *(m)*	Roughness length, Z_o *(m)*	Mean speed exponent α	Gust speed exponent β
1	Open sea, ice, tundra, desert	250	0.001	*0.11*	0.07
2	Open country with low scrub or scattered trees	300	0.03	*0.15*	0.09
3	Suburban areas, small towns, well wooded areas	400	0.3	0.25	0.14
4	Numerous tall buildings, city centres, well developed industrial areas	500	3	*0.36*	0.20

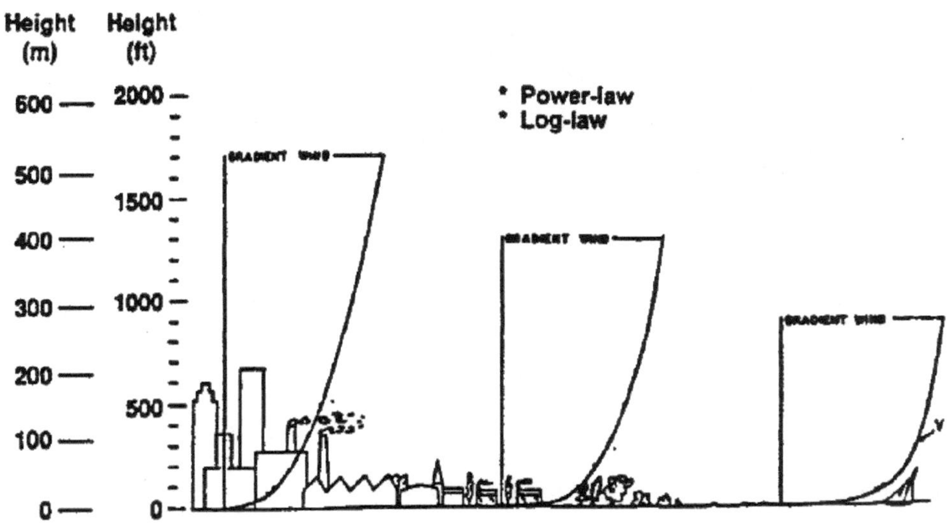

Figure 1. Variation of wind speed with height

Figure 1 shows typical variations of wind speeds above different ground roughness.

Figure 2 shows a typical wind speed variation with time for a 15-minute period. The unsteadiness of the wind is clear. The wind speed can be considered as consisted of two components: mean wind speed and fluctuating component. The fluctuating component, which is known as turbulence, is caused by convective movement (meteorological) and/or ground roughness (mechanical). In the case of boundary layer flow at high wind speed, it is assumed that mechanical turbulence predominates; this is the case of interest to engineers.

The turbulence (mechanical) is higher in rougher terrain than in smoother terrain, e.g. suburban as opposed to flat open terrain and it decreases with increasing height above ground. The turbulence (gust) parameters of importance are:

- turbulence intensity, defined as

$$Iu = \sigma(z)/U(z) \tag{3}$$

where $U(z)$ = mean wind speed at elevation z and σ = standard deviation value of fluctuating velocity.

- gust size (wind snapshot) referred to as integral scale; and
- gust frequency

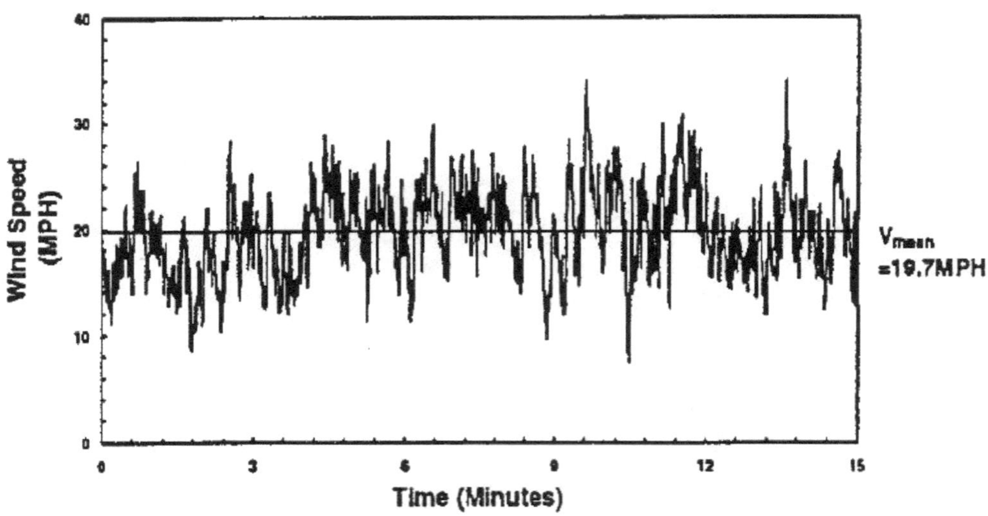

Figure 2. Typical wind speed variation with time

3 Wind structure

Figure 3 shows a typical record of wind velocity obtained from measurements at three different heights of a mast. Stationarity, i.e. the tendency of mean wind speed to stay relatively steady over periods of 10 min to an hour is significant, since it forms the basic idea to wind tunnel testing. The explanation of steadiness lies in the fact that processes generating the mean flow have time scales much, much greater than 1 hour. However, mean speed does vary with time and large fluctuations cover a period of several days.

Some variation in the mean velocity does occur and this is illustrated in Figure 4 by some wind records from a tall tower in Texas. Large fluctuations cover a period of several days, while the speed variation with height is less apparent for some particular days. The variation of wind direction with height is almost nil.

Figure 5 shows the full wind spectrum, as proposed by Van der Hoven (1957). Several distinctive features of the wind speed spectrum have been found from full-scale measurements. Energy appears into two major humps, one at $T = 4$ days, associated with movement of large scale pressure systems and $T = 1$ day, related to the diurnal frequency; the other at $T = 1$ min, associated with turbulence, separated by a gap (T = 10 min to 1 hour), which enables a convenient distinction between "mean wind" and "gusts". The choice of an averaging period for mean wind of 1 hr to 10 min provides fairly stable mean values (thunderstorms may be an exception). It is to be noted that he magnitude of high frequency hump depends on the speed of mean flow.

In addition, it is apparent that wind speed fluctuates randomly at all frequencies. Referring to Fig. 5, the vertical axis represents energy (relative number) at each frequency and the peak is around $n = 0.01$ Hz (period of 100 seconds). Given that the energy reduces drastically above 1 Hz, wind gusts induce negligible dynamic resonance (buffeting) response to buildings or structures with fundamental frequency greater than 1 Hz.

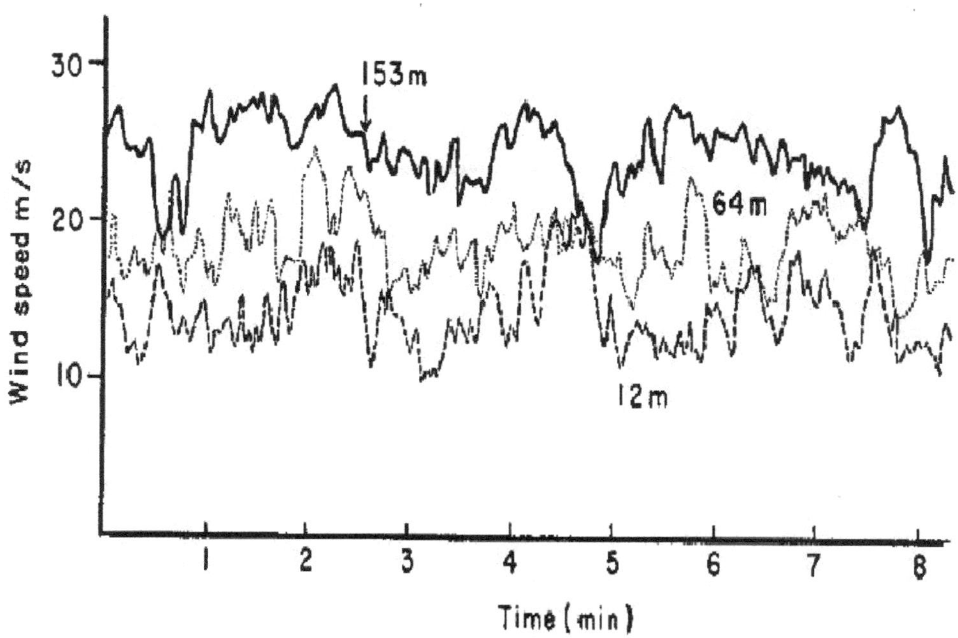

Figure 3. Record of wind speed at three heights on a 153 m mast in open terrain

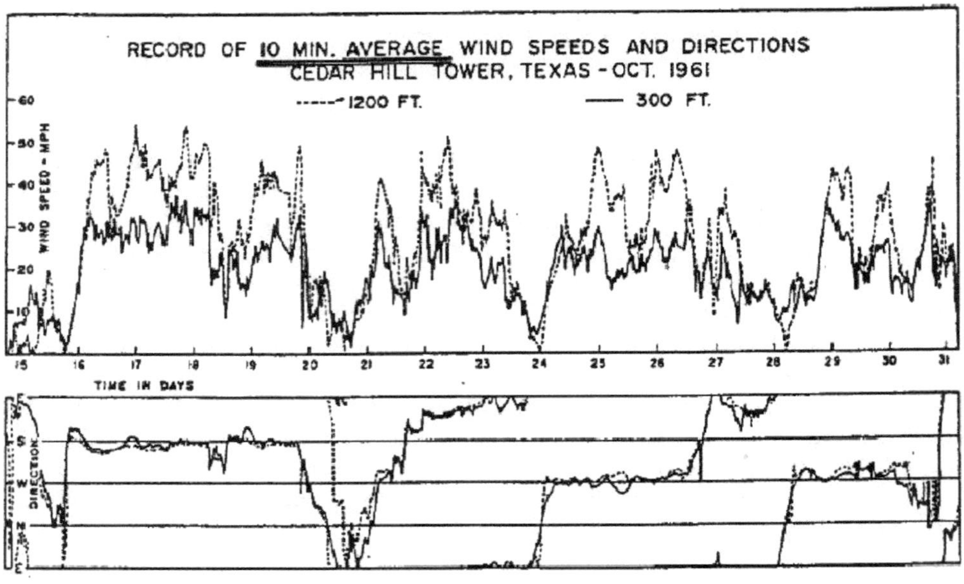

Figure 4. Variation of mean velocity in the atmospheric boundary layer

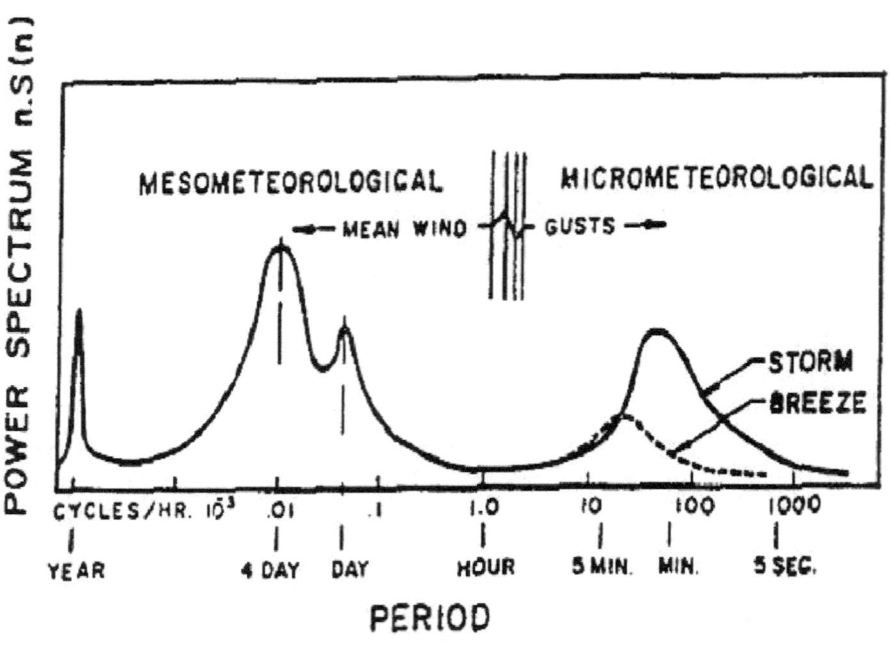

Figure 5. Wind Spectrum, after Van der Hoven (1957)

Figure 6 shows the micrometeorological part of the spectrum, which can be plotted in a smoothed form such as that of Eq. (4), among others, which fits well a series of full-scale data obtained in several locations, as Figure 7 shows.

$$\frac{nS(n)}{k\overline{V}_{10}} = \frac{4.0\left(\frac{nL}{\overline{V}_{10}}\right)^2}{\left[1+\left(\frac{nL}{\overline{V}_{10}}\right)^2\right]^{4/3}} \qquad (4)$$

where

$L \cong 1200$ m

k = surface drag coefficient dependent on the terrain roughness; it has values of about 0.001, 0.005, 0.015 and 0.04 for terrain categories 1-4 (see Table 1) or D-C-B-A respectively

$\overline{V}_{10}$ = mean speed at height of 10 m

Even though mean wind speeds may be lower in a city, gustiness is much higher. In fact, a ratio between the surface drag coefficient in a rough urban area over that in an open terrain is about 10 to 1. Thus in city conditions, the percentage of turbulence in the air speed may be significantly greater.

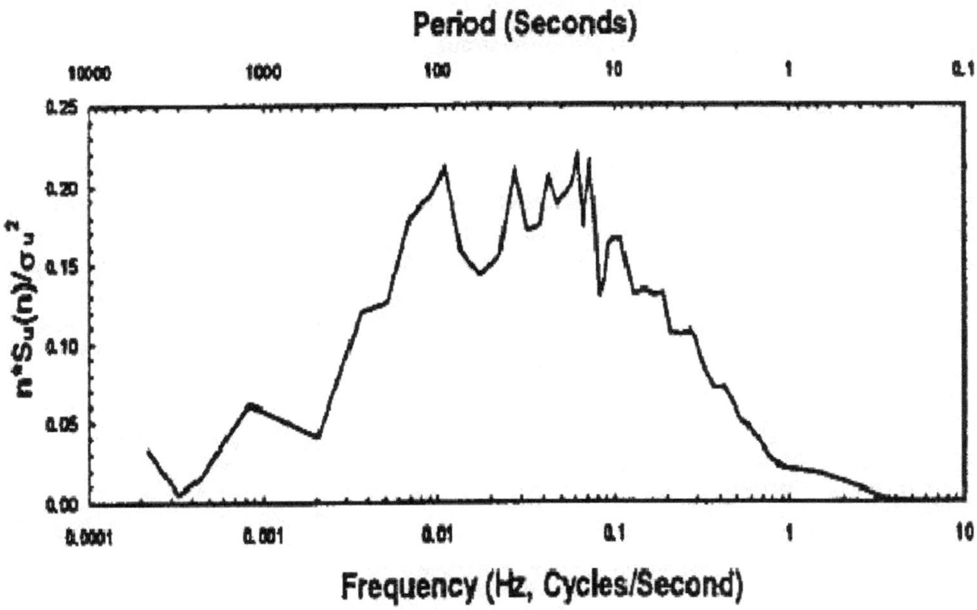

Figure 6. Gust frequency spectrum

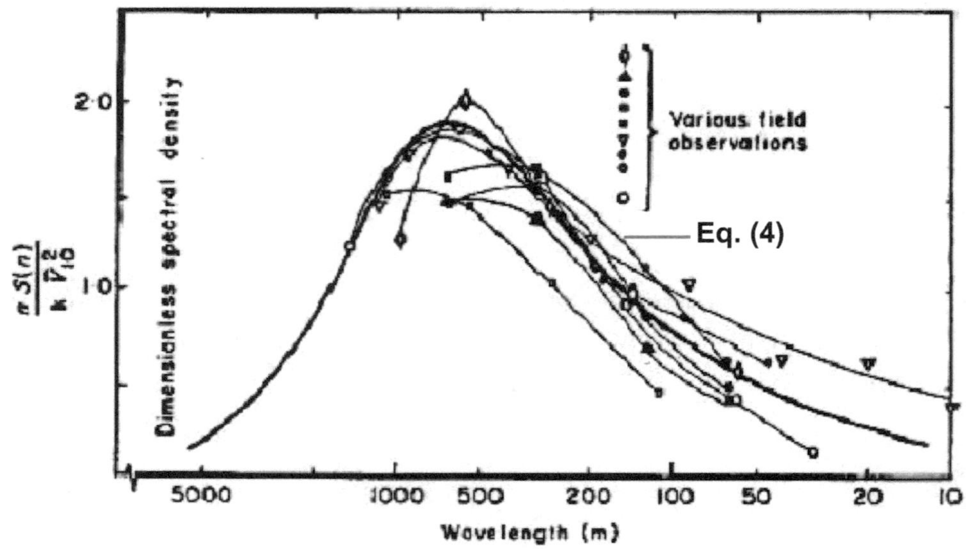

Figure 7. Spectrum of turbulence in strong winds, after Davenport (1967)

4 Upstream exposure

Previous discussion in this chapter addresses mainly upstream exposures with homogeneous roughness. In reality, most practical applications deal with transition cases or mixed upstream exposures, sometimes different for different wind directions as well. Such cases are hard for the design practitioner to handle and codes and standards generally provide little assistance in this regard. Wind speed and turbulence models established recently for such cases are referred to briefly in this section.

4.1 Wind speed model

Recent wind-tunnel investigations confirmed that small-scale roughness changes play an important role on the speed variation above site (Schmid and Bunzli 1995, Zhang and Zhang 2001). Above a well-defined pertinent fetch, the patches may stratify the boundary layer regime into an outer sub-layer or a set of corresponding Internal Boundary Layers (IBL's). The IBL depth growth *g(x)* may be modelled by a 0.8 power law, as mentioned in Garratt (1990), as follows:

$$g(x) \propto z_{o,r}^{0.2} x^{0.8} \tag{5}$$

where $z_{o,r}$ is the greatest of an upstream patch roughness length or its adjacent downstream patch roughness length; x is the distance from the change of roughness to the site. Each segment in the speed profile is dictated by the power law index of the corresponding patch. An illustrative example is shown in Fig. 8 in order to outline the general features of the model.

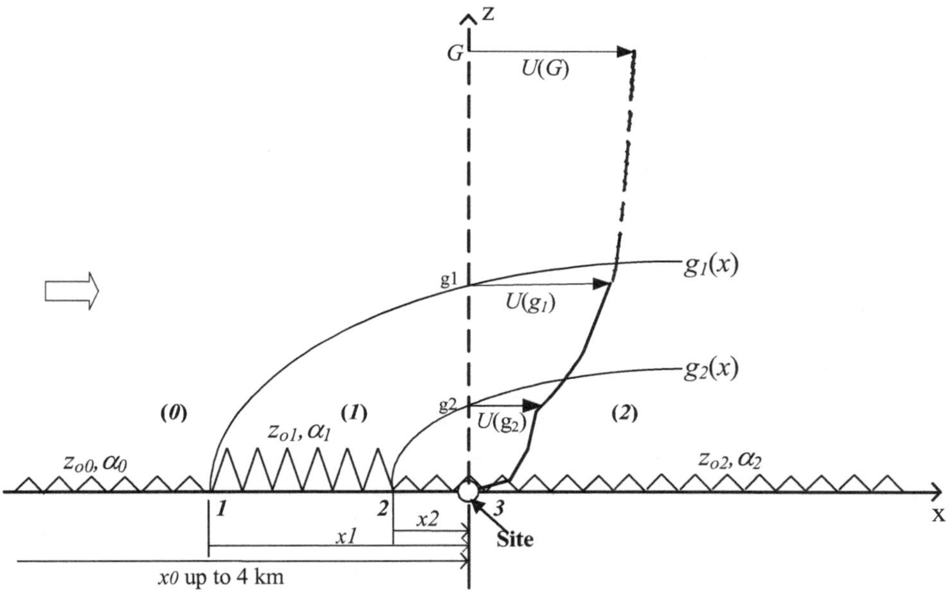

Figure 8. Schematic boundary layer with two IBL's developing above a fetch with two roughness changes

This example has a pertinent fetch of three patches: patch 0 from the 4 km upwind location to the first roughness change location 1; patch 1 from roughness change location 1 to 2; and patch 2 from roughness change location 2 to the site. The entire boundary layer is correspondingly stratified into three sub-layers: outer sub-layer (*0*) in the height range from *G* down to $g_1(x)$, IBL (*1*) from $g_1(x)$ down to $g_2(x)$, and IBL (*2*) from $g_2(x)$ down to the ground. Both $g_1(x)$ and $g_2(x)$ may be modelled with the 0.8 power law (2). The layer depths $g_1(x)$ and $g_2(x)$ break the speed profile at the site into three segments, each of which can be modelled by power-law equation with the patch-respective power-law indices α_o, α_1 and α_2.

The proposed speed model has the form:

Gradient height:

$$g_n(x) = G \quad (n = 0) \tag{6}$$

IBL depth:

$$g_n(x) = 0.5 z_{o,(n,n-1)}^{0.2} x_n^{0.8} \quad (n = 1, 2, \ldots, N) \tag{7}$$

Speed profile segment:

$$U(z) = U(g_n(x))\left(\frac{z}{z_{g_n}}\right)^{\alpha_n} \quad (g_{n+1}<z\leq g_n;\ n = 0, 1,\ldots, N; \tag{8}$$

in which the subscript n denotes the patch number, i.e., patch N is the patch of the site; $g_o(x)$ denotes the gradient height; $g_n(x)$ denotes the depth of the nth IBL; x_n is the distance from the nth roughness change to the site; $z_{o,n}$ and α_n are roughness characteristic values of patch n; and $z_{o,(n,\ n-1)} = z_{o,n-1}$ or $z_{o,n}$, whichever is larger. For homogeneous terrain of no roughness change, $N = 0$, Eq. (7) does not apply and Eq. (8) reduces to the ordinary power law.

In order to use Eqs. (6) – (8), the boundary conditions need to be specified for the pertinent fetch length, the patch information (x_n, $z_{o,n}$, α_n), the gradient height G, and the gradient wind speed $U(G)$. In this study, the pertinent fetch length is assumed as constant 4 km. Patch roughness information (x_n, $z_{o,n}$, α_n), and the gradient height G for the pertinent fetch can be found in Table 2 after Davenport et al. (2000) and ASCE (1999).

Table 2. Roughness classification after Davenport et al. (2000) with power law α values after ASCE (1999)

Roughness Class	z_o (m)	α	G (m)
1. Sea	0.0002	*0.09*	213
2. Smooth	0.005	*0.125*	213
3. OC	0.03	0.15	274
4. Roughly open	0.1	*0.2*	274
5. Suburban (Rough)	0.25	0.25	366
6. Very rough	0.5	*0.3*	366
7. Urban (Closed)	1	0.33	366

Numbers in *italics* are obtained by best fit of data.

The highest roughness grade (Class 8, Chaotic, $z_o > 2$ m) of the original classification (Davenport et al. 2000) is combined within Urban (Closed) in compliance with ASCE 7-02. Values of $U(G)$ are available in wind standards and codes of practice.

Discussion

The application and validation of this new speed model has been discussed in detail by Wang and Stathopoulos (2005). This section includes an example demonstrating how the proposed model fits the actual ESDU (82026) values for a case of inhomogeneous upstream terrain. The example provided in the ESDU document was deemed the most appropriate case to check. The ESDU (82026) example requires to find the mean speed profile at a site downwind of two changes in surface roughness, given the reference speed $U(10)$ = 22 m/s, and the fetch containing three patches, as shown in Figure 9.

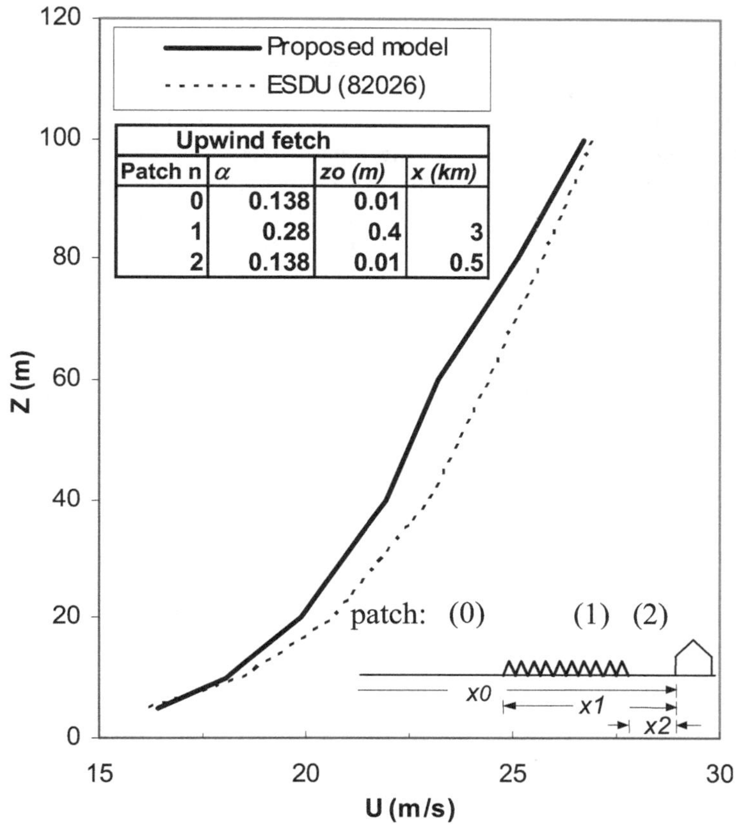

Figure 9. The proposed speed model and the ESDU (82026) model for the ESDU (82026) example case (terrain condition shown), after Wang and Stathopoulos (2005)

ESDU (82026) appears to take a lot more steps than the proposed model in speed profile calculation. It is worth noting that the probability factor, which is taken into account by the original ESDU (82026) data, has been removed in the comparison of Figure 9, which shows that the agreement between the proposed model and ESDU (82026) is reasonable, particularly below 20 m and above 80 m. For intermediate heights, ESDU tends to provide higher values than the proposed model, which thus appears less conservative than ESDU (82026). However, the proposed model agrees better with the full-scale investigation of Letchford et al. (2001) on a geometrically similar fetch configuration under hurricane conditions. It is of interest to note that Letchford et al. (2001) found that the ESDU (82026) transitional speed model may tend to overestimate, by as much as around 20%, the increase in speed at 10 m height induced by an R-S (rough-to-smooth) upstream terrain change. It is also noteworthy that this particular case ensures conditions of neutral atmospheric stability considering the high wind speeds it refers to.

4.2 Turbulence model

Although the detailed derivation of the turbulence model is not presented here, it can be found in Wang and Stathopoulos (2005). However, a comparison with the experimental data for a fetch with a single roughness change (from open country to suburban) is shown in Figure 10. The proposed model agrees well with the experimental data, whereas the ESDU (84030) data overestimates the turbulence intensity at higher heights. However, there is excellent agreement among data, ESDU model and the present study results, as well as the ESDU data for heights less than 30 m, i.e. the height zone of most low buildings

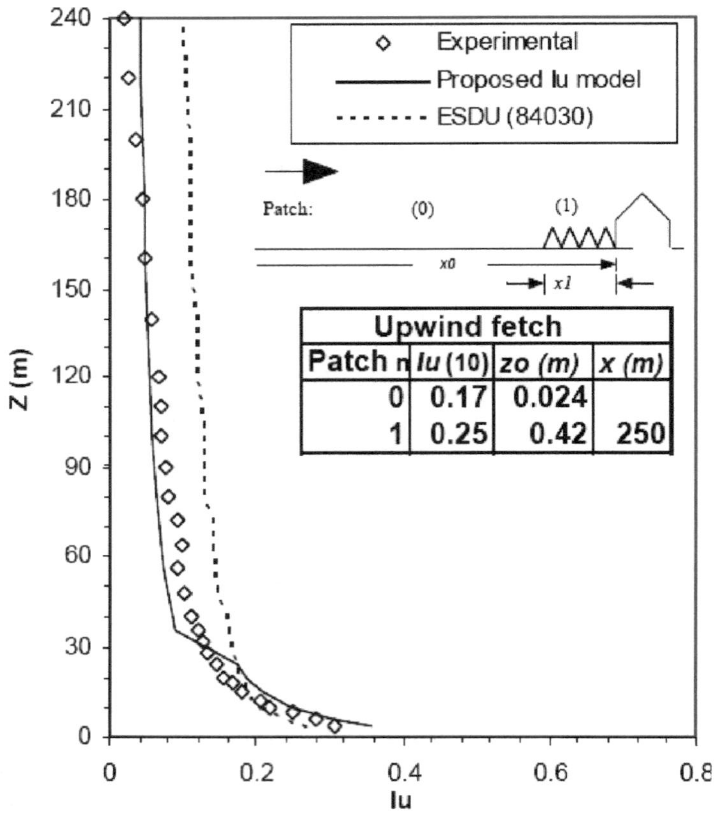

Figure 10. The *Iu* model and wind-tunnel data for a fetch with a single roughness change, and the ESDU (84030) data for a very similar fetch, after Wang and Stathopoulos (2005)

5 Mean wind pressure loads on buildings

Airflow around buildings produces pressure distributions on the building envelope. Knowing these distributions is important in a number of engineering applications. In case of strong oncoming winds, the flow may produce high negative pressures that can cause damage to cladding of tall buildings or roofs of low buildings, particularly near corners and edges. The distribution of air contaminants exhausted from a building may also be affected even for low or moderate winds, the air infiltration will be altered due to the differences in the internal pressures of the building and the intake and exhaust flow rates will be different causing possible problems to the HVAC system performance.

5.1 Uniform flow conditions

The airflow velocity (V) produces a pressure (P) evaluated by the well known Bernoulli's equation, which for horizontal flows can be written as

$$P + 1/2\, \rho V^2 = constant \qquad (9)$$

where the second term is called dynamic pressure and ρ is the air density. This equation is valid for ideal conditions of non-viscous, steady and irrotational flow – see Figure 11 – so it cannot be used in the case of turbulent flows around buildings.

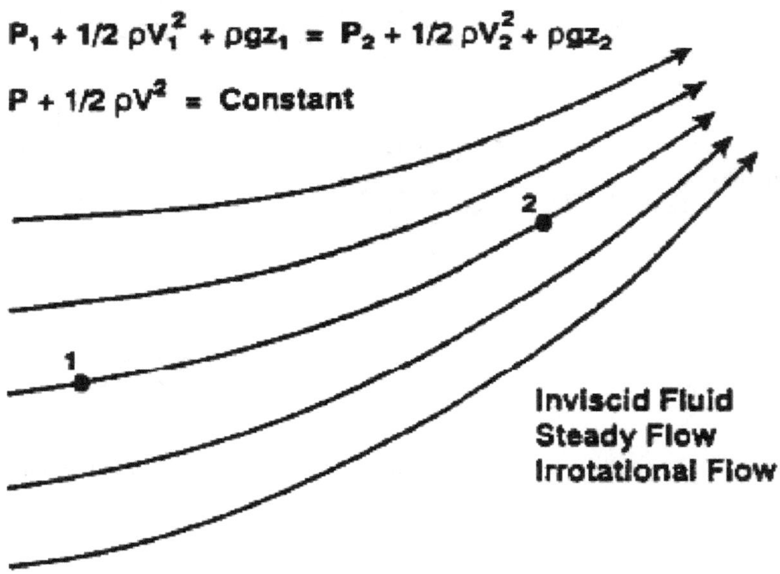

Figure 11. Bernoulli's equation for a streamline

Since most buildings and other civil engineering structures are not streamlined in nature, traditional aerodynamics calls them "bluff bodies". The implication is that wind flow separates and causes the formation of the so-called "wake", see Figure 12 showing the basic characteristics of steady flow around a simple rectangular building or tower. If air was an ideal (non-viscous) fluid, there would not be any separation and Bernoulli's equation would be applicable in all places. However, as Figure 2 demonstrates, the wind flow separates from the body at the two upstream edges and forms two regions: an outer flow, where there is no viscosity effect (except a very thin layer on the front face of the building) and an inner flow, i.e. the wake region. The outer flow is separated by the inner flow by an area of high vorticity, called "shear layer".

Clearly Bernoulli's equation is only valid in the outer flow region and the constant in Eq. (9) can be evaluated by using the pressure (atmospheric) P_o and velocity of the undisturbed free stream flow V_o to get

$$\Delta P = P - P_o = 1/2\, \rho V_o^2 - 1/2 \rho V^2 \tag{10}$$

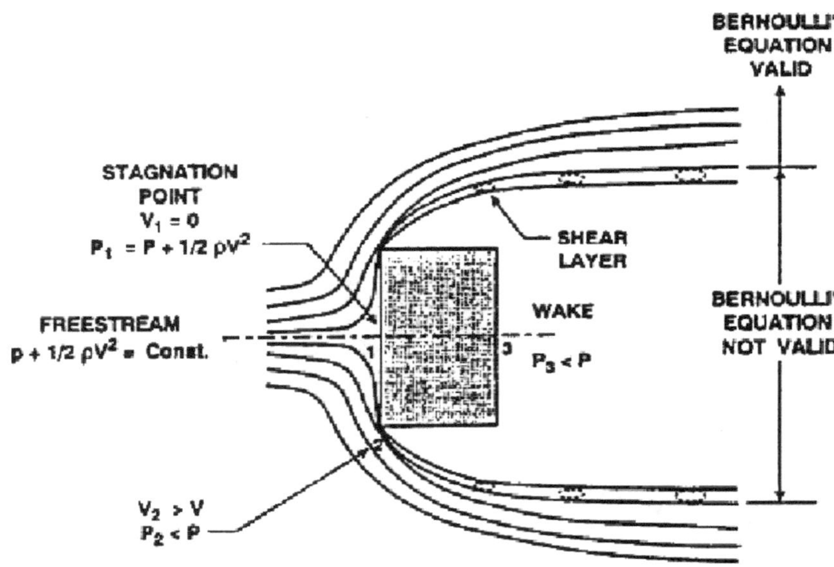

Figure 12. Bernoulli's equation and wind flow around a rectangular building

For ideal conditions of stagnation $V_1 = 0$; $P_1 = P + 1/2\, \rho V^2$
Also, if $V_2 < V$, $P_2 > P$; this implies inward acting pressure (overpressure or simply pressure). However, if $V_2 > V$, $P_2 < P$, i.e. outward acting pressure, known as suction.
Pressures are usually expressed in a dimensionless form to be independent of velocities. This dimensionless form is called pressure coefficient (C_P) and is defined as follows:

$$C_P = \frac{\Delta P}{1/2 \rho V_o^2} \tag{11}$$

in which ΔP is the pressure induced by wind (above or below ambient atmospheric pressure); commonly ΔP is called P.
Eq. (11) can also be rewritten as:

$$C_P = \frac{P - P_o}{1/2\rho V^2} = [1 - (\frac{V}{V_o})^2] \tag{12}$$

Eq. (12) shows that the pressure coefficient can reach a maximum value of +1 at stagnation points where $V = 0$; thus positive values of C_P close to one are observed on sides of building facing the wind. C_P is 0 in the free stream where $V = V_o$ and negative for values of $V > V_o$. Negative values of C_P are observed on roofs and sides of building.

The wake region in Figure 12 is characterized by little pressure gradient. Bernoulli's equation is not applicable but pressure coefficients can also be expressed in dimensionless form:

$$C_{P_w} = \frac{P_w - P_o}{1/2\rho V_o^2} \tag{13}$$

Pressure coefficients in wake are invariably negative. Typical time series of pressure coefficients along with variation of wind speed and their statistics are shown in Figure 13. Definitions of Cp_{mean} and Cp_{peak} are:

Mean pressure coefficient: $$Cp_{mean} = C_P = \frac{\Delta P_{mean}}{1/2\rho V_{mean}^2} \tag{14}$$

Peak pressure coefficient: $$Cp_{peak} = GC_P = \frac{\Delta P_{peak}}{1/2\rho V_{mean}^2} \tag{15}$$

The location of separation points and the geometry of the wake have a substantial influence on the pressure distribution and the total forces on the bluff obstacle. In the case of rectangular cylinder the separation points were dictated by the geometry of the prism. The boundary layer, which builds up on the front surface, fails to flow around the sharp corners boundary layer, which builds up on the front surface, fails to flow around the sharp corners and separates. For other bluff shapes particularly for those with curved surfaces such as wires, chimneys, and circular tanks the separation points are not easy to predict. For a circular cylinder, for example, separation takes place at different positions depending on the magnitude of the viscous forces, which dominate the flow within the boundary layer. The relative magnitude of these viscous forces can be expressed in the form of a dimensionless parameter known as the Reynolds number R_e:

$$R_e = \frac{inertia\ forces}{viscous\ forces} \propto \frac{\rho V_o^2 D^2}{\mu \frac{V_o}{D} D^2} \tag{16}$$

Thus $R_e = \frac{V_o D}{\nu}$ (17)

in which ρ is the density, μ is the dynamic viscosity, and ν is the kinematic viscosity of the air.

Figure 14 shows the variation of pressure coefficients on the surface of a circular cylinder for different values of R_e. Clearly the influence on the side face and the leeward side is significant.

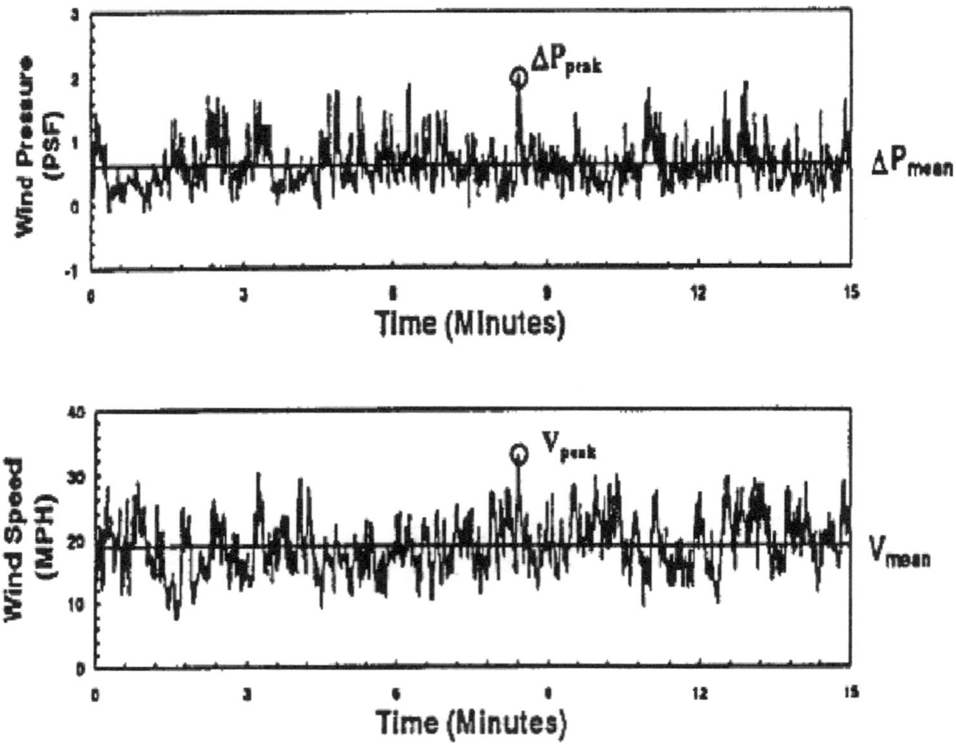

Figure 13. Wind pressure and wind speed traces indicating mean and peak values

The time-averaged aerodynamic forces on structures can be expressed as along wind or drag forces (F_D) and across wind or lift forces (F_L). The latter should not be confused with the upward lift forces acting on horizontal building elements such as roofs. The drag force is normally larger, as far as static loads on buildings is concerned. Both drag and lift forces can be expressed also in terms of coefficient form, as follows:

Drag coefficient $C_D = \frac{F_D}{1/2 \rho V_o^2 h}$ (18)

where h = projected frontal width or height of building

Lift coefficient $C_L = \frac{F_L}{1/2\rho V_o^2 h}$ (19)

For bodies with curved surfaces the drag coefficient depends drastically on Reynolds number, while for square cylinders or buildings with sharp corners on their outline, C_D is almost independent of *Re*. This can be clearly observed in Figure 15, which indicates the variation of drag coefficients for gradually increasing the radius of curvature of building corners as we go from an almost square to a fully circular shape. For the latter, it is interesting to note the variation of C_D with the surface roughness, which affects the location of separation and, consequently, the pressure loads on the surface.

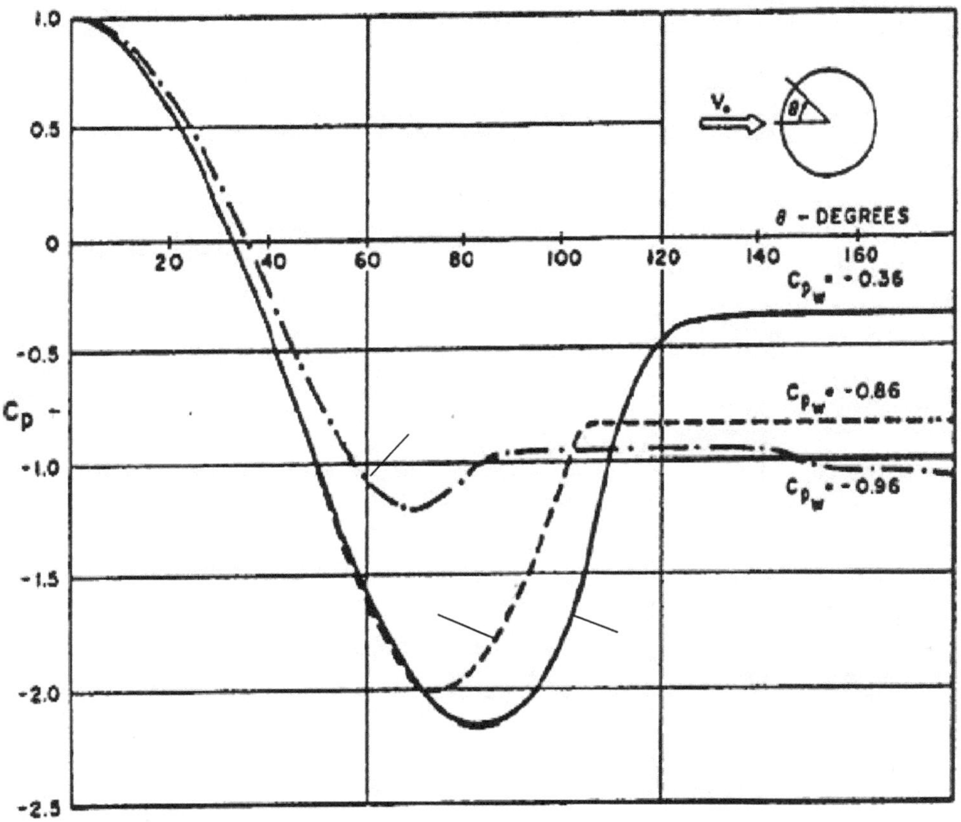

Figure 14. Influence of Reynolds number on the pressure distribution around a circular cylinder

In addition to the previously mentioned factors, the streamwise length of the bluff body plays a significant role on the drag coefficient, as shown in Figure 16. Again, flow characteristics such as the location of re-attachment, if it exists at all, are determinants of the magnitude of pressure-induced drag force. Clearly the size of wake will influence the drag force, so that for two buildings with the same frontal area, that with the longer length, i.e. the narrower wake, will experience the smaller drag force. Turbulence of the oncoming flow causes re-attachment to occur at a relatively shorter length, so turbulence also affects drag forces. This reveals the influence of the upstream exposure to the wind-induced pressures on buildings.

Figure 16 also shows the significant difference between the pressure-induced drag and the skin friction force. The latter is indeed negligible unless the dimensions of a building are excessive.

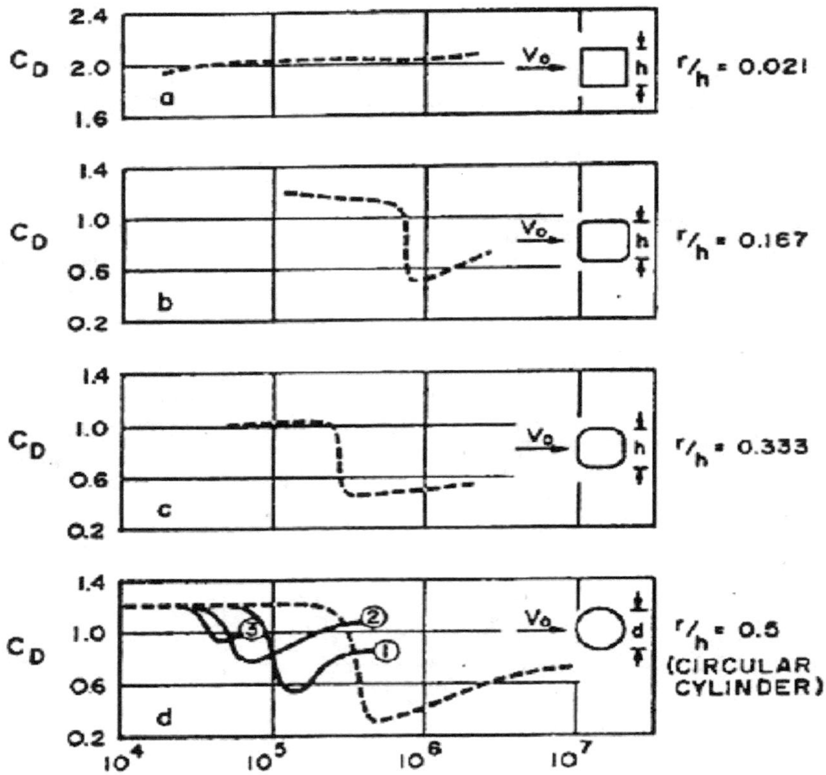

Figure 15. Influence of Reynolds number, corner radius and surface roughness on the values for C_D for cylinders of square and circular sections, 1. $k/d = 0.002$, 2. $k/d = 0.007$, 3. $k/d = 0.020$, where k is the grain size of sand, after Scruton (1971)

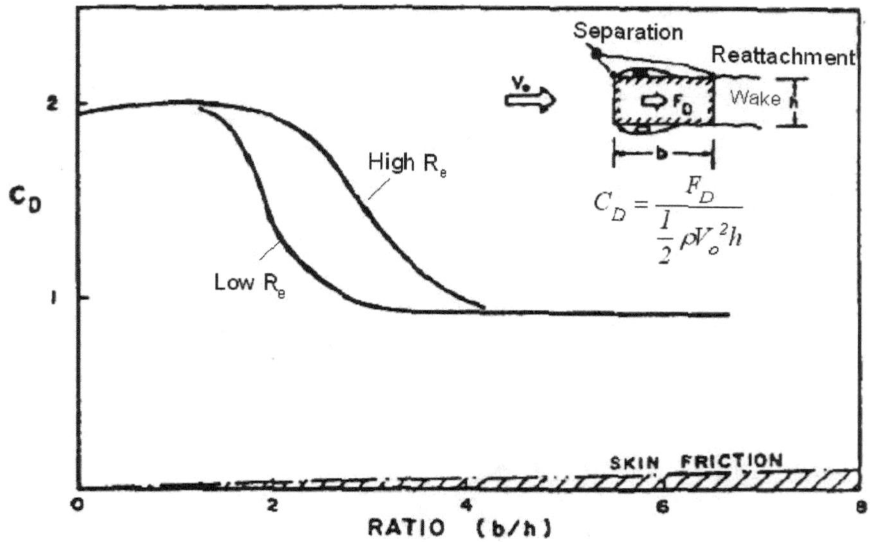

Figure 16. Effect of "after-body" on the drag for rectangular cylinders

5.2 Boundary layer flow conditions

Discussion so far has been restricted to the mean pressures and forces acting on two-dimensional (2D) bodies in uniform flow. Plenty of data is available for such cases but, unfortunately, very little can be used for structural design purposes. This is because such data can be used only for loads of isolated slender members, i.e masts, guy wires and the like. In the case of buildings, loads induced by turbulent boundary layer flow are expected to be quite different from those originating from uniform 2D flow.

Figure 17 shows uniform flow and pressures on a flat plate in a 3D flow situation, for which the interaction of the flow separated over the roof with that separated over the vertical edges is apparent. The pressure of the wake is more or less uniform along the height of the plate with a pressure coefficient in the order of -0.4.

Figure 18 shows exactly the same configuration but the flow now follows the actual boundary layer characteristics. Results are drastically different with the formation of a standing vortex in the front of the structure, creation of a clear down-flow separating from the sides of the plate and lower suctions in the wake, where pressure coefficients are in the order of -0.25. Figure 19 shows the distribution of mean pressure coefficients on an exploded cubical building view. The wind blows perpendicular to the building face and is assumed to have a velocity profile constant with height (uniform flow conditions). In contrast, Figure 20 shows mean pressure coefficient distribution obtained for boundary layer flow conditions reflecting more of reality. The wind is also perpendicular to the building but significant differences exist, as it can be observed easily. This finding makes imperative the testing of model buildings under realistic atmospheric flow conditions in order to obtain data representative of the actual full-scale pressures and forces.

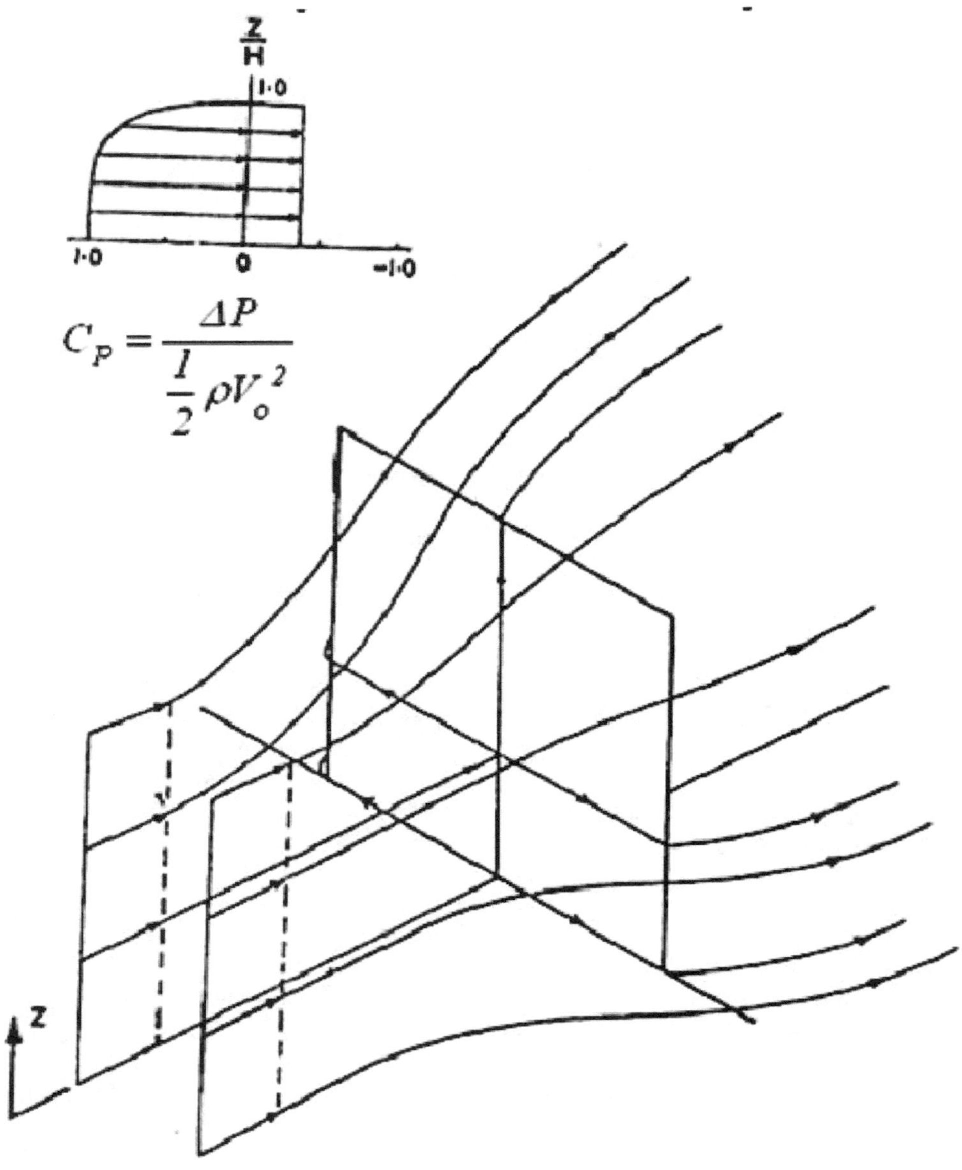

Figure 17. Flow pattern and centre - line pressure distribution - wall of height : width = 1:1, in a constant velocity field, after Baines (1963)

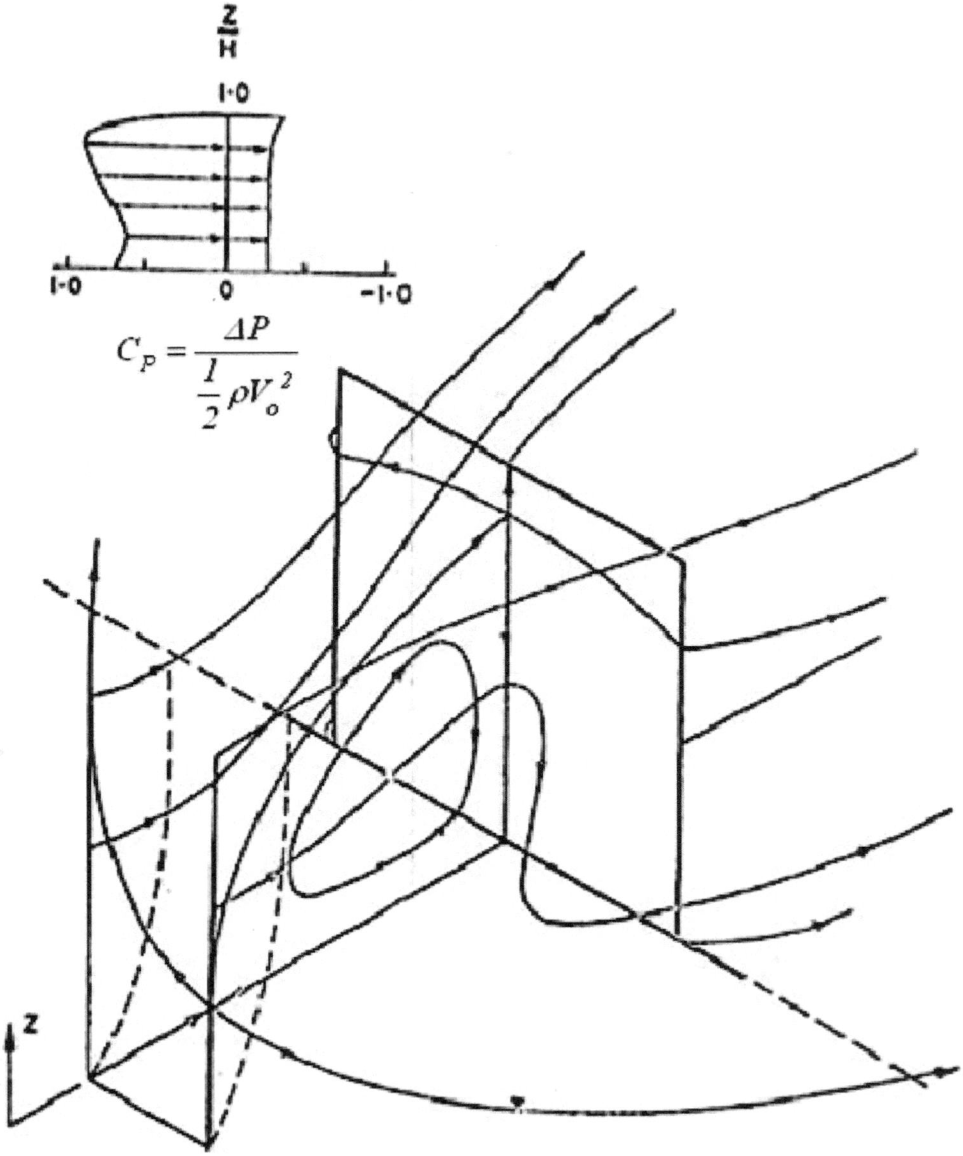

Figure 18. Flow pattern and centre-line pressure distribution - wall of height : width = 1:1, in a boundary-layer velocity field, after Baines (1963)

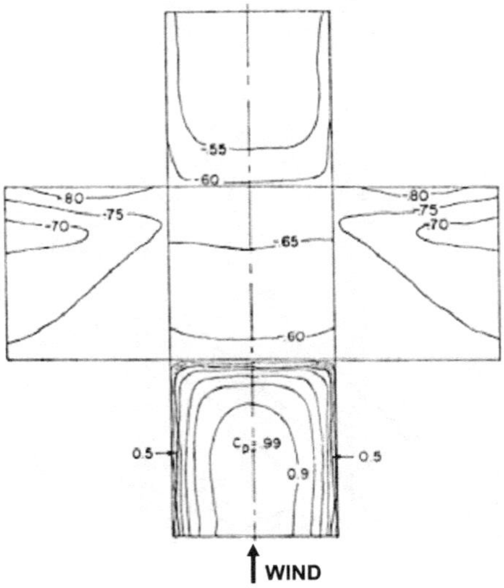

Figure 19. Pressure distribution on a cube in a constant velocity field, after Baines (1963)

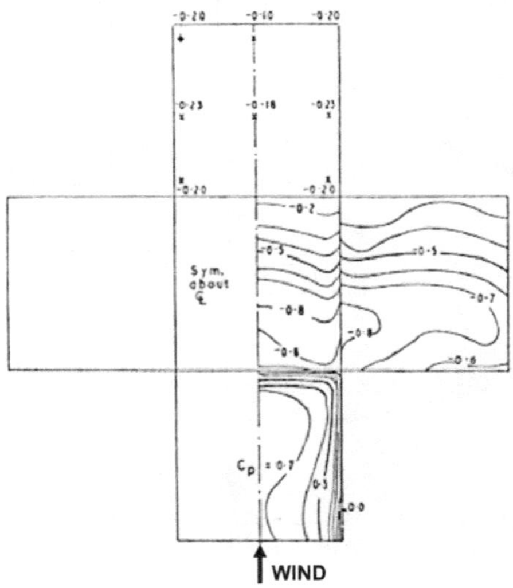

Figure 20. Pressure distribution on a cube in a boundary-layer velocity field, after Baines (1963)

5.3 Common features of pressure distributions on buildings

After testing several building models under turbulent shear flow conditions, the following general characteristics of wind-induced pressure distributions have been noticed:

1) Pressures over the front face are positive but reduce rapidly as the flow accelerates around the sides and upper edge of the face
2) Pressures decrease downwards along the face centre (decreasing velocity in boundary layer); downward flow results in substantial velocities at street level
3) Pressures on the rear face are negative with their absolute value somewhat decreasing downwards
4) Roof and side pressures are mostly negative with very large localized suctions (eaves, corners); see Figure 21 for more details
5) Pressure difference between wake and base of windward face cause horizontal flow through arcades or around corners - very difficult to control, particularly for very tall buildings

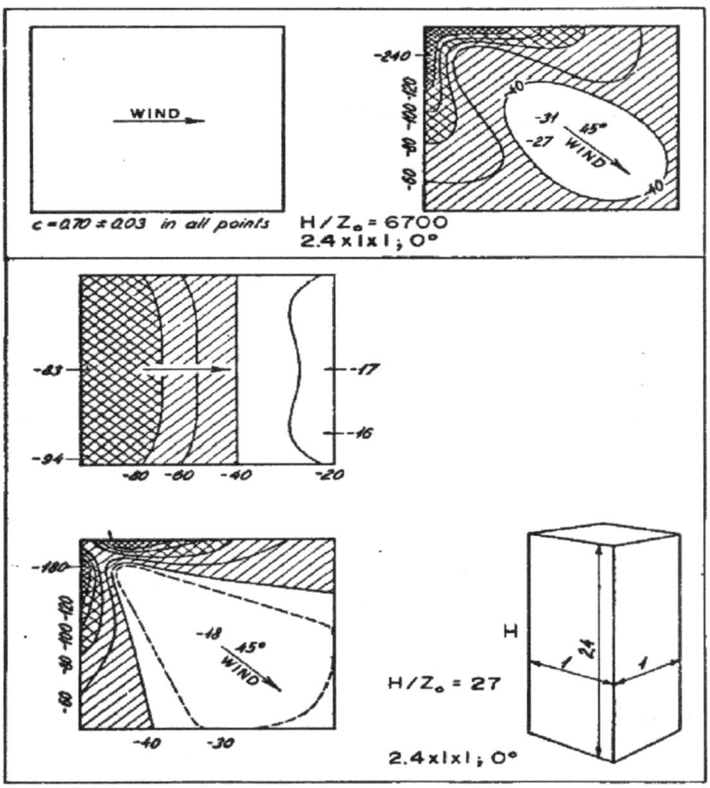

Figure 21. Pressures on horizontal roofs (pressures expressed as a percent of dynamic head at roof level), after Jensen and Franck (1965)

6 Internal pressures

Internal pressures depend on the external pressure distribution, terrain, shape, area and distribution of openings on the façade. In other words, the magnitude of the internal pressure depends primarily on the distribution of vents or openings in relation to the external pressure distribution. In the ideal case of a hermetically sealed building, the internal pressure is not affected by the external wind flow. As shown in Figure 22 a building with dominant venting on the windward side is under positive pressure while building pressures are negative with dominant venting within the wake region.

In a steady state situation, internal pressures can be computed from knowledge of the external pressure distribution and the size, shape and distribution of vents or openings. In this case, the inflow and outflow must balance. The flow through a wall of total open area A, subject to some uniform external pressure P_e and internal pressure P_i is given by:

$$Q = C_D A[\frac{2(P_e - P_i)}{\rho}]^{0.5} \qquad (20)$$

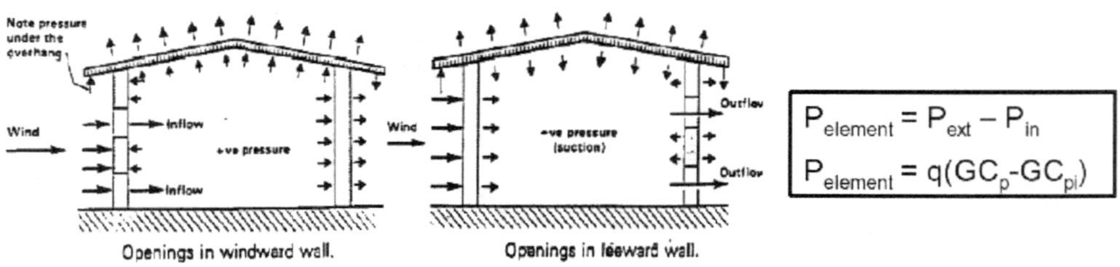

Figure 22. Mean internal pressures in buildings with various opening distributions

where C_D is the discharge coefficient. If the pressures are expressed in terms of coefficients using Eq. (20), then:

$$Q = C_D A V(C_{pe} - C_{pi})^{0.5} \qquad (21)$$

The value of C_{pi} can be computed from a knowledge of the values C_D, A and C_{pe} for each external surface by using the mass balance equation (Aynsley et al. 1977):

$$\Sigma Q = 0 \qquad (22)$$

In a building with two openings, assuming the same leakage characteristics for the inlet and outlet and uniform internal pressure, the following equations for the mean internal pressure coefficient can be derived from the mass balance equation (Liu 1991):

$$C_{pin} = \frac{C_{p2} + \alpha^2 C_{p1}}{1+\alpha^2} \text{ for large openings (windows, doors – turbulent flow)} \quad (23)$$

$$C_{pin} = \frac{C_{p2} + \beta C_{p1}}{1+\beta} \text{ for small openings (cracks – laminar flow)} \quad (24)$$

where
C_{p1} = external pressure coefficient in opening 1
C_{p2} = external pressure coefficient in opening 2
$\alpha = A_1/A_2$;and $\beta = ¼$ for uniform distribution of cracks.

The internal pressure coefficient in buildings with openings in one wall received a lot of attention in the 70's mainly for structural applications (Aynsley et al. 1977; Stathopoulos et al. 1979). Wu et al. (1998) carried out a comparison among various studies for internal pressures.

Stathopoulos et al. (1979) carried out BLWT experiments in order to investigate the impact of various opening configurations on internal pressures for different background leakage values, wall openings and exposures. Figure 23 illustrates the high degree of correlation of the internal pressures. Typical pressures from two pressure taps located at opposite ends of a large building are shown. The traces are virtually identical. This characteristic is quite general for all the various configurations examined and leads to the conclusion that the spatial variation of the internal pressure is minimal. However, recent experiments (Karava et al. 2006) have shown the uniformity of internal pressure is reduced in buildings with significant openings on more than one walls.

Comparison between traces of external and internal pressures not recorded simultaneously is made in Figure 24. The two pressure taps considered are located at the corner of the roof of the building, one on the inside and the other on the outside. The internal pressure appears to be as dynamic as the external pressure but its magnitude is significantly lower.

Figure 25 presents max, min and mean internal pressure coefficients for a large building exposed to an open country terrain for wind perpendicular to the wall with the openings (side wall). Each point on the graph represents a typical value of internal pressure coefficient for one particular configuration, i.e. a particular combination of background porosity and side-wall opening. For the cases shown in Figure 25, mean internal pressure coefficients are generally positive with the exception of the case of high background porosity combined with small or no openings, when they become zero of slightly negative. For wall openings of significant size (more than 20% of the wall area) the internal pressure coefficients depend little, if at all, on the background porosity. The lower the background porosity, the smaller the little, if at all, on the background porosity. The lower the background porosity, the smaller the necessary wall opening before the internal pressure coefficient becomes insensitive to further increase of the openings. Although the unsteady component denoted by Cp_{rms} decreases for cases with very low wall opening ratios (implying that the internal pressure becomes more nearly static for such configurations), the data suggest that internal pressures are normally dynamic in nature, with a typical value of internal gust factor of about two for the open country exposure. It is interesting to note that this is similar to the value of the gust effect factor often taken for overall external loads, suggesting that the reduced internal pressure fluctuations nevertheless include the quasi-steady components, which are encompassing the major part of the structure. The higher external intensities are then primarily local effects.

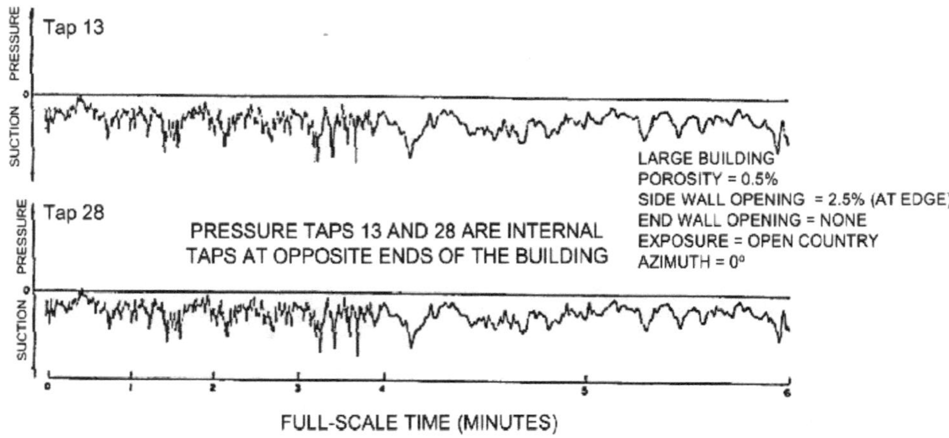

Figure 23. Simultaneous time traces of internal pressures for two different taps, after Stathopoulos et al. (1979)

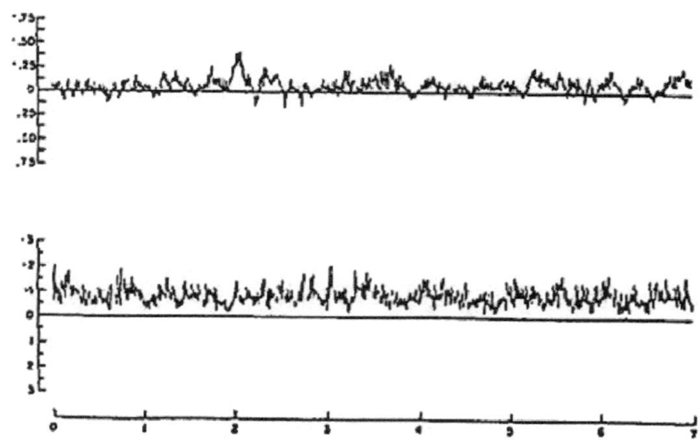

Figure 24. Comparison of internal and external pressures, after Stathopoulos et al. (1979)

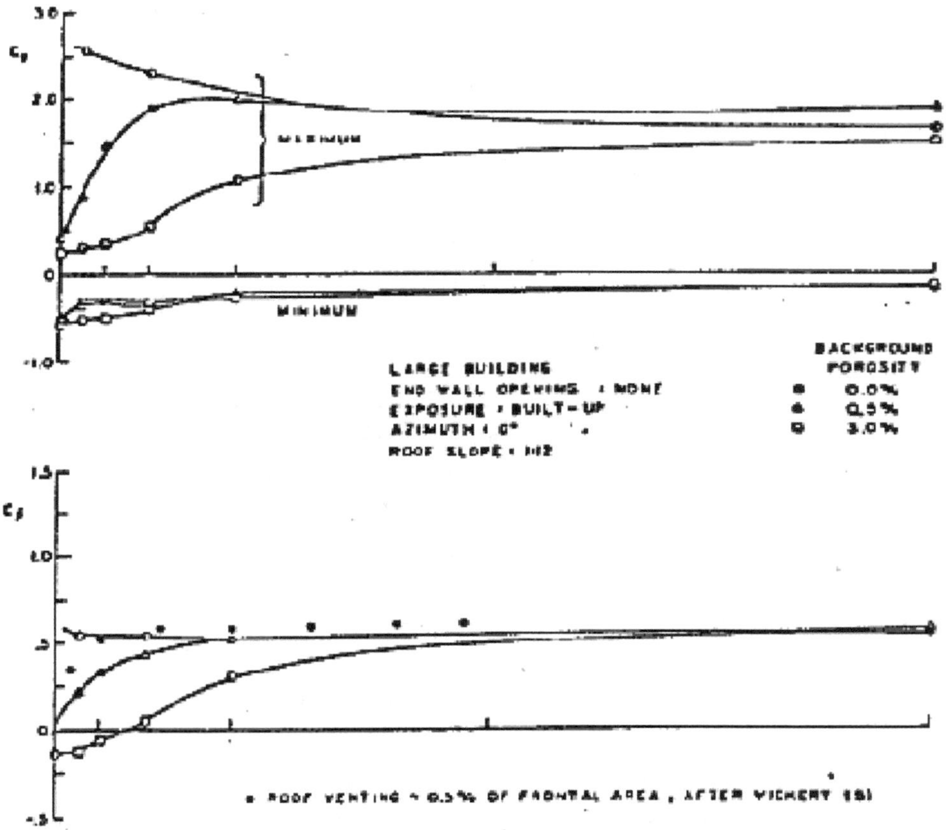

Figure 25. Extreme and mean internal pressure coefficients for various side-wall openings and background porosities - large building, open country exposure, after Stathopoulos et al. (1979)

The impact of a windward wall opening on internal pressure was recently reassessed (Karava et al. 2006) for 0 and 0.5% background leakage and results are presented in Figure 26 for different opening area or windward wall porosity (A_{inlet}/A_{wall}). The internal pressure was measured at different internal taps and it was found to be uniform, as also previously reported by Stathopoulos et al. (1979) and Wu et al. (1998). Wind tunnel results by Aynsley et al. (1977) for a building with windward wall openings and a roof opening equal to 0.5% of the windward wall simulating the background leakage and results by Stathopoulos et al. (1979) for a building with 0.5% uniform background leakage (open country exposure) are also included. The experimental data are compared with the values obtained by Eq. (23) for 0% background leakage. For the case of a single opening, Eq. (20) reduces *to* $Cp_{in} = Cp_1$, which is equal to 0.67. Figure 26 shows good agreement between the experimental results and the theoretical values for 0% leakage. In addition, data obtained for 0.5% background leakage shows a similar trend with that observed in the previous studies. Application of Eq. (23) for 0.5% background leakage will be arguable in this case due to the undetermined character of the flow.

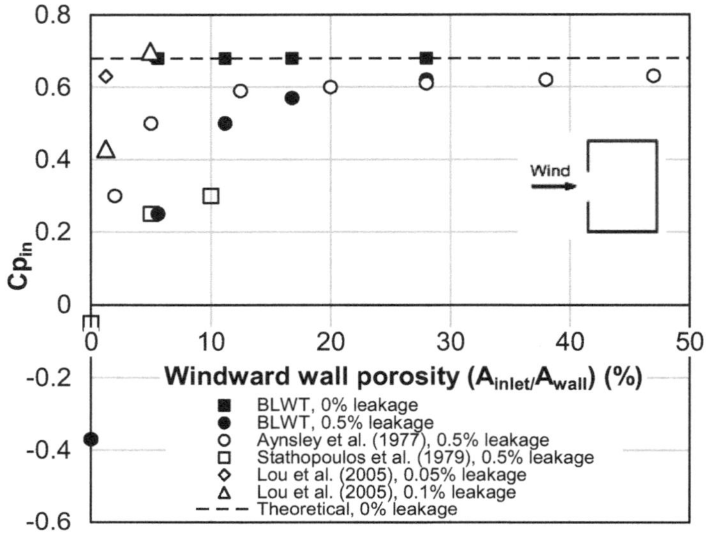

Figure 26. Internal pressure coefficients for single-sided ventilation and different windward wall porosity (A_{inlet}/A_{wall}), after Karava et al. (2006)

The impact of a windward and a side-wall opening of the same area ($A_1 = A_2$) on internal pressure was investigated for 0.5% background leakage and $\theta = 0°$. Measurements were carried out for opening areas up to 10.2 cm^2 (or 22 % windward wall porosity). The external pressure distribution was monitored and found unaffected by the presence of openings on the façade (sealed body assumption) for the range of wall porosity considered.

The variation of internal pressures with the different geometrical and porosity configurations is quite extensive. The determination of appropriate values of internal pressure coefficients for design standards and codes of practice remains a challenge.

7 Conclusion

This chapter introduces the reader to the science of wind engineering, the wind velocity profiles and the wind structure in general. Recently developed analytical models for the description of velocity profiles and wind turbulence are provided. The wind-building interaction is presented and the importance of simulation of atmospheric boundary layer flow is demonstrated. The significance of internal pressures in the design process and the designer's challenge in their evaluation becomes clear.

Acknowledgement

The assistance of Ph.D. student Panagiota Karava in putting this chapter together is gratefully acknowledged and highly appreciated.

References

American Society of Civil Engineers ASCE. (1999). Wind tunnel studies of buildings and structures, ASCE Manuals and Reports on Engineering Practice No. 67, Aerospace Division of the American Society of Civil Engineers, Reston, VA.

American Society of Civil Engineers ASCE. (2003). Minimum design loads for buildings and other structures. ASCE 7-02ASCE, New York, NY.

Aynsley, R. (1999). Unresolved issues in natural ventilation for thermal comfort. In: *Proceedings of 1st International One Day Forum on Natural and Hybrid Ventilation,* Sydney, Australia, International Energy Agency Annex 35 project Technical Paper, IN Annex 35 CD, Ed. Per Heiselberg, Aalborg University, Denmark.

Baines, W.D. (1963), Effects of velocity distribution on wind loads and flow patterns on buildings. *In Proceedings of the Symposium on Wind Effects on Buildings and Structures,* Vol. I, National Physical Laboratories, Teddington, H.M.S.O.

Cermak, J.E.C. (1975). Applications of fluid mechanics to wind engineering – A Freeman Scholar Lecture. *Journal of Fluids Engineering*, March. pp. 9-38.

Davenport, A.D. (1967). Gust loading factors. *Journal of Structural Division,* Proc. ASCE, Vol. 93, No. ST3.

Davenport, A.G., Grimmond, S., Oke, T. and Wieringa, J. (2000). The revised Davenport roughness classification for cities and sheltered country. In *Third Symposium on the Urban Environment. Davis*, California. Aug. 14-18, pp. 7-8.

Engineering Sciences Data Unit, ESDU. Strong winds in the atmospheric boundary layer. Part 1: hourly-mean wind speeds. *Data Item 82026*, Engineering Sciences Data Unit.

Engineering Sciences Data Unit. ESDU.. Longitudinal turbulence intensities over terrain with roughness changes. *Data Item 84030*, Engineering Sciences Data Unit.

Garratt, J.R. (1990). The internal boundary layer – a review. *Boundary-Layer Meteorology*. 50, pp. 171-203.

Jensen, M. and Franck N. (1965). *Model-scale tests in turbulent wind, Part II.* The Danish Technical Press, Copenhagen.

Karava, P., Stathopoulos, T and Athienitis, A. (2006). Impact of Internal Pressure Coefficients on Wind-Driven Ventilation Analysis. *International Journal of Ventilation*. Vol. 5, No. 1, June. pp. 53-66.

Letchford C., Gardner, A., Howard, R. and Schroeder, J. (2001). A comparison of wind prediction models for transitional flow regimes using full-scale hurricane data. *Journal of Wind Engineering and Industrial Aerodynamics*, 89, pp. 925-945.

Liu, H. (1991). *Wind engineering – A handbook for structural engineers.* Prentice-Hall, New Jersey.

Lou, W-J, Yu, S-C and Sun, B-N. (2005). Wind tunnel research on internal wind effect for roof structure with wall openings. In: Choi, C.K., Kim, Y.D., Kwak, H.G. (Eds.*), Proceedings of 6th Asia - Pacific Conference on Wind Engineering (APCWE – VI)*, Seoul, Korea, Sept. 12-14, pp. 1606-1620.

Scruton, C. (1971). Steady and unsteady wind loading of buildings and structures. *Phil. Trans. Roy. Soc.* London, A 269. pp. 353-383.

Schmid, H.P. and Bunzli, B. (1995). The influence of surface texture on the effective roughness length. *Quarterly Journal of Royal Meteorological Society*, 121, pp. 1-21.

Stathopoulos, T., Surry, D. and Davenport, A.G. (1979). Internal pressure characteristics of low-rise buildings due to wind action. In: *Proceedings of the 5th International Wind Engineering Conference*, Vol. 1, Forth Collins, Colorado USA, July.

Van der Hoven, I. (1957). Power spectrum of wind velocities fluctuations in the frequency range from 0.0007 to 900 Cycles per hour. *Journal of Meteorology*, 14, pp. 1254-1255.

Wang, K. and Stathopoulos, T. (2005). Exposure model for wind loading of buildings. In: *4th European-African Conference on Wind Engineering*. Prague. Czech Republic. July 11-15, 2005.

Zhang, X. and Zhang, R.R. (2001). Actual ground-exposure determination and its influences in structural analysis and design. *Journal of Wind Engineering and Industrial Aerodynamics*, 89, pp. 973-985.

Wu, H., Stathopoulos, T. and Saathoff, P. (1998). Wind-induced internal pressures revisited: low-rise buildings. In: *Proceedings of Structural Engineers World Congress,* San Francisco, CA, USA.

Wind Loading on Buildings: Eurocode and Experimental Approach

Chris Geurts, Carine van Bentum

TNO Built Environment and Geosciences, Delft, The Netherlands

Abstract. This chapter deals with the application of European wind loading code EN 1991-1-4 as well as experimental techniques to determine wind loads on buildings. The general outline of EN 1991-1-4 is presented, and the procedure how to come to a wind load on structures is presented. Situations for which wind tunnel or full scale experiments are useful are described and the basic principles for these experiments are given.

1 Introduction

Wind loads on structures, such as buildings, bridges, masts, but also on parts of these structures, are derived from wind loading standards. Structural engineers within Europe will soon be obliged to use the Eurocode System for the calculation of structures. Wind loads are given within this system in EN 1991, Actions on Structures, part 1-4, Wind Loads. This code covers a wide range of building shapes and dimensions. However, many cases still exist for which the code gives no, or a very unsatisfactory, answer. For such cases, wind tunnel experiments, or in special cases, full scale experiments may lead to an answer. This paper describes the principles and some backgrounds for these subjects:

First, the main properties for EN 1991-1-4 are given. Secondly, the main requirements and boundary conditions for wind tunnel research, with a special interest to tests to obtain wind loads for specific structures in the design stage, are given. Thirdly, an outline of full scale testing is given, including the pros and cons of this technique, and its applicability to the design of building structures and building parts.

2 Wind loading in EN 1991-1-4

2.1 The Eurocode System

In 1975, the Commission of the European Community decided on an action programme in the field of construction. The objective of the programme was the elimination of technical obstacles to trade and the harmonisation of technical specifications. In 2010, in all CEN countries, all national standards on the design of building structures will be replaced by the Eurocodes. All Eurocodes, however, include a National Annex, to specify values for which the Eurocode leaves national choice open. Without National Annex, and without translation in the official language of the country considered, the EN's can not be used.

The Eurocode system is based on performance based design. This means that the action effects and the resistance of a structure are treated separately. Action effects are independent of structural material, unless the material itself is a source for an effect (e.g. temperature effects). The effects determine the level of the strength (performance) that needs to be fulfilled by the structure. The Eurocode series consists of 10 series of documents: EN 1990 to EN 1999, where EN 1991 deals with the actions.

- EN 1990 'Basis of Design' specifies the general principles for classification of actions on structures including environmental impacts and their modelling in verification of structural reliability [1]. EN 1990 defines characteristic, representative and design values used in structural calculations.
- EN 1991 'Actions on Structures' specifies the characteristic values of the actions on structures. EN 1991 is divided in 10 volumes, each specifying a specific action. The wind loading is specified in EN 1991-1-4.

Then, 6 Eurocodes specify the calculation of the strength of structures with a specific material:

- EN 1992: 'Design of concrete structures'
- EN 1993: 'Design of steel structures'
- EN 1994: 'Design of composite steel and concrete structures'.
- EN 1995: 'Design of timber structures'.
- EN 1996: 'Design of masonry structures'.
- EN 1999: 'Design of aluminum alloy structures'

The Eurocode series is completed with:

- EN 1997: 'Geotechnical design' gives rules how to calculate the strength of foundations and other geotechnical structures.
- EN 1998: 'Design of structures for earthquake resistance' gives both action effects by earthquakes and design rules to determine the resistance of structures to earthquakes.

Every Eurocode consists of a number of documents. Overall, the Eurocode system consists of 58 documents, which will be used in 28 countries (at least). All these documents, for all countries, require National Annexes, so in the end the total set of regulations will be a large number of documents. It is expected that in the next generation of these Eurocodes, this will be an important issue.

2.2 Field of application

EN 1991-1-4 specifies natural wind actions for the structural design of building and civil engineering works for each of the loaded areas under consideration. This includes the whole structure or parts of the structure or elements attached to the structure, e. g. components, cladding units and their fixings, safety and noise barriers.

The field of application of EN 1991-1-4 is limited to buildings and civil engineering works with heights up to 200 m and to bridges having no span greater than 200 m. EN 1991-1-4 is dealing mainly with the wind loading on the gross amount of structures. It therefore gives limited information on special actions, such as: torsional vibrations, bridge deck vibrations from transverse wind turbulence, cable supported bridges and vibrations where more than the fundamental mode needs to be considered. Although in principle EN 1991-1-4 should be the only document specify-

ing wind loads, guyed masts and lattice towers, as well as lighting columns are treated in separate EN-codes.
A special group of wind actions on structures are the aero-elastic effects, where the response of a structure interacts with the wind flow. These effects are given in EN 1991-1-4 in an informative annex.

2.3 Wind loading model

The wind induced force acting on a structure or a structural component F_w can be determined by three procedures in EN 1991-1-4. The first procedure uses:

$$F_w = c_f\, c_s c_d q_p(z_e) A_{ref}$$

The second is using vectorial summation over individual structural elements, by:

$$F_w = c_s c_d \sum c_f q_p(z_e) A_{ref}$$

The third procedure uses summation of pressures on sides of the structure:

$$F_w = c_s c_d \sum c_{pe} q_p(z_e) A_{ref} \quad \text{(external pressures)}$$

$$F_w = c_s c \sum c_{pi} q_p(z_i) A_{ref} \quad \text{(internal pressures)}$$

$$F_w = c_{fr} q_p(z_e) A_{ref} \quad \text{(external pressures)}$$

In which:

F_w is the wind induced force;
c_p is the pressure coefficient for the effect under consideration;
c_s is a size factor, taking the lack of correlation of the wind pressures on a building into account;
c_d is the dynamic factor, taking the effects of resonance into account;
q_p peak dynamic pressure;
ρ is the density of air.

The characteristic value of the wind loading should be multiplied with appropriate partial safety factors to arrive at the design values of the wind load. Furthermore, the combinations of loads to be taken into account should be specified. This is not given in EN 1991-1-4, but is described in EN 1990 'Basis of Design'. It will not be treated in detail in this paper, since the values to be used are subject of the National Annex of EN 1990, and may therefore differ in the different countries.

2.4 Peak velocity pressure

General. The specification of the wind actions on buildings first requires a specification of the wind which is taken into account. The primary parameter in the determination of wind actions on structures is the characteristic peak velocity pressure q_p. The peak velocity pressure accounts for the effect of the mean wind velocity and a turbulence component. The peak velocity pressure is influenced by the regional wind climate, local factors (e.g. terrain roughness and orography) and the height above terrain. The peak dynamic pressure is found by:

$$q_{\mathrm{p}}(z) = [1 + 7 \cdot I_{\mathrm{v}}(z)] \cdot \frac{1}{2} \cdot \rho \cdot v_{\mathrm{m}}^2(z) = c_{\mathrm{e}}(z) \cdot q_{\mathrm{b}} \text{ with:}$$

$$v_{\mathrm{m}}(z) = k_{\mathrm{r}} \ln(\frac{z}{z_0}) \cdot c_{\mathrm{o}}(z) \cdot v_b \text{ and } q_b = \frac{1}{2} \rho \mathrm{v}_b^2$$

In which:

v_b is the basic wind velocity;
c_0 is the orography factor;
k_r is the terrain factor;
z is the height above ground;
z_0 is the roughness length;
ρ is the mass density of air;
$I_v(z)$ is the turbulence intensity which is the ratio between the standard deviation and the mean value of the wind speed;
q_b is the basic velocity pressure.

The height z at which the dynamic pressure is determined depends on the load effect to be calculated, and on the definition of the reference height for the pressure and force coefficients.

Basic wind velocity. The basic wind velocity v_b is defined in EN 1991-1-4 from the fundamental basic wind velocity. The fundamental basic wind velocity is the 10 minute mean wind velocity with a return period of 50 years. This wind velocity is obtained by a statistical analysis of measurements at meteorological stations. The values are not given in EN 1991-1-4, but in the National Annexes. Some countries give areas for which a certain value of the basic wind velocity applies. Other countries provide a map with isolines of values for the basic wind velocity, and a procedure to interpolate between these lines. There are and will be many discontinuities of this wind speed along national borders. In the future, when drafting of a next generation of the Eurocode on Wind Actions, a joint effort to come to a European wind map is strongly recommended.

Roughness length, terrain categories and terrain factor. In EN 1991-1-4, 5 terrain categories are defined, with given values for the roughness length z_0. These terrain categories are illustrated in Figure 1. Below a minimum height z_{min} the wind speed at height $z = z_{min}$ should be used. EN 1991-1-4 leaves national choice for the definition of the roughness classes. A range of 5 classes is probably too much for many countries.

Terrain category	*Illustration*	z_0 [m]	z_{min} [m]
0: Sea, coastal area exposed to the open sea		0.003	1
I: Lakes or area with negligible vegetation and without obstacles		0.01	1
II: Area with low vegetation such as grass and isolated obstacles (trees, buildings) with separations of at least 20 obstacle heights		0.05	2
III: Area with regular cover of vegetation or buildings or with isolated obstacles with separations of maximum 20 obstacle heights (such as villages, suburban terrain, permanent forest)		0.3	5
IV: Area in which at least 15 % of the surface is covered with buildings and their average height exceeds 15 m	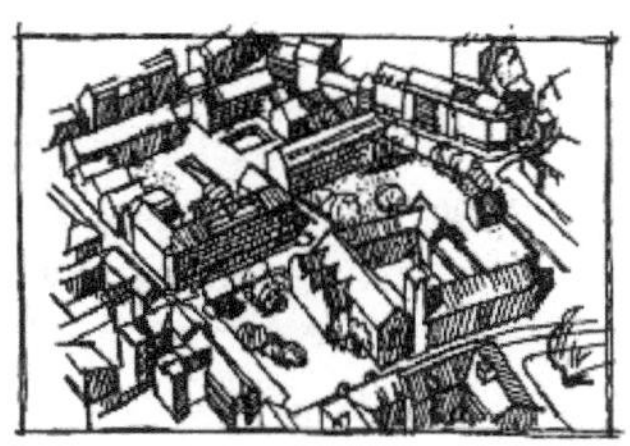	1.0	10

Figure 1: Terrain categories in EN 1991-1-4

The terrain factor takes the effect of terrain roughness on the mean wind profile into account. It is given by:

$$k_r = 0.19\left(\frac{z_0}{z_{0,II}}\right)^{0.07} \quad \text{where } z_{0,II} \text{ is } 0.05 \text{ m (terrain category II).}$$

Exposure factor. The exposure factor c_e is a measure for the dependence of the wind effects over height on the roughness of the terrain, peaks in the wind velocity and orography (hills, cliffs, ...etc.). If orography plays no role ($c_0 = 1$), the exposure factor is given by:

$$c_e = \left(k_r \ln\frac{z}{z_0}\right)^2 (1 + 7I_v(z))$$

Values for the exposure factor are given in Figure 2 for the terrain categories in EN 1991-1-4.

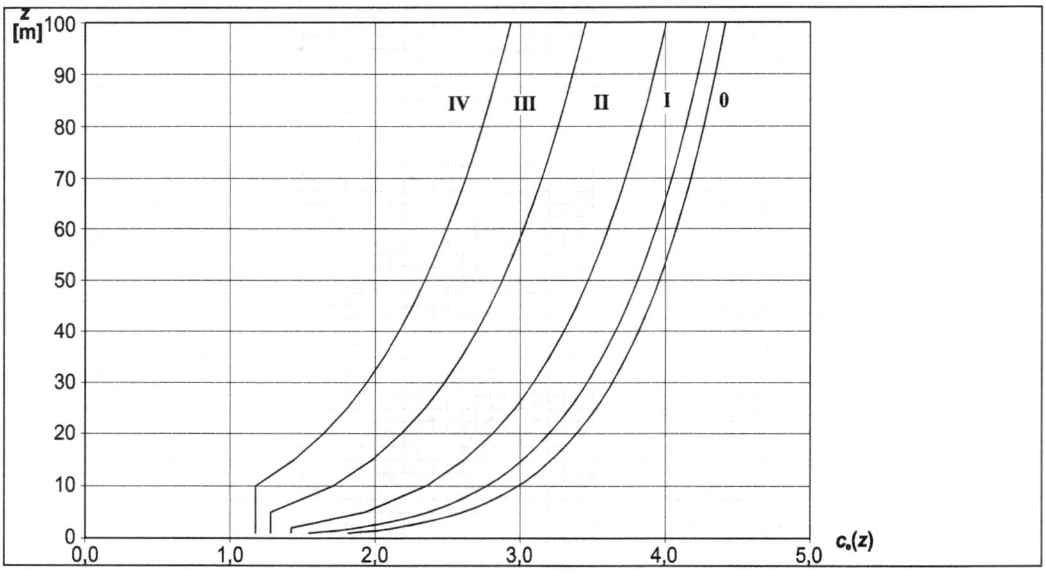

Figure 2: Illustrations of the exposure factor $c_e(z)$ for c_o=1.0

Local effects of the surroundings on wind velocity. The exposure factor as described above does not include the effects of orography or nearby buildings. EN 1991-1-4 gives provisions how to account for these effects. The influence of sloping terrain is incorporated in EN 1991-1-4 as the orography factor c_0 for the wind velocity. This factor depends on size and dimensions of the hill, and on the position of the building considered on the slope. Figure 3 shows the principle of this factor. The mean wind velocity is accelerated depending on the length and height of the slope.

High rise buildings, especially when these are positioned solitary, may lead to an increase of the wind velocities at low heights. This is the cause of problems occurring at pedestrian level, but may also lead to an increased loading for nearby lower buildings. EN 1991-1-4 gives a very rough rule to account for this effect. If a building is more than twice as high as the average height h_{ave} of the neighbouring structures then, as a first approximation, the design of any of those nearby structures may be based on the peak velocity pressure at height z_n ($z_e = z_n$) above ground, see Figure 4.

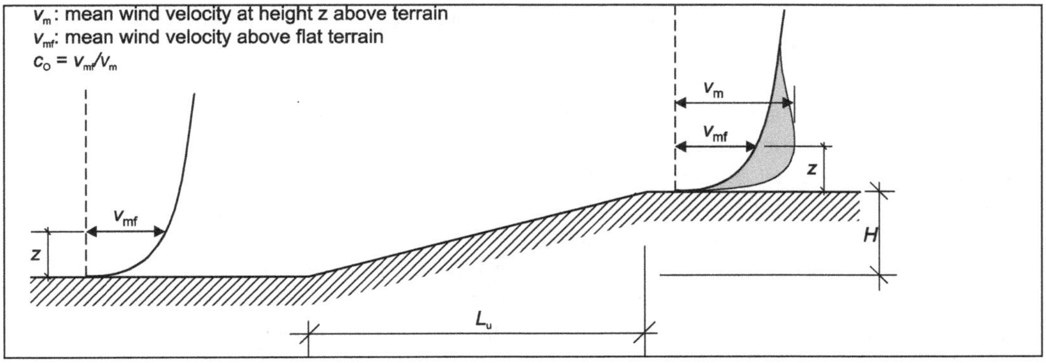

Figure 3: Schematic view of wind over a hill

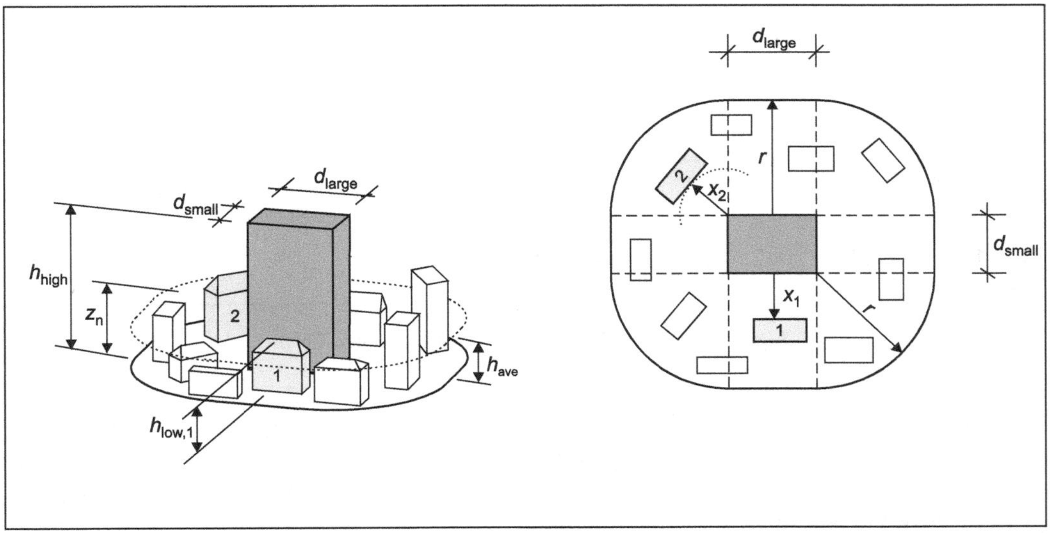

Figure 4: Sketch of parameters relevant to estimate the effect of nearby high buildings

2.5 Pressure and force coefficients

Wind forces and wind induced pressures are found by the multiplication of the peak dynamic pressure with an aerodynamic coefficient. External and internal pressure coefficients, force coefficients and friction coefficients are specified in EN 1991-1-4. The values of these coefficients are specified in such a way, that application of these values leads to loads with a return period of 50 years. Most coefficients in EN 1991-1-4 are based on wind tunnel studies, sometimes dating back to the 1950's. An overview of the backgrounds of the pressure and force coefficients in EN 1991-1-4 is given in [Geurts et. al, 2001]. Where the codes do not specify values, wind tunnel experiments, or sometimes full scale experiments, may be an alternative. In section 3 and 4 of this paper, the backgrounds on the application of wind tunnel and full scale experiments to determine the aerodynamic coefficients are described in more detail.

External pressures. The wind loads on buildings are found by defining first the distribution of the wind induced pressures on the walls and roofs. Based on these distributions, combinations of external and / or internal pressures determine the overall wind loading on the structure under consideration. Internal pressures are described later. The external pressure coefficients c_{pe} for buildings and parts of buildings depend on the size of the loaded area A, which is the area of the structure, that produces the wind action in the section to be calculated. The external pressure coefficients are specified for loaded areas A of 1 m^2 and 10 m^2 as $c_{pe,1}$, for local coefficients, and $c_{pe,10}$, for overall coefficients, respectively. The relation between these values is given in Figure 5.

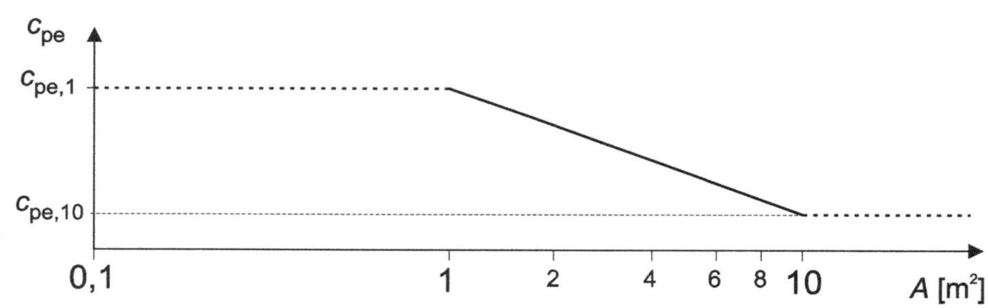

For 1 m^2 < A < 10 m^2 $\quad c_{pe} = c_{pe,1} - (c_{pe,1} - c_{pe,10}) \log_{10} A$

Figure 5: Relation between local and global coefficients

External pressures on walls. The pressure coefficients for 10 m^2 and larger are primarily needed when the overall structure of a building is designed. The wind loads on the overall structure are determined mainly by the external pressures on the vertical walls. The values depend on the position on the walls, defined by zones A, B, C, D and E, see Figure 6.

The lack of correlation of the wind pressures between the windward and leeward side (a maximum at windward side does not occur at the same instance as the maximum on leeward side) may be taken into account by applying a reduction factor to the overall wind loading. In this case, the wind force on building structures is determined by application of the pressure coefficients c_{pe} on zones D and E (windward and leeward side). For buildings with $h/d \geq 5$ the reduction factor applied is 1. For buildings with $h/d \leq 1$, the resulting force is multiplied by 0.85. The rules described above may be subject to changes through the National Annex.

Table 1 Values for pressure coefficients on vertical walls

Zone	**A**		**B**		**C**		**D**		**E**	
h/d	$c_{pe,10}$	$c_{pe,1}$	$c_{pe,10}$	$c_{pe,1}$	$c_{pe,10}$	$c_{pe,1}$	$c_{pe,10}$	$c_{pe,1}$	$c_{pe,10}$	$c_{pe,1}$
5	-1.2	-1.4	-0.8	-1.1	-0.5		+0.8	+1.0	-0.7	
1	-1.2	-1.4	-0.8	-1.1	-0.5		+0.8	+1.0	-0.5	
≤ 0.25	-1.2	-1.4	-0.8	-1.1	-0.5		+0.7	+1.0	-0.3	

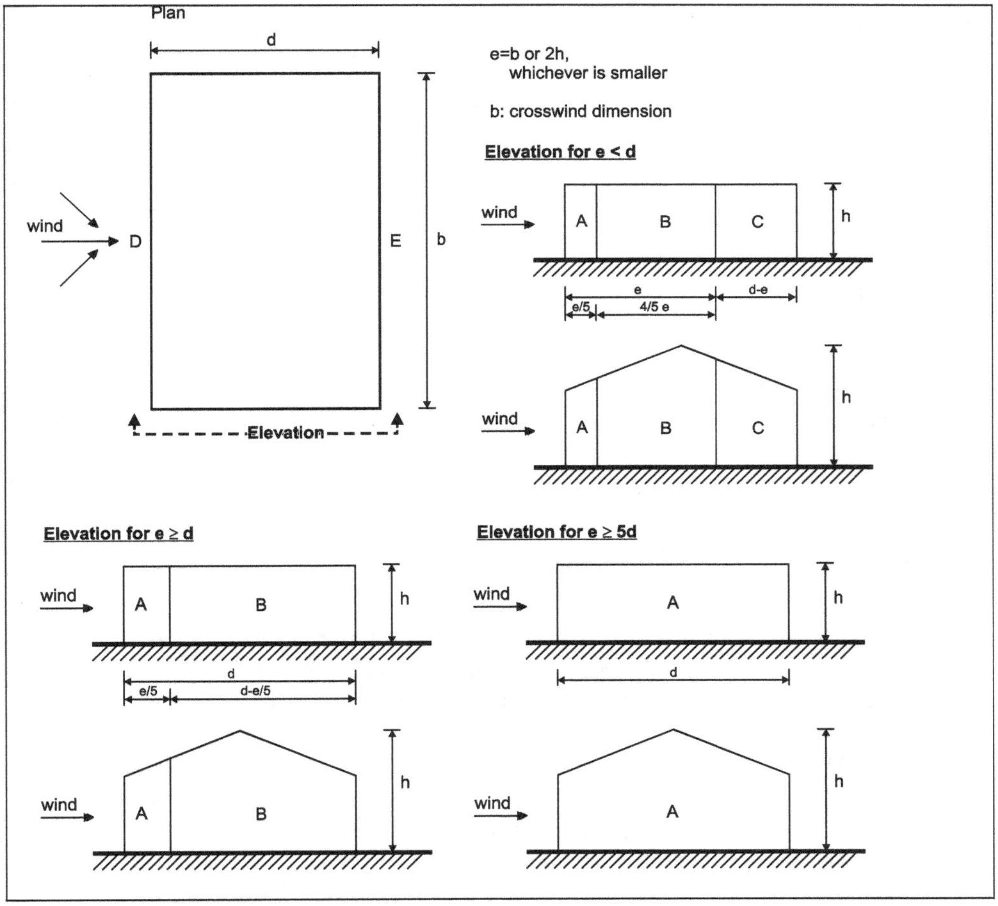

Figure 6 Zones of external pressures on vertical walls

External pressures on roofs. EN 1991-1-4 also specifies external pressure coefficients for flat and pitched roofs, and curved structures. The pressure coefficients are all given for specified wind directions, and for standard, rectangular ground plans. As an example, in Figure 7, the wind zones on flat roofs are given.

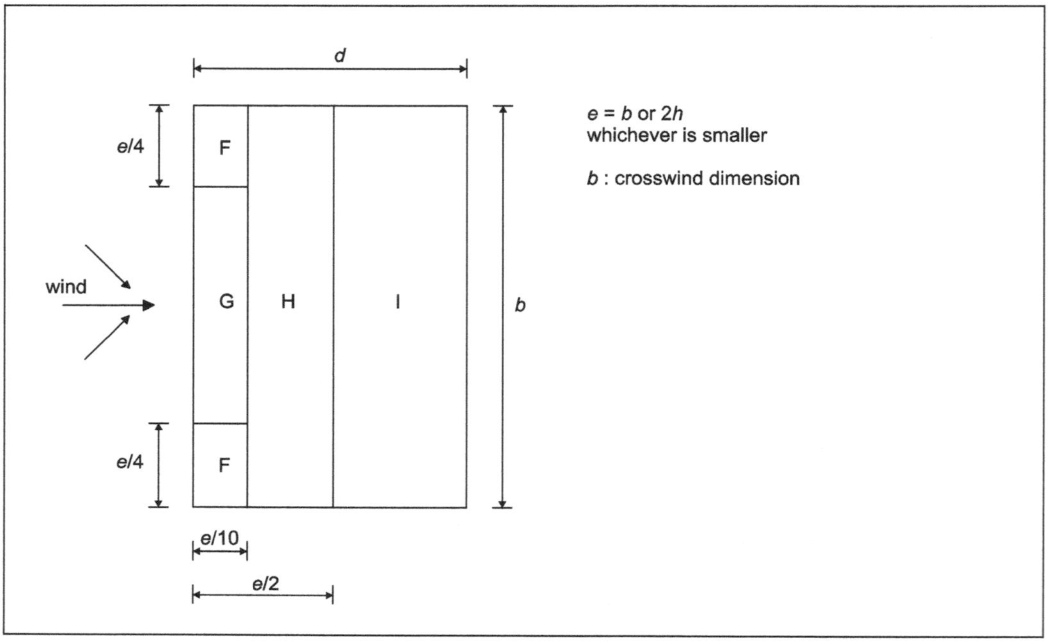

Figure 7: Wind zones on flat roofs

Reference height The definition of pressure coefficients should always include the reference height relative to which corresponding peak dynamic pressure the loading is calculated. For roofs, the reference height is often taken as ridge height. The reference heights, z_e, for windward walls of rectangular plan buildings (zone D) depend on the aspect ratio h/b and are always the upper heights of the different parts of the walls.

Local pressures. The pressure coefficients $c_{pe,1}$ are relevant for the design of cladding and roofing. The area that is to be considered in the design of a structural component depends e.g. on the properties of the component and on the stiffness of the cladding elements. It is assumed that a loaded area smaller than 1 m^2 can appropriately be represented by the value of $c_{pe,1}$. For larger areas, the wind loading will decrease with loaded area, until 10 m^2, for larger areas, the values stay constant. It is up to the designer of the cladding and loading to decide on which area is to be considered. Note that there may be different areas found for fixings, the supporting structure and the cladding elements.

Table 2: Values for the pressure coefficients on flat roofs.

Roof type		**Zone**							
		F		G		H		I	
		$c_{pe,10}$	$c_{pe,1}$	$c_{pe,10}$	$c_{pe,1}$	$c_{pe,10}$	$c_{pe,1}$	$c_{pe,10}$	$c_{pe,1}$
Sharp eaves		-1.8	-2.5	-1.2	-2.0	-0.7	-1.2	± 0.2	
With Parapets	h_p/h=0,025	-1.6	-2.2	-1.1	-1.8	-0.7	-1.2	± 0.2	
	h_p/h =0,05	-1.4	-2.0	-0.9	-1.6	-0.7	-1.2	± 0.2	
	h_p/h =0,10	-1.2	-1.8	-0.8	-1.4	-0.7	-1.2	± 0.2	
Curved Eaves	r/h = 0,05	-1.0	-1.5	-1.2	-1.8	-0.4		± 0.2	
	r/h = 0,10	-0.7	-1.2	-0.8	-1.4	-0.3		± 0.2	
	r/h = 0,20	-0.5	-0.8	-0.5	-0.8	-0.3		± 0.2	
Mansard Eaves	α = 30°	-1.0	-1.5	-1.0	-1.5	-0.3		± 0.2	
	α = 45°	-1.2	-1.8	-1.3	-1.9	-0.4		± 0.2	
	α = 60°	-1.3	-1.9	-1.3	-1.9	-0.5		± 0.2	

Pressures inside buildings. The design of roofing, cladding and internal walls requires that the internal pressures inside buildings are known. The internal pressure coefficient, c_{pi}, depends on the size and distribution of the openings in the building envelope. The openings of a building include small openings such as: open windows, ventilators, chimneys, etc. as well as so-called background permeability such as air leakage around doors, windows, services and through the building envelope. The background permeability is typically in the range 0.01% to 0.1% of the face area.

The calculation of the internal pressures in EN 1991-1-4 depends on the fact whether a building has dominant faces ore not. A face of a building should be regarded as dominant when the area of openings at that face is at least twice the area of openings and leakages in the remaining faces of the building considered.

For a building with a dominant face the internal pressure should be taken as a fraction of the external pressure at the openings of the dominant face. When the area of the openings at the dominant face is twice the area of the openings in the remaining faces, then:

$$c_{\mathrm{pi}} = 0.75 \cdot c_{\mathrm{pe}}$$

When the area of the openings at the dominant face is at least 3 times the area of the openings in the remaining faces, then:

$$c_{\mathrm{pi}} = 0.90 \cdot c_{\mathrm{pe}}$$

Where c_{pe} is the value for the external pressure coefficient at the openings in the dominant face.

For buildings without a dominant face, the internal pressure coefficient is a function of the ratio of the height and the depth of the building, *h*/*d*, and the opening ratio μ for each wind direction θ, which should be determined from:

$$\mu = \frac{\sum \text{area of openings where } c_{\text{pe}} \text{ is negative or - 0.0}}{\sum \text{area of all openings}}$$

This applies to façades and roof of buildings with and without internal partitions. Where it is not possible, or not considered justified, to estimate μ for a particular case then c_{pi} should be taken as the more onerous of +0.2 and -0.3. This will be the most common case.
The reference height z_i for the internal pressures is equal to the reference height z_e for the external pressures on the faces which contribute by their openings to the creation of the internal pressure. If there are several openings the largest value of z_e is used to determine z_i.

2.6 Dynamic response

Slender structures may be prone to dynamic response to the wind. EN 1991-1-4 specifies the structural factor c_sc_d which takes the effect on wind actions from the non-simultaneous occurrence of peak wind pressures on the surface together with the effect of the vibrations of the structure due to turbulence into account. This structural factor c_sc_d may be separated into a size factor c_s and a dynamic factor c_d.

The detailed procedure for calculating the structural factor c_sc_d is given as:

$$c_{\text{s}}c_{\text{d}} = \frac{1 + 2 \cdot k_{\text{p}} \cdot I_{\text{v}}(z_{\text{e}}) \cdot \sqrt{B^2 + R^2}}{1 + 7 \cdot I_{\text{v}}(z_{\text{e}})}$$

where:

z_e is the reference height;

k_p is the peak factor defined as the ratio of the maximum value of the fluctuating part of the response to its standard deviation;

I_v is the turbulence intensity;

B^2 is the background factor, allowing for the lack of full correlation of the pressure on the structure surface;

R^2 is the resonance response factor, allowing for turbulence in resonance with the vibration mode.

The size factor c_s takes into account the reduction effect on the wind action due to the non-simultaneity of occurrence of the peak wind pressures on the surface and may be obtained from:

$$c_{\text{s}} = \frac{1 + 7 \cdot I_{\text{v}}(z_{\text{e}}) \cdot \sqrt{B^2}}{1 + 7 \cdot I_{\text{v}}(z_{\text{e}})}$$

The dynamic factor c_d takes into account the increasing effect from vibrations due to turbulence in resonance with the structure and may be obtained from:

$$c_{\mathrm{d}} = \frac{1 + 2 \cdot k_{\mathrm{p}} \cdot I_{\mathrm{v}}(z_{\mathrm{e}}) \cdot \sqrt{B^2 + R^2}}{1 + 7 \cdot I_{\mathrm{v}}(z_{\mathrm{e}}) \cdot \sqrt{B^2}}$$

These procedures shall only be used if the structure is a vertical or horizontal structure, like a building or a bridge, and when only the along-wind vibration in the fundamental mode is significant, and this mode shape has a constant sign. The contribution to the response from the second or higher alongwind vibration modes is negligible.

EN 1991-1-4 provides two calculation procedures for B^2 and R^2. For information, EN 1991-1-4 gives safe estimates of $c_s c_d$ for a variation of structural systems, e.g. Figure 8.

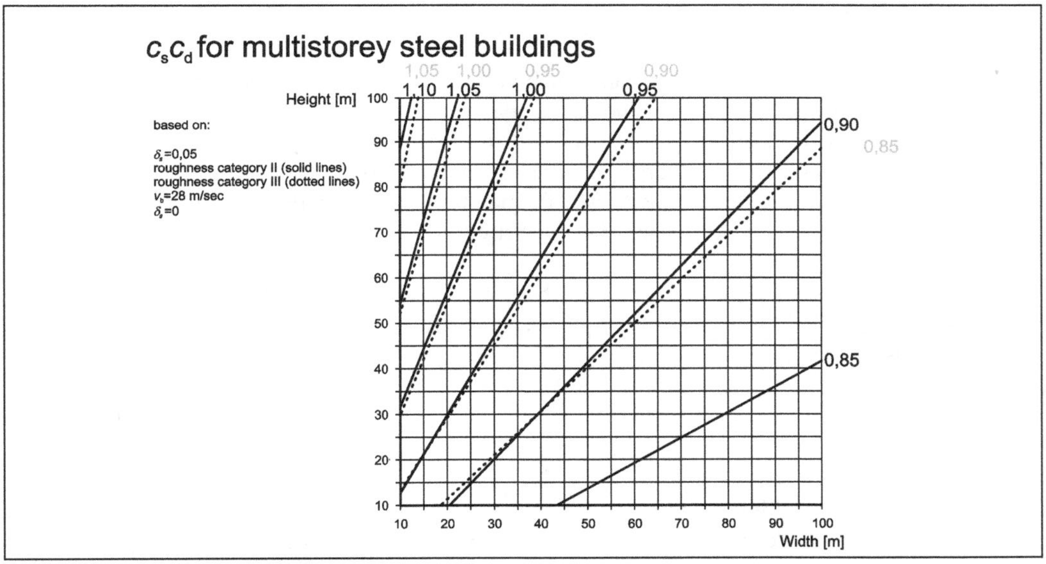

Figure 8: Example of $c_s c_d$ for multi-storey steel buildings

Vibration levels in buildings. The standard deviation $\sigma_{a,x}$ of the characteristic along-wind acceleration of the structural point at height z should be obtained using:

$$\sigma_{\mathrm{a,x}}(z) = \frac{c_{\mathrm{f}} \cdot \rho \cdot b \cdot I_{\mathrm{v}}(z_{\mathrm{e}}) \cdot v_{\mathrm{m}}^2(z_{\mathrm{e}})}{m_{1,\mathrm{x}}} \cdot R \cdot K_{\mathrm{x}} \cdot \Phi_{1,\mathrm{x}}(z)$$

where:

c_f is the force coefficient;

ρ is the air density;

b is the width of the structure;

$I_v(z_e)$ is the turbulence intensity at the height $z = z_e$ above ground;

$v_m(z_e)$ is the mean wind velocity for $z = z_e$;
R is the square root of resonant response, see above;
K_x is a non-dimensional coefficient;
$m_{1,x}$ is the along-wind fundamental equivalent mass;
$n_{1,x}$ is the fundamental frequency of along-wind vibration of the structure;
$\Phi_{1,x}(z)$ is the fundamental along-wind modal shape.

Peaks of the acceleration levels can then be found by applying a peak factor to this value of $\sigma_{a,x}$.

2.7 Wind – structure interaction effects.

Slender and flexible structures may, under certain circumstances, interact with the wind field. This may lead to excitations of the structure. These effects are partially covered in EN 1991-1-4, in informative annexes.

Vortex shedding. Vortex-shedding occurs when vortices are shed alternately from opposite sides of the structure. This gives rise to a fluctuating load perpendicular to the wind direction. Structural vibrations may occur if the frequency of vortex–shedding is the same as a natural frequency of the structure. This condition occurs when the wind velocity is equal to the critical wind velocity. Typically, the critical wind velocity is a frequently occurring wind velocity indicating that, besides the ultimate stresses, also fatigue, and thereby the number of load cycles, may become relevant.
The critical wind velocity for the bending vibration mode *i* is defined as the wind velocity at which the frequency of vortex shedding equals a natural frequency of the structure or a structural element and is given as:

$$v_{\mathrm{crit,i}} = \frac{b \cdot n_{\mathrm{i,y}}}{St}$$

where:
b is the reference width of the cross-section at which resonant vortex shedding occurs and where the modal deflection is maximum for the structure or structural part considered; for circular cylinders the reference width is the outer diameter;
$n_{i,y}$ is the natural frequency of the considered flexural mode *i* of cross-wind vibration;
St Strouhal number.

The Strouhal number *St* is specified for different cross-sections in EN 1991-1-4. Typical values are 0.18 for circular cross sections, and a range of values from 0.06 to 0.15 for typical sharp-edged sections, see Figure 9.

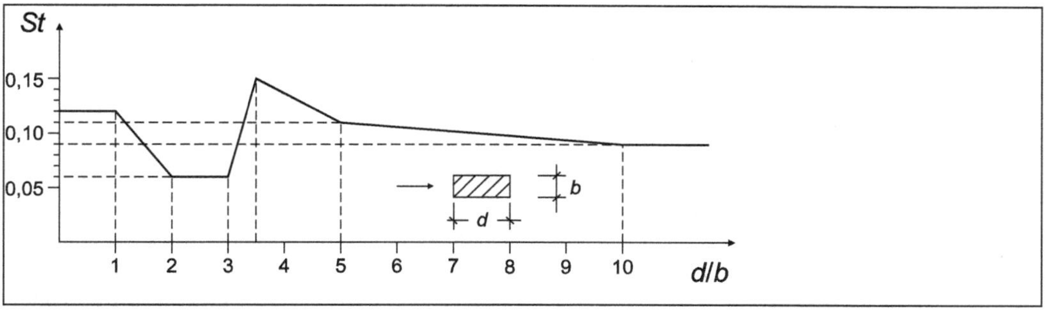

Figure 9: Strouhal number (St) for rectangular cross-sections with sharp corners

The susceptibility of vibrations depends on the structural damping and the ratio of structural mass to fluid mass. This is expressed by the Scruton number Sc, which is given as:

$$Sc = \frac{2 \cdot \delta_s \cdot m_{i,e}}{\rho \cdot b^2}$$

where:

δ_s is the structural damping expressed by the logarithmic decrement;

ρ is the air density under vortex shedding conditions;

$m_{i,e}$ is the equivalent mass m_e per unit length for mode i;

b is the reference width of the cross-section at which resonant vortex shedding occurs.

Two different approaches for calculating the vortex excited cross-wind amplitudes are given in EN 1991-1-4. These two approaches lead to different outcomes for Scruton numbers occurring in practice. These two methods are not described in detail here. The National Annex will have to decide upon which procedure to uses. It may also specify other procedures than the ones proposed in EN 1991-1-4.

Galloping. Galloping is a self-induced vibration of a flexible structure in cross wind bending mode. Non circular cross sections including L-, I-, U- and T-sections are prone to galloping. Ice may cause a stable cross section to become unstable. Galloping oscillation starts at a special onset wind velocity v_{CG} and normally the amplitudes increase rapidly with increasing wind velocity. The onset wind velocity of galloping, v_{CG}, is:

$$v_{CG} = \frac{2 \cdot Sc}{a_G} \cdot n_{1,y} \cdot b$$

where:

Sc is the Scruton number;

$n_{1,y}$ is the cross-wind fundamental frequency of the structure; approximations of $n_{1,y}$ are given in EN 1991-1-4;

b is the width of the structure;

a_G is the factor of galloping instability; if no factor of galloping instability is known, $a_G = 10$ may be used. Typical values are given in EN 1991-1-4.

Divergence and flutter. Divergence and flutter are instabilities that occur for flexible plate-like structures, such as signboards or suspension-bridge decks, above a certain threshold or critical wind velocity. The instability is caused by the deflection of the structure modifying the aerodynamics to alter the loading. Both divergence and flutter should be avoided. EN 1991-1-4 gives procedures below to provide a means of assessing the susceptibility of a structure in terms of simple structural criteria. To be prone to either divergence or flutter, the structure satisfies all of the three criteria given below. The criteria should be checked in the order given (easiest first) and if any one of the criteria is not met, the structure will not be prone to either divergence or flutter:

1. The structure, or a substantial part of it, has an elongated cross-section (like a flat plate) with b/d less than 0,25. Here b is the crosswind dimension, and d is the alongwind dimension of the cross section. Note that for bridge decks, the notations b and d have swapped (e.g. see Figure 12).
2. The torsional axis is parallel to the plane of the plate and normal to the wind direction, and the centre of torsion is at least $d/4$ downwind of the windward edge of the plate, where b is the inwind depth of the plate measured normal to the torsional axis. This includes the common cases of torsional centre at geometrical centre, i.e. centrally supported signboard or canopy, and torsional centre at downwind edge, i.e. cantilevered canopy.
3. The lowest natural frequency corresponds to a torsional mode, or else the lowest torsional natural frequency is less than 2 times the lowest translational natural frequency.

Bridges. EN 1991-1-4 covers bridges of constant depth and with cross-sections as shown in Figure 10 consisting of a single deck with one or more spans. Wind actions for other types of bridges (e.g. arch bridges, bridges with suspension cables or cable stayed, roofed bridges, moving bridges and bridges with multiple or significantly curved decks) may be defined in the National Annex.

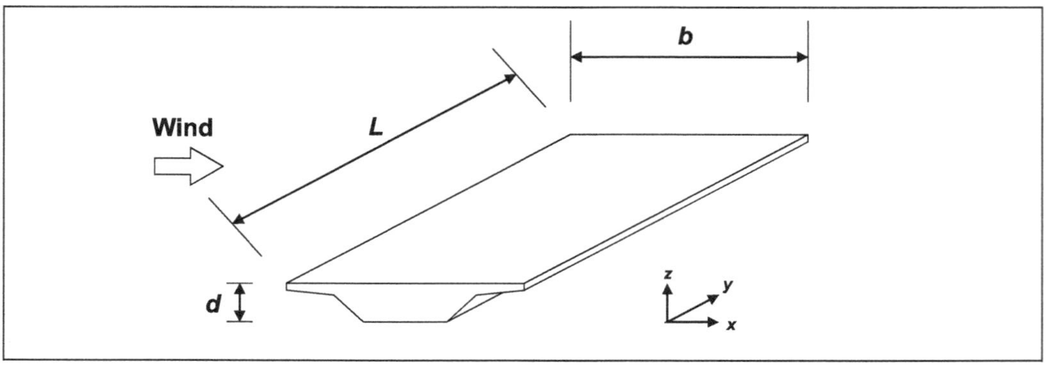

Figure 10: Bridges of constant depth

2.8 National Annex.

There is a large number of clauses in EN 1991-1-4 for which national choice is allowed. The most relevant items are the specification of the wind climate and terrain categories. Others are some pressure and force coefficients, the choice of procedures for dynamic response and vortex excitation. Every country should publish a National Annex in which the values for these clauses should be specified. Without NA, the Eurocodes can not be used. Some of the choices are obligatory, such as the definition of the wind climate. In some other cases, it may be enough to state that a note stays for information only. The National Annex should also provide guidance on the use of the annexes.

3 Application of Wind Tunnel Experiments

3.1 Introduction

Experiments have been essential in the development of current design procedures for wind loads on structures. Design coefficients in codes and guidelines are almost without exception based on wind tunnel experiments. Wind tunnel experiments are also used as alternative for codes of practice in cases outside the scope of these codes, or when it is assumed necessary to obtain the wind loading more precisely. This section deals with the principles and boundary conditions for wind tunnel experiments, as a tool to find wind loads on buildings.
In a wind tunnel, the wind, the building, its surroundings, and in particular cases its behaviour are modeled on scale. It is possible to measure wind velocities, pressures, forces, moments and accelerations. For wind loading studies, measured data are transferred into nondimensional coefficients, such as pressure coefficients. These coefficients can be defined in various ways, using a range of reference wind speeds, and defining different statistical properties.

3.2 Field of application for wind tunnel experiments

Wind tunnel research to determine the wind loading on buildings and building parts is recommended when buildings:

- have shapes that are significantly different from those given in building codes. This may be related to the shape of the plan, but also to the height of the structure;
- are situated in a complex environment, causing interaction effects (leading to reduced and or increasing wind loads). This includes cases where a planned building exists of more than one independent structure (e.g. two towers on a joint lower base building).

Wind tunnel experiments are also used for validation of other methods, e.g. Computational Wind Engineering, or for fundamental research, including the search for data to apply in the next generation of guidance and codification documents.

Wind tunnel experiments are not used for the estimation of:

- internal pressure coefficients;
- friction coefficients;
- the effect of pressure equalisation;

- the dynamic forces on slender structures with limited stiffness, such as cables, bridge decks, flexible roof coverings.

It may be expected that, because of the importance of wind tunnel results in our codes, a general code exists on how to carry out wind tunnel tests. Guidelines on wind tunnel research should give information on how to prepare, to set up, to carry out and to analyse wind tunnel research. A number of guidelines have been developed recently to help those involved with wind tunnel experiments to carry out, analyse and apply wind tunnel data. The most extensive guideline is published by ASCE. Others are those published by WTG (in German) or the guideline of BLWTL, and recently, in the Netherlands, CUR Recommendation 103 (in Dutch).

In this text, a brief overview of all these aspects is given. The reader is referred to the literature list for more detailed information.

3.3 Wind tunnel technique

An atmospheric boundary layer wind tunnel consists at least of the following elements (see also Figure 13):

- One or more ventilators to develop moving air;
- Devices to 'straighten' the flow coming from the ventilator, before it enters the test section. These devices usually contain a contraction, to accelerate the flow, and one or more honeycombs and directional vanes to make the flow low turbulent;
- A working section which is usually adjustable. It contains the model of the building under consideration, and specific features to generate the flow in the atmospheric boundary layer. These features are, seen from upstream:
    - A barrier, or step, at the entrance of the tunnel to generate large scale turbulences;
    - An array of spires;
    - A large fetch of roughness elements to generate a boundary layer flow;
    - A test section, with a turn table, on which the model is placed.
- An outlet of the flow.

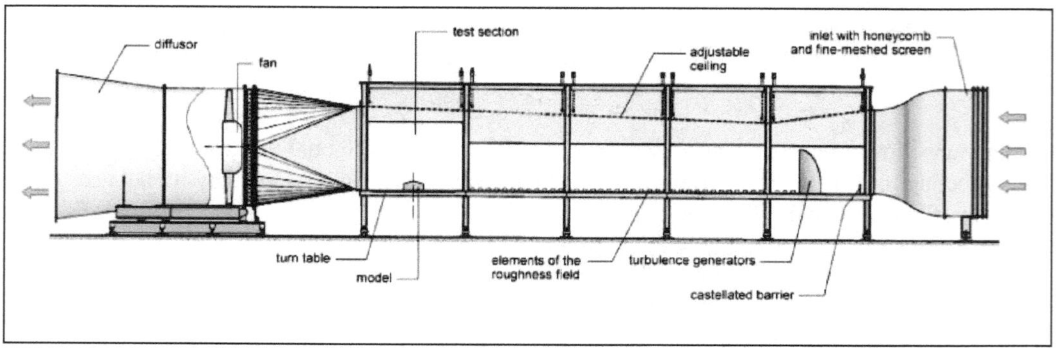

Figure 11: Open section wind tunnel of the Ruhr University in Bochum, Germany

Depending on the type of wind tunnel, the flow is recirculated (closed section tunnel), or connected to the outside world (inside or outside), for an open section tunnel. Open section tunnels can have the fan places before or behind the test section.
Variations in wind tunnel technique may be adjustments to be made to the ceiling, the positions of the roughness elements. At the Boundary Layer Wind Tunnel Lab in London, Ontario, adjustable roughness elements have been installed. The wind tunnel of BRE, in the UK, has walls in the test sections which are partially open, to minimize blockage effects.

3.4 Modelling techniques

Modelling the building and surroundings. A wind tunnel test is carried out on a scale model of the building and surroundings. This scale model is usually custom made for the wind tunnel research. The model has to be suited to mount specific instruments, and is of a geometric scale, that meets the relevant scaling requirements. The geometric scale λ_g is defined as:

$$\lambda_g = \frac{L_{WT}}{L_{FS}}$$

(The index WT stands for Wind Tunnel, the index FS stands for Full Scale)

The geometric scale needs to be specified in the test report. When choosing the geometric scale, the maximum allowed blockage, the amount of detailing required, the effect of nearby surrounding obstacles are relevant. The following minimum demands are specified:

Blockage: The blockage ratio has to be given in the test report. Also, the effects of blockage on the results, and corrections applied. A blockage of 5% is a maximum below which no corrections are needed. In other cases the choices made should be specified in the test report. A special way to treat blockage is the application of slotted walls at the test section. This has been applied at BRE in the UK, see Figure 12.

Figure 12: Interior of BRE wind tunnel, with slotted walls at the side and ceiling of the test section.

Surrounding buildings and obstacles: The effect of obstacles in the direct environment needs to be taken into account in connection with the principle that all loads that are assumed to work on a structure during its lifetime have to be considered. To determine the wind loads on high rise buildings and its components, the following situations are considered (see Figure 13):

1. A wind tunnel test where the direct surroundings are modelled as known at the time the research is carried out. Both local loads and the overall loads on the load bearing structure are determined. If known, future developments should be considered in this test.
2. A second test, where the surrounding buildings on the turn table are taken away. Alternatively, the surrounding buildings may be reduced to a lower height in full scale.

Figure 13: Left: Building with surroundings; Right: Building on empty turn table

Details: The amount of detail of the wind tunnel model depends on the objective of the wind tunnel test. Small details are less relevant for the overall loads. The amount of detail modelled needs to be specified in the test report. Besides the roughness of facades, all significant differences between full scale shape and wind tunnel model should be motivated. This includes features such as parapets, roof overhangs, installations, rounded corners etcetera.

Roughness of surrounding terrain: The roughness length z_0 of the upstream fetch is scaled according to the Jensen Number $Je = h/z_0$. For the determination of wind loads on buildings and components (to check the ultimate limit states), application of a lower value for z_0 is a conservative choice. This leads to the following minimum demand:

$$\frac{z_{0WT}}{z_{0FS}} \leq \lambda_g$$

Usually, in wind tunnel tests, the same roughness length is applied for all wind directions. The value of the roughness length should be chosen which gives the most conservative results. Usually, this is the lowest value. When the wind tunnel results are used to check the serviceability limit state (accelerations of the building), applying a lower roughness length may lead to an underestimation of the fluctuating component of the wind loads. This may lead to an underestimation of the vibration levels. The effect however is usually small.

Modelling the atmospheric boundary layer. The wind in the atmospheric boundary layer varies in time and space. It depends on the terrain roughness, the local wind climate, and on variations in temperature. Usually, the effects of temperature are assumed negligible, when studying wind loads. Relevant are the proper simulation of the wind speed with height (the wind profile), and the turbulent characteristics.
The profile of the mean wind velocity is modeled by applying the Jensen law, or by applying an appropriate exponent in a power law profile. The Jensen law demands that the value of the roughness length in the wind tunnel is geometrically scaled from the full scale value. When the power law is applied, the exponent of the power law should be similar to the value expected in full scale. The shape of the profile is determined on the features installed in the wind tunnel to generate a boundary layer flow with appropriate turbulence. Modern ABL Wind tunnels have a range of profile characteristics available on request.
Minimum demands are specified to the profile of the mean wind with height, and to the specification of the turbulent components. Besides the geometric scale scaling λ_g the wind velocity scale λ_v and time scale λ_t are relevant. These are defined as follows:

$$\lambda_v = \frac{v_{WT}}{v_{FS}}$$

$$\lambda_t = \frac{T_{WT}}{T_{FS}}$$

The frequency *f* is the inverse of time *T*. For the frequency scale this yields:

$$\lambda_f = \frac{f_{WT}}{f_{FS}} = \frac{T_{FS}}{T_{WT}} = \frac{1}{\lambda_t}$$

Wind velocity: The wind velocity which is applied in the wind tunnel, has to fulfill the minimum demands specified by the Reynolds number *Re* and Strouhal number *St*. The Reynolds number is defined as follows:

$$\mathrm{Re} = \frac{vL}{\nu_a}$$

Where:
L is the width of the structure;
ν_a is the kinematic viscosity of air, equal to $1.5.10^{-5}$ m^2/s;
v is the mean wind velocity.

The pressure and force coefficients may depend on Reynolds number. This may be the case for buildings or building parts with rounded shapes. Methods to take Reynolds effects into account are applying various wind speeds in the experiment, or applying roughened surfaces of the building. The wind tunnel institute should report the corrections applied to account for this effect, including the value of *Re* in full scale, the method of scaling this value to the wind tunnel and the method of taking *Re* into account.

Strouhal number: The Strouhal number is defined as follows:

$$St = \frac{fL}{v}$$

To scale frequency, time, length and wind speed appropriately, the Strouhal number is equal in full scale and in the wind tunnel:

$$St_{WT} = St_{FS}, \text{ or: } \frac{1}{\lambda_f} = \lambda_t = \frac{1}{\lambda_v} \lambda_g$$

This scaling demand is relevant to determine:

- the length of the samples in the wind tunnel;
- the sample frequency;
- the model properties when applied in the high frequency force balance.

A wind velocity scaling in the order of 1 to 5 is common for the (West) European wind climate. The wind tunnel velocity has to be given in the test report.

Boundary layer height: The height of the boundary layer in the wind tunnel, at the measurement section, should be high enough that the measured properties represent the full scale situation well. Usually this is achieved when the boundary layer height in the wind tunnel is at least twice the model height.

Reference height: The reference wind velocity at a reference height h_{ref} has to be specified. The position of this reference with velocity v_{ref} has to be chosen so, that the test results lead to reliable predictions of the wind loading. This reference height may be taken equal to the height of the building, when the building is lower than twice the average height of the surrounding buildings. For buildings which are at least three times the average height of the surrounding buildings, the reference height may be chosen between 2/3 and the total building height. This reference height needs to be specified in the test report.

Turbulence: The turbulent characteristics can be represented by the turbulence intensity, the spectral density functions and the correlation lengths in the flow. Turbulence intensity and spectral density functions can be represented in nondimensial form. The requirements is simply that the model scale and full scale values (when presented in non-dimensional form) should be the same. The correlation lengths are represented by so called integral length scales. These are scaled down by the geometrical scale. In most wind tunnels the above demands are not met simultaneously. For wind loading studies, a lower level of turbulence than required, usually leads to higher loads, and is therefore conservative, so the turbulence intensity in the scaled wind climate in the wind tunnel needs to be smaller than or equal to the value in full scale:

$$I_{WT}\,(h_{ref}) \leq I_{FS}\,(h_{ref})$$

3.5 Measurements

As a general demand for measuring velocities, pressures and moments, instrumentation should be applied for which the calibration results are known. In the test reports the measurement techniques applied and the instrumentation used are specified, along with the accuracy of these instruments.

Measuring the simulated boundary layer. To determine the properties of the wind in the wind tunnel, pitot-tubes and hot wire anemometry are appropriate. Hot wire anemometry allows to measure the mean and fluctuating properties of the wind profile. Also, the spectral density of the wind fluctuations are determined. The simulation applied for the atmospheric boundary layer has to be reported. This documentation has to be available on request. The measurement techniques should be specified. The test report needs to specify how the relevant demands are fulfilled.

Measurement of pressures. Pressures are measured as pressure differences between the building surface and a reference pressure. These measurements may be used to determine the local loads on facades and roofs, or to determine the overall loads on the load bearing structure. The minimum demands for pressure measurements are given below. The following minimum demands should be fulfilled:

- The position of the reference pressure is chosen so, that this pressure is independent of wind direction or changes made to the model;
- The frequency-response characteristics of the pressure measurement equipment need to be available on request.

The following additional demands are relevant for local loads:

- A selection of locations is made, at which increased local loads are expected. Usually, these locations are near extremities of facades and roofs;
- Pressure measurements have to be carried out with sufficiently high frequency, that the extreme loads are determined which correspond to the loaded area A_{ref}, as defined in the building codes. The sample frequency is determined according to the Strouhal number.

When applying pressure measurements to determine the wind loads on the overall load bearing structure, additional demands to those given above, are given:

- All surfaces are provided with a large number of pressure measurement points. The number and position of these pressure taps depends on the shape and dimensions of the building (see e.g. Figure 14), and should be motivated in the test report;
- The contribution of friction is not taken into account, when using pressure measurements to determine the overall forces and moments. The friction has to be taken into account by application of the appropriate rules of the building codes;
- When integrating pressures to obtain forces, the individual pressures should be simultaneous to keep the time information. This yields new time series of forces and moments.

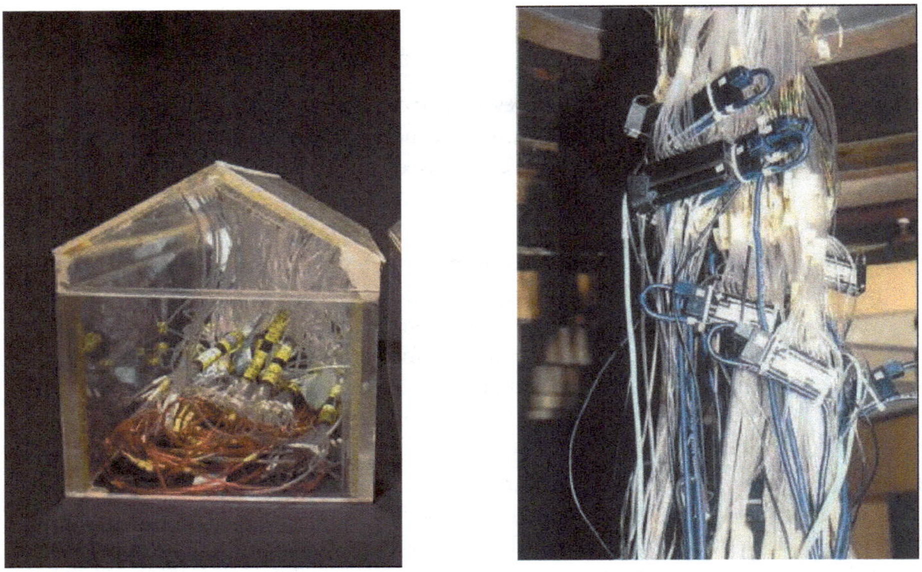

Figure 14: Pressure transducers inside a model (left), or under the wind tunnel turn table (right) (picture left from Benoit Parmentier, WTCB, Belgium)

Forces and moments from static balance measurements. A so called static force balance is able to measure mean values of forces and moments only. This method gives three values for force; F_x, F_y and F_z, and three values for the moment M_x, M_y, M_z. When applying a static force balance measurement, the effect of wind friction is assumed to be implicitly taken into account.

Forces and moments from high frequency force balance measurements. A dynamic, or high frequency force, balance is applied to measure time series of the wind loading. When applying a high frequency force balance measurement, the effect of wind friction is assumed to be implicitly taken into account. Measurements with a high frequency force balance do not include the effect of resonance of the structures. The natural frequency of the model applied should be at least 2 times the value of the highest frequency of interest for the measurement. This frequency should be determined using the Strouhal number. This should be motivated in the test report.

Sample length, number of samples and sample frequency. To determine the statistical properties (mean, maximum, minimum and root-mean square values), one or more time series should be measured, for every wind direction and for every measurement channel, per configuration examined.

When only mean values are of interest, one time series is needed per wind condition, with a length chosen so, that a longer time series does not give another mean value. When applying extreme value analysis, more than one time series is needed with a certain length *T*. Every time series is measured with the same sample frequency. The following minimum demands are relevant, related to the extreme value analysis, which is applied:

- The time series are of the same length;
- There is no overlap in the time series (time series are independent);
- The time series each represent a full scale duration of at least 60 seconds, which should be determined using the Strouhal number. There are various theories in the literature on this demand, check e.g. (ASCE, 1999) or (Cook, 1986) for more information.
- At least 24 time series are required per wind direction;
- The sample frequency is at least twice the value of the frequency that is of interest for the wind effect studied. This frequency should be determined using the Strouhal number.

The measurement method applied should be described in the test report, together with the number of time series, sample frequency and sample length.

Analysis of measured data. There is a wide range of analysis methods available to obtain structural loads from wind tunnel experiments. These can roughly be divided in two. When only time averaged information is available, analysis of the results may be based on the quasi-steady assumption. This means that fluctuations in the pressures can adequately be represented by the fluctuations in the oncoming flow. In that case, mean pressure or force coefficients may be used to calculate the extreme wind loads. This assumption is not valid for all situations covered.
The second group of analysis methods deals with extreme value analysis. In that case, a minimum number of time series with sufficient length should be available, of the property needed, e.g. pressures or forces. The peaks in of each time series are analysed according to an extreme value method. A consistent theory on this analysis method has been derived by Cook and Mayne, and values obtained with this theory have been applied in the Eurocode. The detailed background of these procedures is described in the literature, start with (Cook, 1986), and use the reference list of his book.

4 Application of Full Scale Experiments

4.1 Introduction

Wind tunnel experiments have limitations which may be solved by experiments in full scale. Wind tunnel research means scaling down the sizes of building and flow properties. In those cases where these small scales become important, full scale data provide necessary information. Such cases include the pressure difference over permeable façade and roof systems, and the local loads on small areas. A second reason where full scale experiments are still usefull is when wind tunnel or other simulations need to be validated and verified. A third group of 'full scale measurements' consists of observations of severe motions of structures under wind loading, sometimes with catastrophic results. These observations are never set-up as an experiment on an object of study, however, many lessons have been learned from e.g. the movie of the Tacoma Narrows Bridge collapse, or the films that where made of cable swinging.

Full scale experiments are limited with respect to:

- the availability of sufficient wind conditions;
- the large amount of time needed to collect sufficient data to cover the items of interest
- the effects of surrounding terrain;
- variations in pressure tap set up or building configuration;
- instrumentation;
- costs involved.

Full scale experiments are generally not suited to predict the wind loading for a specific building in a specific surrounding. That is the domain of wind tunnel experiments or numerical models. Also, parametric studies are better carried out in laboratory circumstances. In this section, a general description is given of the techniques involved when using full scale experiments to obtain wind loading information. Some examples are described of experiments of which many information can be obtained in the open literature.

4.2 Full scale experimental Techniques

The wind loading on structures is expressed in various ways, like pressures, forces, moments but also as accelerations or deformations. In all cases, the relation between these aspects is expressed in coefficients, relative to the wind velocity. Experiments to investigate wind-induced pressures in full scale, are generally set up according to the schedule in Figure 15.
Examinations of the wind field (expressed in figure 17 with u,v,w) are made to analyse the reference conditions during the measurement of the wind-induced pressures. Wind-induced pressures are measured with differential pressure transducers (T). One side of the pressure transducer is connected by flexible tubing (t) with a pressure tap (p) in the facade; the other side is connected to a reference pressure (R). Analogous to this picture, instead of pressures, forces or accelerations can be measured. Data are recorded using an analogue-to-digital converter (ADC). Data analysis is usually done afterwards.

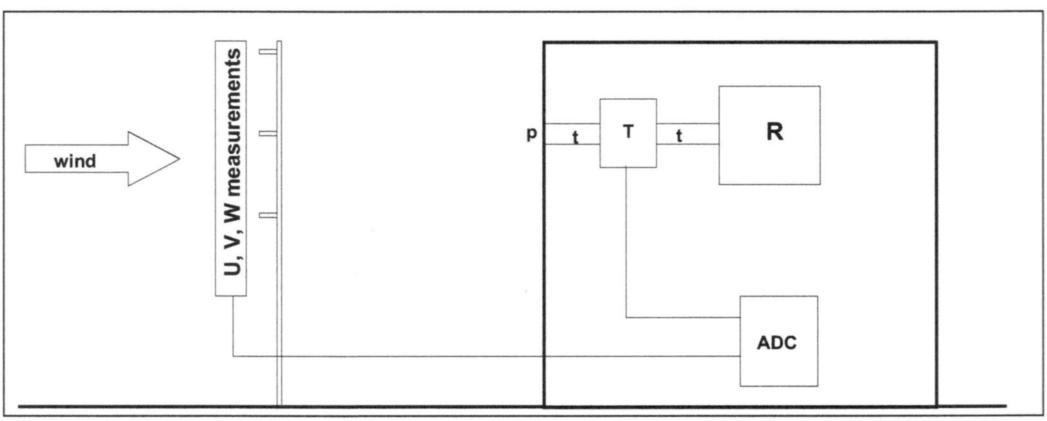

Figure 15: Schedule of pressure measurements in full scale

Wind field measurements. The mean and turbulent quantities of the undisturbed upstream wind are measured outside the direct influence of the test building (i.e. outside the recirculation and acceleration zones of the flow). However, the instruments must be close enough to the building to measure flow characteristics that are representative for the measured wind loads. A distance about three times the building height upstream is sufficient to fulfill these requirements. The measurement height should be at least the roof height of the test building. Building height is used in many experiments and is a frequently used reference in building standards, so it is easy to compare results from different references.

Measurements of the flow closer to the building and measurements downstream the test building are influenced by the building itself. Measurements of the flow with the instrumentation on top of the test building are only possible at a height, outside the zone where separation and acceleration of the flow occurs. This is in the order of at least one to two times the building height above the building. This is however not very easy to build and maintain.

Reference velocity measurements can be made in full scale with relatively cheap cup-vanes and/or more expensive ultrasonic anemometers. These instruments should be placed free from the mast, on which they are mounted. The World Meteorological Organisation gives a guideline for the distance of the instruments to the tower: '*...wind sensors should preferably be located on top of a solitary mast. If side mounting is necessary, the boom length should be at least 3 times the mast width...*'.

In past full scale experiments, where wind velocities and wind-induced pressures have been measured, different set ups have been used. Masts are put on top of the test building, on a neighbouring higher building, upstream at a location, free from local obstacles. In some experiments more measurement positions are used.

In full scale it is usually not possible to analyse the pressures on all sides of a building and for all wind directions with only one position where the reference velocity is measured, since the measurements will be in the influence area of the building itself for some wind direction. A possible solution is to install more then one reference positions for the wind velocity. Another solution is to place the building on a turn table, to obtain the wind effects required for any wind direction available. Such a special experiment was designed at Texas Tech University (TTU). The TTU building is placed on a large turn table, to obtain measurements independent of wind direction. This is however only possible only for relative small scale experiments (e.g. low rise buildings), under laboratory conditions. The costs involved are a relevant issue.

For the design of structures, the wind loading with a mean return period typically in the range between 15 and 100 years are relevant. Those wind speeds usually are characterized by a wind profile which is not effected by thermal convection (i.e. neutral atmosphere). Full scale experiments usually have a much shorter period. However, it is important to perform these measurements under atmospheric circumstances similar to those under extreme conditions. Periods with very low wind velocities are usually non-neutral. Therefore, threshold velocities are applied, which are typically in the order of 5 to 10 m/s.

Wind tunnel measurements can be turned on and off on request, but in full scale, the wind can not be delivered on request. Automatic data acquisition procedures are necessary, to obtain the data without the need for persons to interact with the measurements. Many full scale experiments require several years of data acquisition before sufficient data are available.

Measurement of wind induced pressures. Measurements of pressures in full scale are based on the same principle as in the wind tunnel: Differential pressure transducers measure the difference between the pressure on the surface and a reference pressure, preferably the ambient pressure. The techniques used are similar, however the practical consequences differ.

Instantaneous wind-induced pressures for mean wind velocities up to 25 m/s are in a range between 50 and 1000 Pa. Differential pressure transducers are available in several types: capacity cells, strain gauge or inductive transducers. In recent experiments, transducers are used of the strain gauge type. Pressure transducers are mounted flush on the facade or connected with pressure taps by flexible tubing. Pressure transducers tend to drift in time. Full scale measurements are usually done over a long period and data acquisition is done automatically. Drift may therefore be an issue during the measurement period, in contrary to wind tunnel experiments. Therefore it is recommended to install an automatic calibration in the measurement configuration. A schematic diagram of this technique is provided in Figure 16.

When the pressures are measured, the ports 1 and 2 are open and 3 is closed. After or before a measurement the zero-pressure calibration is obtained by closing 1 and opening 3 automatically, so that both sides of the transducer are connected to the reference pressure.

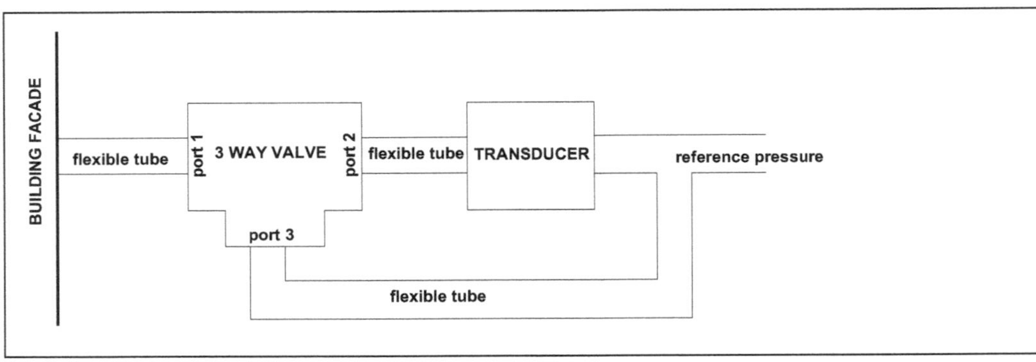

Figure 16: Pressure tap configuration in a full scale experiment

Levitan (1993) gives an overview of systems used in past full scale experiments and designs the reference pressure system for the TTU building. He gives three types of reference pressure systems: atmospheric static pressure systems, internal pressure systems and constant pressure systems.

In full scale tests on low-rise buildings in an undisturbed flow field, reference pressures can be obtained from the atmospheric static pressure upstream of the building. The pressure measured is p_s-p_a. Several methods have been used, like ground cavities, static pressure tubes on directional vanes, or fixed static pressure probes.

For buildings in a built-up area and for high-rise buildings, an undisturbed measurement of p_a is practically impossible. Internal pressures are used, the measured pressure is p-p_i. The internal pressure is not equal to the ambient pressure and corrections of the measured value of wind-induced pressures are needed.

In a constant pressure system, the reference pressure is provided by an insulated tank, which is placed in a protected location. The measured pressure is p_s-p_{tank}. A slight change in temperature of the air inside the tank influences the pressure coefficient unacceptably. The pressure in the tank is usually not equal to the ambient pressure.
Pressure transducers are placed behind the surface, connected by flexible tubing to a pressure tap in the surface. Pressure taps have a diameter in the order of 6 to 10 mm in full scale. Rain penetration can be a problem, especially for roof taps. This can be overcome using a buffer, where the water is saved, which is applied in the TTU building.
In full scale, the frequency response of the tubing system is usually no problem up to 10 Hz, when tubes with a maximum length of 1 metre are used.

Measurements of forces and accelerations. Full scale experiments to the wind loading on buildings often include a direct measurement of the forces in the structure, or accelerations at the top levels of the building. Such measurements are most common on high rise buildings. In the 1960's and 1970's, quite a number of such measurements have been performed. These measurements have lead to empirical values for structural damping and natural frequencies in building codes. Many experiments have been accompanied by pressure measurements.
Usually, high rise buildings are situated in complex surroundings. The measurement of an undisturbed reference velocity could be a problem in such case. Often, the reference velocity is measured on top of the same building.

Data acquisition. It depends on the objectives of the research project, which sample frequency and record length is required. A high sample frequency restricts the number of channels or the record length, because of limitations in data storage and file handling. Typical sample frequencies used in full scale are between 5 and 20 Hz. This is usually sufficient to catch the relevant time scale for local loads for relatively small surfaces.
Full scale record lengths are typically between 10 and 60 minutes. Full scale data recording is often done automatically. When the wind velocity exceeds a predefined limit, a record starts. Nowadays, computers are no restriction to the amount of data that can be stored. Earlier experiments, until the late 1990's, however, often were not suited to store all data.

Data analysis. The analysis of data from full scale measurements usually starts with an evaluation of the quality of the data measured. Mean wind speed, and wind direction are usually relevant, but also a check on the stationarity of the data is required.
The time series is called stationary if the statistical characteristics are independent of the starting point of the time series. A stationary signal is assumed representative for the whole time domain. For stationary time series, the mean and standard deviation do not change with time. There is a range of criteria for stationarity, which are more or less severe. Tests to determine whether a signal is stationary are given in (Bendat et.al., 1986). In full scale, stationary signals hardly occur.

4.3 Possible risks, problems etcetera

Full scale measurements are very time consuming en relatively expensive. Many problems may occur during the running of the experiments. Problems which are known to occur are insects or sand creeping into the pressure taps and closing the active sides. Also, condensation of moist air can be a problem. Human behaviour may also play a role. Some of the risks involved are described briefly below:

Change of terrain properties. Experiments, such as the cases described in chapter 3, are in place for a relatively long period. During this period, new buildings can be planned and built in the direct vicinity, or cities may become larger, so that the roughness of the upstream fetch increases. Also, seasonal variations may be important, e.g. the presence of snow, or the difference between summer and winter vegetation. such variations should be monitored.

Occurrence of wind directions. The wind in a full scale experiment cannot be predicted in advance. For wind loading studies, normally wind field data above a certain threshold are needed. Usually a range of wind directions is of interest. It requires time to have enough data of sufficient wind speed level, at all required wind directions. Placing the building on a turn table may help, but in that case, the upstream terrain conditions differ per measurement.

Water, sand and insects. Pressures are measured using pressure taps, usually holes in the order of 6 to 10 mm, in the face of the building studied. These taps are connected to the pressure transducers by flexible tubing. These holes may catch drops of rain, or humid air may condensate in the flexible tubes. Also, insects may find it an interesting place to creep in. Also, fine sand particles, e.g. in snow, may settle in the pressure tap.

Reference pressure. The reference pressure is very sensitive to the position and way it is measured, and often basis for discussion. When there is a flat upstream fetch, the reference pressure could be measured flush with the terrain. In complex terrain, it is hardly possible to find a nicely undisturbed position. A comparative wind tunnel research would help establishing a measure of the error made. When the project is focussing on pressure fluctuations, or on pressure differences, the reference pressure is less important.

Human behaviour. Human beings can in various ways influence the measurements. Of course, people directly involved may make mistakes. Other experiences include theft of instrumentation, unplanned influencing the measurements by the inhabitants of the test building.

Lightning. Outdoor measurements of wind field characteristics usually require a position of the equipment well above the surrounding structures. This makes such a measurement vulnerable for lightning. This should be accounted for, by installing a safety system, or connecting the mast to an existing safety system.

4.4 Examples

In this section, some examples, which are extensively described in the free literature, are described very briefly. For details, the reader is referred to the reference list.

Commerce Court Building, Canada. During the 1970's, the Commerce Court Building, a 57 storey (239 m high) building in Toronto, had been used as a full scale experiment to investigate wind induced pressures and responses. These measurements have been compared with wind tunnel tests and have served as a valuable basis for further development of both experimental techniques and building codes.
Pressures have been measured at 32 locations around the exterior of the building on four levels. An internal pressure was used as reference. Wind measurements where made on a mast on the roof at 285 metres height. Sample frequency was 20 Hz, for both wind data and pressures. Both mean, rms and peak pressures were measured and analysed. W.A. Dalgliesh has published a number of papers on this project.

Texas Tech University:. The Wind Engineering Research Field Laboratory (WERFL) is a major facility for conducting wind engineering research. This facility is used to conduct full-scale measurements of wind and its effects on buildings. It consists of a 160-ft high meteorological tower, a rotatable 30 x 45 x 13 ft test building, and a 10 x 10 x 8 ft data acquisition room. Figure 7 shows the facility with the tower and test building. The meteorological tower is equipped with wind-speed anemometers, wind direction and temperature sensors, and barometric pressure and relative humidity measurement instruments. These instruments provide a complete wind characterization at the field site. The test building can be rotated in a full circle on a track that is embedded in a concrete pad. The wind-induced pressures at different locations on the roofs and walls of the building are recorded by pressure taps. The pressure, wind speed and wind direction data are recorded automatically by a network of 15 computers to monitor and automatically acquire when specified conditions are met. All data is acquired at 30Hz. Wind Engineering plans to use a high-performance, high-speed networking backbone to make data from the Wind Engineering Research Field Laboratory available in near real-time across the Texas Tech campus and over the internet.

Figure 17: From left to right: Commerce Court Building, Texas Tech University building, Eindhoven University of Technology building

Eindhoven University of Technology. Since 1994, measurements of wind and wind induced pressures have been performed on the campus of Eindhoven University of Technology. Eindhoven is situated in the south of the Netherlands, surrounded by flat terrain. Most of the experiments have been carried out on the main building of the university. This building has dimensions height

$H = 44.6$ metres, width $W = 167$ metres and depth $D = 20$ metres. The building has an exact north-south orientation so that the long facades are the western and eastern facade. The fetch of the building in the prevailing (western) wind directions consists of trees up to 12 metres height and low-rise buildings (maximum height 10 metres), with a few taller buildings in between. Studies have been performed and published on the fluctuating pressures on building facades (Geurts, 1997), Driving rain (Van Mook, 2001) and pressure equalization (Suresh Kumar, 2002, Li et al, 2004). This experiment is still running.

5 Concluding remarks

This chapter has given relevant principles and insights for the use of the Eurocode, of wind tunnel experiments and full scale data to obtain wind loads on structures. The Eurocode, as any other code, can never be representative of all structures to be built. Therefore, experiments, such as wind tunnels and full scale data will be necessary in future.

Since there is a range of demands, often conflicting, when setting up an experiment, there are still discussions going on about the optimal way to perform wind tunnel experiments. When wind tunnel experiments are commissioned, expert judgement of the modeling and analysis of the results is still an important issue. Experience shows that differences in resulting values of 20 % between wind tunnels are not uncommon, although methods applied, using the same measurements, may all meet the relevant demands.

Full scale testing has been the basis for validation of wind tunnel modelling since the beginning of wind engineering. The Journal of Wind Engineering and Industrial Aerodynamics and the Journal of Wind and Structures frequently contain papers on full scale measurements. Full scale measurements all represent a single case (under local circumstances, for the local wind climate and for the specific building under study). These case studies are very important for validation of wind tunnel testing, and, even more important, are very useful for studies into local effects, which can not be covered by wind tunnel data.

6 Bibliography

Akins, R.E, Cermak, J.E. (1976). *Wind pressures on buildings, Report ENG72-04260-A01, ENG76-03035*, Colorado State University, USA.

ASCE (1999). Wind Tunnel Studies of Buildings and Structures, Manual of practice no. 67, American Society of Civil Engineers.

Beeck, J. van, Corieri, P., Parmentier, B., Deszo, G. (2004). Full-scale and wind tunnel tests of un steady pressure fields of roof tiles of low rise buildings. In: *Proceedings of the Cost C14 International Conference on Urban Wind Engineering and Building Aerodynamics, VKI, Rhode St.-Genese*, pp. D1.1.

Bendat, J.S., Piersol, A.G. (1986). *Random data: Analysis and measurement procedures, second edition (revised and expanded),* John Wiley and Sons

CEN (2005). EN 1991-1-4: Eurocode 1: Actions on structures – Part 1-4: General Actions – wind actions

Cook, N.J. (1986): *The Designer's Guide to Wind loading of Building Structures, Part 2 Static Structures*, Butterworths,

CUR (2005). Windtunnelonderzoek voor de bepaling van ontwerp-windbelasting op (hoge) gebouwen en onderdelen ervan. *CUR Aanbeveling 103* (in Dutch)

Dalgliesh, W.A. (1970). Experience with wind pressure measurements on a full-scale building, In: *Wind Loads on Buildings and Structures, NBS building series 30, Proceedings of Technical Meeting concerning Wind Loads on Buildings and Structures* (ed. R.D. Marshal)

Dalgliesh, W.A. (1975). Comparison of model/full-scale wind pressures on a high-rise building, *Journal of Industrial Aerodynamics, no. 1*, pg. 55-66

Dalgliesh, W.A., Templin, J.T., Cooper, K.R. (1979). Wind tunnel and full-scale building surface pressures, Wind Engineering, In: *Proceedings of the Fifth International Conference*, Fort Collins, USA, pp. 553-565

Dalgliesh, W.A., (1982). Comparison of model and full scale tests of the commerce court building in Toronto, In: *Wind tunnel modelling for civil engineering applications, Int. workshop, Gaithersburg*, Cambridge University Press, 575-589

Dalgliesh, W.A., Cooper, K.R. Templin, J.T,. (1983). Comparison of model and full-scale accelerations of a high-rise building, *Journal of Wind Engineering and Industrial Aerodynamics*, no. 13, pg. 217-228

Geurts, C. (1997). *Wind-induced pressure fluctuations on building facades*, PhD thesis, Eindhoven University of Technology, 1997, 256 pg.

Geurts, C.P.W. (1997). Naturmessungen und Modellsimulation des Windfeldes über einer Vorstadt, In: *Windkanalanwendungen für die Baupraxis, WTG Berichte Nr. 4*, pp. 7-25 (In German)

Geurts, C.P.W. et.al. (2001). Transparency of pressure and force coefficients, In: *Proceedings of 3EACWE*, pp. 165-172

Geurts, C.P.W., Bouma, P.W., Aghaei, A. (2005). Pressure equalisation of brick masonry walls, In: *Proceedings of 4EACWE, Prague*, published on CD

Holmes, J.D. (1976). Pressure fluctuations on a large building and along wind structural loading, *Journal of Industrial Aerodynamics 1*, pp. 249-278

Holmes, J.D. (1995). Methods of fluctuating pressure measurement in wind engineering, In: *A state of the art in Wind engineering, Davenport sixtieth birth anniversary volume*, pp. 26-46

Holmes, J.D. (2001). *Wind loading of Structures,* Spyon Press London, New York

Hunt, J.C.R., Kawai, H., Ramsey, S.R., Pedrizetti, G., Perkins, R.J. (1990). A review of velocity and pressure fluctuations in turbulent flows around bluff bodies, *Journal of Wind Engineering and Industrial Aerodynamics*, vol. 35, pp. 49-85

Kawai, H. (1983). *Wind pressure on a tall building*, PhD Thesis, (In Japanese)

Kiefer, H., Plate, E. (2001). Local wind loads on buildings in Built-up Areas – comparison of full scale results and wind tunnel data. In: *Proceedings of 3EACWE, Eindhoven*, pp. 91-98

Letchford, C.W., Sandri, P., Levitan, M.L., Mehta, K.C. (1992). Frequency response requirements for fluctuating wind pressure measurements, *Journal of Wind Engineering and Industrial Aerodynamics*, no. 40, pg. 263-276

Letchford, C.W., Iverson, R.E., McDonald, J.R. (1993). The application of the quasi-steady theory to full scale measurements on the Texas Tech building, *Journal of Wind Engineering and Industrial Aerodynamics*, no. 48, pp. 111-132

Levitan, M.L., Mehta, K.C. (1992). Texas Tech field experiments for wind loads, I: Building and pressure measurement system, II: Meteorological Instrumentation and terrain parameters, *Journal of Wind Engineering and Industrial Aerodynamics*, no. 41-44, pg. 1577-1588

Levitan, M.L. (1993). *Analysis of reference pressure systems used in field measurements of wind loads*, PhD. thesis, Texas Tech University 156 pg.

Li, C., De Wit, M. (2004). Pressure equalization: The full scale experiment in Eindhoven. In: *Proceedings of the Cost C14 International Conference on Urban Wind Engineering and Building Aerodynamics, VKI, Rhode St.-Genese*

Littler, J.D. (1993). An assessment of some of the different methods for estimating damping from full-scale testing, In: *Wind Engineering, 1st IAWE European and African Conf. On Wind Engng.* pp.209-219

Matsui, G., Suda, K., Higuchi, K. (1982). Full-scale measurement of wind pressures acting on a high-rise building of rectangular plan, *Journal of Wind Engineering and Industrial Aerodynamics* 10, pg. 267-286

NBS (1982). *Wind tunnel testing for Civil Engineering Applications*

Ng, H.H.T. (1988). *Pressure measuring system for wind-induced pressure on building surfaces*, Master's Thesis, Texas Tech University

Ng, H.H.T., Mehta, K.C. (1990) Pressure measuring system for wind-induced pressures on building surfaces, *Journal of Wind Engineering and Industrial Aerodynamics* no 36, pp. 351-360

Ohkuma, T., Marukawa, T., Niihori, Y., Kato, N. (1991) Full-scale measurement of wind pressures and response accelerations of a high-rise building, *Journal of Wind Engineering and Industrial Aerodynamics*, no 38, 185-196

Okada, H., Ha, Y-C. (1992). Comparison of wind tunnel and full scale pressure measurement tests on the Texas Tech Building, *Journal of Wind Engineering and Industrial Aerodynamics* 41-44, pp. 1601-1612

Parmentier, Benoit, Schaerlaekens, Steven, Vyncke, Johan (2001). Net pressures on the roof of a low-rise building Full-scale experiments. In: *Proceedings of 3EACWE*, Eindhoven, pp. 471- 478.

Parmentier, Benoit, Schaerlaekens, Steven, Vyncke, Johan (1999). A Belgian research program to determine the net forces on rooftiles, In: *Proceedings of the 10th ICWE conference*, Copenhagen.

Parmentier, B., Hoxey, R., Buchlin, J.M., Corieri, P. (2002). The assessment of full-scale experimental methods for measureing wind effects on low rise buildings, In: *Proceedings of the first workshop of Cost C14 'impact of wind and storm on city life and built environment'*, Nantes, 2002, pp. 91-103 (includes a large reference list on the full scale measurements on low-rise buildings)

Peterka, J.A. (1982). Selection of local peak pressure coefficients for wind tunnel studies of buildings, *Journal of Wind Engineering and Industrial Aerodynamics*, 13, pp. 477-488

Richardson, G.M., Surry, D. (1992). The Silsoe Building: a comparison of pressure coefficients and spectra at model and full-scale, *Journal of Wind Engineering and Industrial Aerodynamics*, vol. 41-44, pp. 1653-1664

Rijkoort, P.J. (1983). *A compound Weibull model for the description of surface wind velocity distributions report WR 83-13,* KNMI

Robertson, A.P., Glass, A.G. (1988). *The Silsoe Structures Building - its design, instrumentation and research facilities, Div. Note DN 1482, AFRC Inst. Engng Res.*, Silsoe, 59 pg.

Sharma, R. (1996). *The influence of internal pressure on wind loading under tropical cyclone conditions*, PhD thesis, University of Auckland

Sill, B., Cook, N.J., Fang, C. (1992). The Aylesbury Comparative Experiment: A final report, *Journal of Wind Engineering and Industrial Aerodynamics*, 41-44, pp. 1553-1564

Simiu, E., Scanlan, R.H. (1996). *Wind effects on structures, third edition* Wiley and Sons, New York

Snaebjornson, J.T. (2001). Spectral characteristics of wind induced pressures on a model- and a full-scale building. In: *Proceedings of 3EACWE*, Eindhoven, pp. 333-340

Snaebjörnsson, J.T., Geurts, C.P.W. (2006). Modelling surface pressure fluctuations on medium-rise buildings, *Journal of Wind Engineering and Industrial Aerodynamics*, 94, pp. 845-858

Stathopoulos, T. (1979). *Turbulent wind action on low rise Structures*, PhD Thesis, Univ. of Western Ontario

Suresh Kumar, K., Stathopoulos, T. and Wisse, J.A. (2002). Field Measurement Data of Wind Loads on Rainscreen Walls, *Journal of Wind Engineering and Ind. Aerodynamics*.

Tieleman, H.W., Surry, D., Mehta, K.C. (1996). Full/model-scale comparison of surface pressures on the Texas Tech experimental building, *Journal of Wind Engineering and Industrial Aerodynamics,* no. 61, pp. 1-23

University of Western Ontario Boundary Layer Wind Tunnel Laboratory. *Wind tunnel testing, a general outline*

Van Mook, F. (2002). *Driving rain on building envelopes*, PhD Thesis, Eindhoven University of Technology (check: http://fabien.galerio.org/teksten/fjrvanmook2002.html)

Wind Technologische Gesellschaft (1995). WTG Merkblatt über Windkanalversuche in der Gebäudeaerodynamik.

Xie, Irwin, (1999). Wind load combinations for structural design of tall buildings, In: *Wind Engineering into the 21st Century*, Balkema, pp. 163-168.

Introduction to the Prediction of Wind Loads on Buildings by Computational Wind Engineering (CWE)

Jörg Franke

Department of Fluid- and Thermodynamics, University of Siegen, Germany

Abstract. In this chapter the numerical prediction of wind loads on buildings as a branch of Computational Wind Engineering (CWE) is introduced. First the different simulation approaches are described with their corresponding basic equations and the necessary turbulence models. The numerical solution of the systems of equations is sketched and the most important aspects and their influence on the computational results are highlighted.

1 Introduction

Computational Wind Engineering (CWE) is the usage of Computational Fluid Dynamics (CFD) for the solution of problems encountered in wind engineering. Typical application examples are the prediction of wind comfort, pollution dispersion and wind loading on buildings, which is the main topic of this chapter. The loading is a result of the pressure distribution on the building or structure in general. The variation of the pressure is determined by the flow field around the structure which itself depends on the shape of the structure and its immediate surroundings, and on the approach flow characteristics. In structural engineering this dependence is described with the first three links of the wind load chain (e.g., Dyrbye and Hansen, 1997). The first link determines the regional wind climate of the site from meteorological data. The second link describes the conversion of these data in the profile of the wind at lower heights, which is determined by the terrain surrounding the structure. The transformation of the wind profile into the pressure distribution forms the third link. These last two links are increasingly examined by means of CFD and reviews for the simulation of the flow over complex terrain and the computation of pressures are available (e.g., Stathopoulos, 1997; Stathopoulos, 2002; Bitsuamlak et al., 2004). With these application reviews available the present chapter tries to focus on the basics of CFD and therefore addresses novices to this field of wind engineering. The presented material is of very general nature, biased by the author's experience.

1.1 Outline of a CFD simulation

Like in all other applications of CFD, knowledge on several ingredients of the problems to be evaluated is required. First of all the user has to have knowledge of the area of application, here wind loading on buildings. Secondly, she or he must be aware of the assumptions made in describing the physics by a mathematical model. And finally the influence of the numerical approximations on the solution should be known.

In Figure 1 a typical flow chart of the numerical solution of an engineering problem by means of CFD is shown. First one has to decide which mathematical equations should be used to describe the physical problem. For these equations boundary and initial conditions are necessary, which are ideally available from measurements. Next one has to decide about the domain in which one wants to compute the flow field. The size of this computational domain is determined by the knowledge of the flow conditions on the boundaries and by the available resources of computer hardware, man power and time. Inside the computatonal domain and on its boundaries a grid then has to be generated, which determines the discrete locations at which the flow is computed. On this grid the basic system of partial differential equations is discretized with the aid of several numerical approximations and transformed into a non linear algebraic system of equations that can be solved by the computer. After analyzing the computed solution one has to decide whether it is necessary to modify one or more of the previous choices, e.g. to use different equations or boundary conditions, a larger or smaller domain, a different mesh or different numerical approximations. These choices depend on the intended outcome of the simulation, indicated by the terms *verification* and *validation* in the arrow with broken lines. While verification deals with mathematical aspects of the solution, validation denotes the comparison of the simulation results with experimental data. If the solution is regarded as appropriate, concerning the physical and numerical parameters, it is used for further analysis, interpretation and finally the solution of the initial engineering problem.

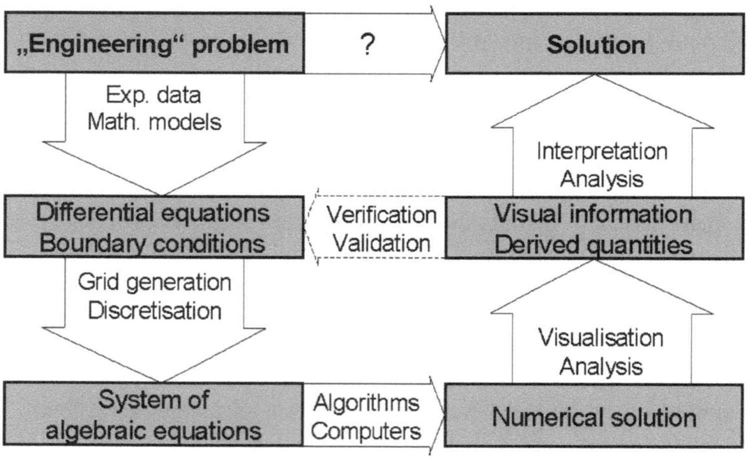

Figure 1. Flow chart of a CFD analysis (after Schäfer, 1999).

In every step of the solution process errors and uncertainties are introduced by the several approximations that are made. These can be most generally grouped into two broad categories.

- Approximation of reality by a system of partial differential equations with corresponding initial and boundary conditions.
- Approximation of a solution to this initial-boundary value problem by numerical computation.

The user of a CFD code must choose between different approximations, in the following termed models. With regards to the above two categories the most important models specific for CFD are the

- models for the physics (turbulence model, initial and boundary conditions, geometry) and the
- models for the numerics (discretisation in space and time, iterative convergence of solution).

An appropriate choice of the user can only be based on knowledge of the physical and mathematical basics of the different models. This knowledge is most efficiently transferred by Best Practice Guidelines which are available for industrial CFD in general (Casey and Wintergerste, 2000) and also for loading predictions (Tamura et al., 2006) and other aspects of CWE (Mochida et al., 2002; Scaperdas and Gilham, 2004; Bartzis et al., 2004; Tominaga et al., 2004; Franke et al., 2004; VDI, 2005; Yoshie et al., 2005). When entering the field of CWE it is highly advisable to consult these guidelines and recommendations.

1.2 Outline of the chapter

As stated above, this chapter is mainly intended as a basic introduction for an audience which is new in the field of CWE and wants to use CFD for wind loading predictions. To that end it is structured in the following way. In section 2 the different simulation approaches are described with their corresponding basic equations. Emphasis is put on the physical meaning of the solutions obtained. Furthermore the unknown quantities that require physically based approximations are introduced. The most common models for these quantities, known as turbulence models, are described for each simulation approach. Section 3 presents a short overview of the solution of the corresponding equations by means of the Finite Volume method, which is the most common approach in commercial CFD codes. Again the influence of the necessary numerical approximations on the solution are discussed. Their principal influence is the same with every other numerical approximation, e.g. Finite Difference or Finite Element methods. In section 4 the computational domain, the boundary conditions and the computational grid are discussed and conclusions are drawn in section 5.

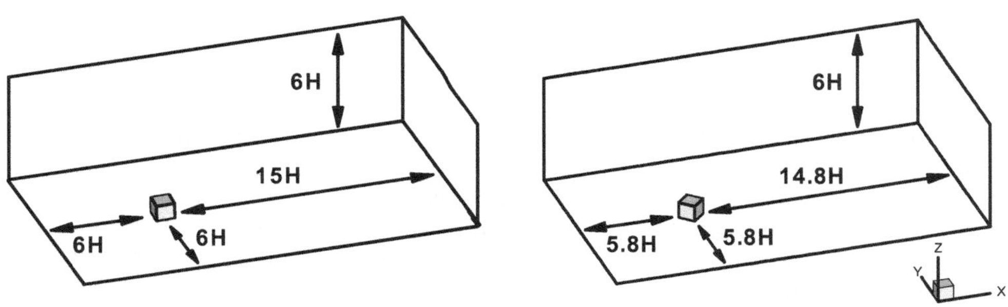

Figure 2. Computational domains for the flow simulation around the Silsoe cube. Left: 0° approach flow. Right: 45° approach flow.

No thorough review of CFD applications for wind loading is made, but simulation results of other authors are only cited when judged suitable to illustrate certain aspects of a model. The presentation of own simulation results serves the same purpose. These results are all for the flow around the Silsoe cube, for which many full scale (Hoxey et al., 2002), wind tunnel (Richards et al., 2005) and CFD results (Richards et al., 2002; Wright and Easom, 1999; Wright and Easom, 2003) are available. The computational domains for the two approach flow cases studied are shown in Figure 2. The height of the cube is H = 6m. The reference velocity of the approach flow is U_{ref}= 10m/s at $z = H$. The surrounding homogeneous terrain is described by a hydrodynamic roughness height of z_0 = 0.01m.

2 Simulation approaches

In this section the different modeling approaches of the flow physics which are most common in CWE are presented. First the basic equations for all simulation approaches are introduced and then their simplifications for the different approaches. The simplifications of the equations require the introduction of models for the turbulence. The most often used turbulence models for the respective approaches are introduced. Flow modeling close to walls is described separately in the final subsection.

2.1 Basic equations for isothermal flows

Wind effects on structures are mainly of interest in situations of high wind velocities. Therefore thermal effects are normally neglected in structural engineering (e.g. Simiu and Scanlan, 1996). In addition to that the influence of the Coriolis force on the lowest part of the Atmospheric Boundary Layer (ABL) in which the structures are immersed is also neglected as it leads only to minor changes in the mean wind direction with height (e.g. Holmes, 2001). The ABL can then be described by the well known conservation equations for mass and momentum with constant fluid properties of density ρ and dynamic viscosity μ.

$$\frac{\partial u_j}{\partial x_j} = 0 \tag{2.1}$$

$$\frac{\partial u_i}{\partial t} + \frac{\partial u_i u_j}{\partial x_j} = -\frac{\partial P}{\partial x_i} + \nu \frac{\partial}{\partial x_j}\left(\frac{\partial u_i}{\partial x_j} + \frac{\partial u_j}{\partial x_i}\right) \tag{2.2}$$

u_i are the velocity components, $P=p/\rho$ with the static pressure p, and $\nu=\mu/\rho$ is the kinematic viscosity.

The continuity equation (2.1) and momentum equation (2.2) together with initial and boundary conditions can in principle be directly used to compute the flow in the ABL. The turbulent nature of the ABL leads to several requirements for the direct solution of (2.1) and (2.2), which is known as Direct Numerical Simulation (DNS). The main requirement is that all relevant length scales of the flow have to be resolved with a DNS. In Figure 3 the typical distribution of the kinetic energy of the turbulence, $E(K)$, is shown as function of the wave number $K=2\pi/l$. The spectrum extends over a wide range of wave numbers and is generally interpreted as representing a superposition of turbulent eddies with characteristic length scales of $l \sim 1/K$. With a DNS all these eddies must be resolved. Therefore the computational domain has to be large enough to contain the largest scales (small K). On the other hand the computational grid used inside the computational domain has to be fine enough to resolve the smallest scales (large K). In addition also the time step used in the necessarily unsteady computation has to be small enough to resolve the evolution of the flow. Due to these requirements the application of DNS to problems in wind engineering, i.e. the direct solution of equations (2.1) and (2.2), is not feasible within the foreseeable future.

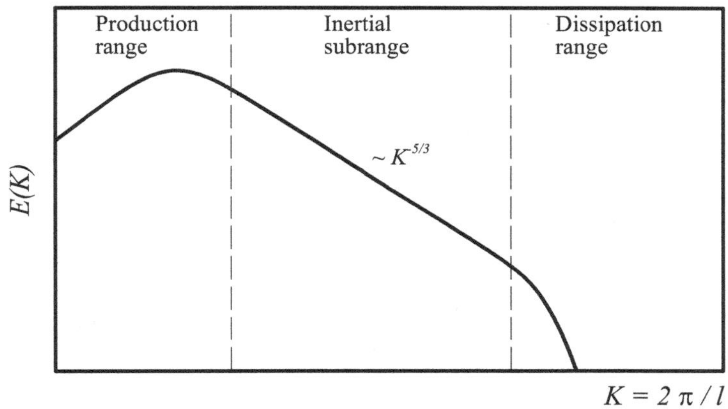

Figure 3. Turbulent energy spectrum.

Equations (2.1) and (2.2) therefore have to be modified prior to their numerical solution. The different modifications leading to different simulation approaches are described in the following.

2.2 Large Eddy Simulation (LES)

With the Large Eddy Simulation (LES) approach the turbulent flow in the ABL is still treated in four dimensions, i.e. unsteady and three dimensional in space. Contrary to DNS not the entire spectrum is resolved but only scales up to a cut-off wave number $K_c \sim 1/\Delta$, which is defined by the so called filter width Δ, see Figure 4. Therefore a coarser grid together with a larger time step size can be used in LES than in DNS, reducing the computational costs substantially. Fröhlich and Rodi (2002) succinctly call LES a "poor man's DNS" and state that the price to pay for the reduction of scales is the usage of a model for at least the unresolved scales' influence on the resolved scales.

The resolved flow variables are formally defined by applying a low-pass filter in space to the basic variables. E.g. the resolved velocity components are defined as (e.g. Sagaut, 2002)

$$\overline{u}_i(x,t) = \int_{-\infty}^{\infty} G(x - x')u_i(x',t)\,dx' , \quad \int_{-\infty}^{\infty} G(x')dx' = 1 \quad , \tag{2.3}$$

using a constant filter width Δ in an unbounded domain. In Equation (2.3) G is the filter function or filter kernel, which depends on the filter width Δ and is normalized to preserve constants. The unresolved velocity component is

$$u_i''(x_j,t) = u_i(x_j,t) - \overline{u}_i(x_j,t) \quad . \tag{2.4}$$

The three dimensional filter function is generally obtained from one dimensional filter functions via

$$G(\boldsymbol{x} - \boldsymbol{x}') = \prod_{j=1,3} G(x_j - x_j') \quad . \tag{2.5}$$

The discussion in the following will therefore be mainly restricted to one dimensional filter operations which can be transformed to 3D filter operations with (2.4). One classical filter function is the box or top hat filter G_B.

$$G_B(x - x') = \begin{cases} 1/\Delta & \text{for} \quad |x - x'| \le \Delta/2 \\ 0 & \text{for} \quad |x - x'| > \Delta/2 \end{cases} \tag{2.6}$$

Figure 3 shows the filtered energy spectra resulting from the application of the box filter, $E_B(K)$, and from the spectral cut-off filter, $E_S(K)$, which removes all contributions to the Fourier transformed flow variables that are greater than the cut-off wave number K_c (e.g., Sagaut 2002). While the spectral filter therefore leads to a sharp cut-off in wave number space, the box filter leads to a gradual decay of the kinetic energy when approaching the cut-off wave number K_c. The sharp cut-off of the spectral filter makes it well suited for the discussion of the basics of the LES approach in wave number space and will therefore be always used in the following figures. From Figure 4 it can be seen that scales larger than the filter width Δ are resolved, while scales smaller than Δ are not resolved. Furthermore the two directions of the energy transfer across the cut-off wave number are shown. While in the mean there must be a net transfer of turbulent kinetic energy from the large scales to the small scales, instantaneously one finds also a transfer of energy from the small towards the large scales. This backward transfer of energy is normally called backscatter.

The formal introduction of the filter operation (2.3) does not contain any information about the computational grid. Except for the fact that the filter width Δ cannot be smaller than the grid width it can be chosen independently of the grid width. In order to resolve as many scales as possible on a given grid, Δ is normally set equal to the grid width or twice the grid width. In most practical applications the filter function G does not even appear explicitly in the computational code but is only used to explain the concepts of LES, like above. This approach is known as implicit LES in contrast to explicit LES, where a finer grid than the filter width is chosen and the resolved scales are obtained by actually applying the filter operation (2.3) with a specific filter function G. In CWE implicit filtering is used exclusively and only the filter width is computed from the grid widths.

The relation between the filter and the grid width leads to further complications in the practical application of LES. The filter operation (2.3) is only valid for constant Δ, as the filter function then only depends on the distance $x - x'$. For implicit LES this implies that also the grid width has to be constant. For explicit LES one could theoretically use a constant Δ together with variable grid widths. As Δ has to be at least as large as the largest grid width this would however lead to a bad resolution in the regions with small grid widths. Therefore a variable filter width should be used in practical applications.

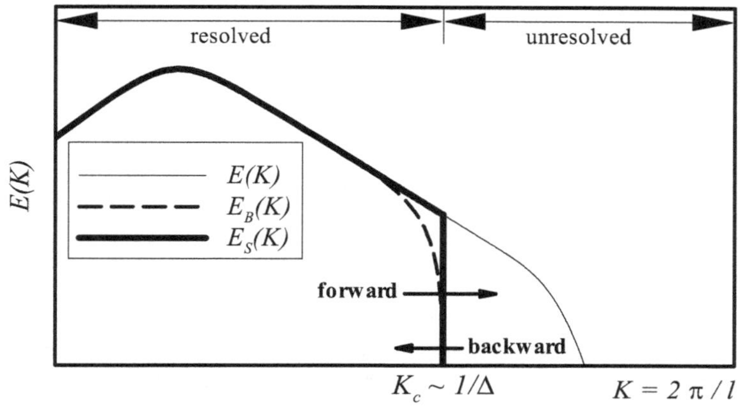

Figure 4. Turbulent energy spectrum $E(K)$ and filtered spectra $E_B(K)$ (box filter) and $E_S(K)$ (spectral cut-off).

Another problem of the filter operation (2.3) is the usage of an unbounded domain, which is never the case in structural engineering, where loads on buildings placed on the earth's surface have to be computed. Therefore a more general definition of the filtering is required, which takes spatial variations of the filter width and a bounded domain into account. Equation (2.7) redefines the resolved variable for this case.

$$\overline{u}(x,t) = \int_a^b G(x,x')u(x',t)dx' \tag{2.7}$$

With this definition of the filtering space differentiation and filtering do no longer commute. The difference between the filtered partial space derivative and the space derivative of the filtered flow variables constitutes the spatial commutation error.

$$\left[\frac{\partial u}{\partial x}\right] \equiv \overline{\frac{\partial u}{\partial x}} - \frac{\partial \overline{u}}{\partial x} = -\int_a^b u(x',t)\left(\frac{\partial G(x,x')}{\partial x'} + \frac{\partial G(x,x')}{\partial x}\right) dx' + \left(u(x',t)G(x,x')\right)_{x'=a}^{x'=b} \tag{2.8}$$

The first term on the right hand side of Equation (2.8) shows the influence of the variable filter width $\Delta(x)$ within the entire domain, while the second term contains the influence of the domain's boundary. The physical content of Equation (2.8) is the description of the fact that the definition of the resolved variable changes with the spatial variation of the filter width. E.g. in the case of a reduction of the filter width away from the domain boundary, a part of the resolved variable becomes unresolved due to the smaller filter. This is described by the first term on the right hand side of (2.8). The second term reflects the problem that the unfiltered solution has to be known up to $\Delta(x)/2$ outside of the domain to obtain the filtered values at the domain boundary according to definition (2.7). Commutation errors are therefore only absent for the ideal case of a uniform filter width in an unbounded domain, i.e. for a convolution filter. In wind engineering applications this is never the case and one always has to deal with the commutation error.

For the general definition of the filter operation, (2.7), the system of equations that describe the evolution of the resolved variables is obtained by filtering equations (2.1) and (2.2). Taking the spatial commutation errors into account, the basic equations of the LES approach are

$$\frac{\partial \overline{u}_j}{\partial x_j} = -\left[\frac{\partial u_j}{\partial x_j}\right] , \tag{2.9}$$

$$\frac{\partial \overline{u}_i}{\partial t} + \frac{\partial \overline{u}_i \overline{u}_j}{\partial x_j} = -\frac{\partial \overline{P}}{\partial x_i} + \nu \frac{\partial}{\partial x_j}\left(\frac{\partial \overline{u}_i}{\partial x_j} + \frac{\partial \overline{u}_j}{\partial x_i}\right) - \frac{\partial \tau_{ij}}{\partial x_j} - \left[\frac{\partial u_i u_j}{\partial x_j}\right] - \left[\frac{\partial P}{\partial x_i}\right] + \nu\left[\frac{\partial u_i}{\partial x_k \partial x_k}\right] . \tag{2.10}$$

This system of equations contains more unknowns than there are equations. This so called closure problem is caused by the spatial commutation errors and the subfilter or subgrid stress tensor τ_{ij}, which results from the replacement of the averaged product of the velocity components with the product of the averaged components in the momentum equation (2.10) and describes the interaction of the resolved and the unresolved scales. It is defined as

$$\tau_{ij} = \overline{u_i u_j} - \overline{u}_i \overline{u}_j \quad . \tag{2.11}$$

While the expression subfilter stress tensor is more appropriate in the context of explicit filtering, the expression subgrid stress tensor is often used, due to historical reasons (Fröhlich and Rodi, 2002). In the following the latter denomination will be used.

To render the system of equations solvable, models for the spatial commutation errors and the subgrid stress tensor have to be used. The spatial commutation errors are normally simply neglected. In the context of explicit filtering that can be motivated by the observation that the spatial commutation error is of second order in the filter function, i.e. $O(\Delta^2)$, when filtering is defined after a change of coordinates (Ghosal and Moin, 1995; see also Sagaut, 2002). However, the spatial variation of the filter width should also be smooth and slow (Geurts, 2004), corresponding to a small variation in grid widths.

After the removal of the spatial commutation errors from (2.9) and (2.10) the remaining closure problem is the subgrid stress tensor. With a subgrid scale model τ_{ij} is expressed in terms of the resolved variables and the system of equations is closed. The most common subgrid models that are used in wind engineering applications are briefly introduced in the following.

The Smagorinsky model. Smagorinsky (1963) introduced the first subgrid scale model which is still widely used in CWE. With this model the subgrid stresses are approximated like the viscous stresses.

$$\tau_{ij} - \frac{1}{3}\tau_{kk}\delta_{ij} \approx -2\nu_{sgs}\overline{S}_{ij} , \quad \overline{S}_{ij} = \frac{1}{2}\left(\frac{\partial \overline{u}_i}{\partial x_j} + \frac{\partial \overline{u}_j}{\partial x_i}\right) \tag{2.12}$$

While $1/3\ \tau_{ij}\ \delta_{ij}$ is added to the resolved pressure and does not require further modeling, the subgrid scale (SGS) viscosity ν_{sgs} still has to be determined. Contrary to the molecular viscosity the SGS viscosity is no fluid property but depends on the local subgrid scales. Based on dimensional analysis, ν_{sgs} is proportional to a length scale and a velocity, both characteristic of the local subgrid scales. For the length scale the filter width is used and the velocity is computed from the magnitude of the resolved rate of strain tensor, yielding

$$\nu_{sgs} = C_s^2 \Delta\left(\Delta\left|\overline{S}\right|\right) = \left(C_s\Delta\right)^2\left|\overline{S}\right|, \quad \left|\overline{S}\right| = \sqrt{2\overline{S}_{ij}\overline{S}_{ij}} \quad . \tag{2.13}$$

The Smagorinsky constant C_s is usually taken as $C_s = 0.1$. Referring to Figure 4, the Smagorinsky model only provides forward transfer of kinetic energy and is therefore purely dissipative.

The Bardina scale similarity model. The scale similarity model of Bardina et al. (1983) was the first model to acknowledge the fact that the strongest interaction takes place between the smallest resolved scales and the largest unresolved scales.

$$\tau_{ij} \approx \overline{\overline{u}_i\overline{u}_j} - \overline{\overline{u}}_i\overline{\overline{u}}_j \tag{2.14}$$

This model therefore requires a second application of the filtering operator. The model shows a good correlation between the exact and modeled SGS stresses and is capable of providing the backward transfer of kinetic energy from the unresolved to the resolved scales, see Figure 4. However, it is not dissipative enough, underestimating the transfer of kinetic energy from the resolved scales to the unresolved. Therefore it is often combined with the purely dissipative Smagorinsky model, yielding the mixed Smagorinsky-Bardina model.

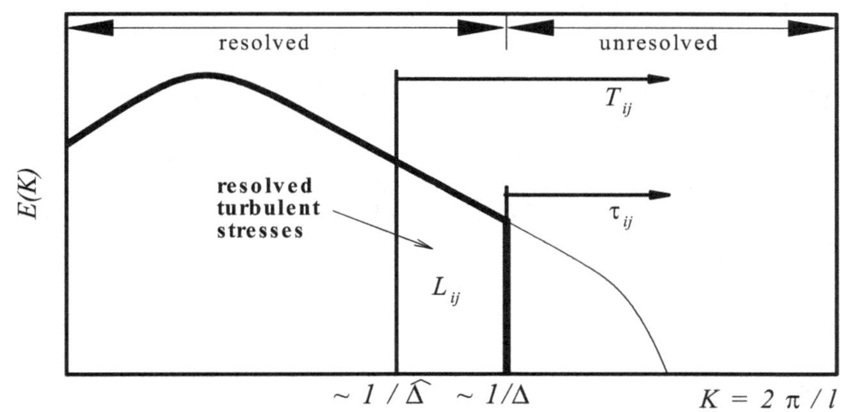

Figure 5. Principle of the dynamic modeling approach (after Fröhlich and Rodi, 2002).

The dynamic modeling approach. Dynamic modeling as introduced by Germano et al. (1991) can be combined with nearly any subgrid scale model. It formally takes the aforementioned strong interaction between the smallest resolved and largest unresolved scales into account by introducing a second filter width $\hat{\Delta} > \Delta$. The specific SGS model is then applied for the two SGS stress tensors T_{ij} and τ_{ij}, resulting from filtering with $\hat{\Delta}$ and Δ, respectively. In Figure 5 the positions where these subgrid stress tensors apply in the wave number space are shown. Also shown is the region that corresponds to the resolved subgrid stresses for the filter level $\hat{\Delta}$, L_{ij}. It is clearly visible that L_{ij} can be computed from T_{ij} and τ_{ij}.

$$L_{ij} = \widehat{\overline{u}_i\overline{u}_j} - \hat{\overline{u}}_i\hat{\overline{u}}_j = T_{ij} - \hat{\tau}_{ij} \tag{2.15}$$

With the aid of the identity (2.15) any "constant" in the common model applied for T_{ij} and τ_{ij} can then be determined as local, instantaneous function of the flow field. The most common model in CWE is the dynamic Smagorinsky model, which uses (2.13) at both filter levels. Due to the dynamic procedure this model is capable of computing the backward energy transfer from small to large scales. In the context of the Smagorinsky model this means that the subgrid scale viscosity can become negative. As a negative viscosity can destabilize the computation, the SGS viscosity is normally bounded so that the sum of molecular and SGS viscosity is not negative. In addition the dynamically computed constant is not allowed to exceed a maximum value.

The dynamic modeling approach is able to detect laminar regions with no SGS activity and does not need any further modeling assumptions close to walls, see section 2.5. Its general use in wind engineering has already been recommended by Murakami (1998).

Other modeling approaches and applications. Besides the two models described in more detail and the dynamic modeling approach there are many more models available for the subgrid scales. The most complete survey is provided by Sagaut (2002). First of all the models based on one or two additional transport equations are also available for the LES approach. These can be combined with the dynamic approach. A completely different view on SGS modeling is followed with the Monotonically Integrated Large Eddy Simulation (MILES) approach in which the built in dissipation of certain numerical approximations is used to model the subgrid scales (see, e.g., Grinstein and Fureby, 2004).

In general LES is said to be the most promising approach in CWE. For wind loading predictions the principle advantage of the method over the steady RANS approach described in the following section is that as a time dependent approach it can be used to predict pressure fluctuations and therefore also extreme pressures. This has been done by Nozawa and Tamura (2002) for the case of a low rise building. The computed maximum pressures compare in general well with corresponding measurements. Nozawa and Tamura (2003) also examined the pressure fluctuations on a high rise building model and found good agreement, as did Ono et al. (2006) for the flow over a low rise building with 45° approach flow direction. For further applications of LES see the reviews by Murakami (1998) and Stathopoulos (2002).

2.3 Steady Reynolds Averaged Navier Stokes (RANS) simulation

The still most common simulation approach is based on the steady Reynolds Averaged Navier Stokes (RANS) equations. With this approach the time history of the flow variables can no longer be computed but just their time average. This means that the turbulent kinetic energy spectrum shown in Figure 3 is not resolved at all, but has to be modeled.

The time averaged velocity components for example are formally defined as

$$\langle u_i \rangle(\boldsymbol{x}) = \lim_{T\to\infty} \frac{1}{2T} \int_{-T}^{T} u_i(\boldsymbol{x},t)dt \quad , \tag{2.16}$$

and the corresponding fluctuations are

$$u_i'(\boldsymbol{x},t) = u_i(\boldsymbol{x},t) - \langle u_i \rangle(\boldsymbol{x}) \quad . \tag{2.17}$$

Application of the time average (2.16) to the basic equations (2.1) and (2.2) yields the RANS equations.

$$\frac{\partial \langle u_j \rangle}{\partial x_j} = 0 \tag{2.18}$$

$$\frac{\partial \langle u_i \rangle \langle u_j \rangle}{\partial x_j} = -\frac{\partial \langle P \rangle}{\partial x_i} + \nu \frac{\partial}{\partial x_j}\left(\frac{\partial \langle u_i \rangle}{\partial x_j} + \frac{\partial \langle u_j \rangle}{\partial x_i} \right) - \frac{\partial \langle u_i' u_j' \rangle}{\partial x_j} \tag{2.19}$$

Contrary to the general spatial filter applied in the LES approach, time averaging does commute with the spatial derivatives and hence does not lead to commutation errors. However, the replacement of the averaged product of velocity components with the product of their averages in the momentum equation (2.19) again leads to a new term, the Reynolds stress tensor

$$\langle u_i' u_j' \rangle = \langle u_i u_j \rangle - \langle u_i \rangle \langle u_j \rangle \quad . \tag{2.20}$$

The Reynolds stress tensor describes the interaction of the fluctuations with the mean flow and constitutes the only closure problem in the RANS approach. This tensor has to be modeled in terms of the average flow variables which is done with turbulence models. For turbulence modeling different approaches are available in the context of the steady RANS equations. The most widely used approaches are presented in the following.

Two equation models. The most common turbulence models used in CWE are models that solve two additional transport equations to determine the Reynolds stress tensor (2.20). For linear two equation models the Reynolds stresses depend linearly on the turbulent kinetic energy k and the mean velocity gradients, like with the Smagorinsky model (2.12) used in the LES approach.

$$\langle u_i' u_j' \rangle - \frac{2}{3} k \delta_{ij} = -2\nu_t \langle S_{ij} \rangle, \quad k = \frac{1}{2} \langle u_i' u_i' \rangle, \quad \langle S_{ij} \rangle = \frac{1}{2}\left(\frac{\partial \langle u_i \rangle}{\partial x_j} + \frac{\partial \langle u_j \rangle}{\partial x_i} \right) \tag{2.21}$$

This is known as Boussinesq hypothesis and reduces the modeling to the determination of k and the turbulent viscosity ν_t, which is a function of the local turbulence in the flow field. From dimensional analysis it follows that the turbulent viscosity can be expresses as a velocity and a length scale, both characteristic of the turbulence. These two quantities are determined by two additional transport equations, one for k and one for its dissipation rate ε, or its specific dissipation $\omega \sim \varepsilon/k$. In CWE the standard k-ε model (Launder and Spalding, 1972) is still often used. For this model the corresponding transport equations are

$$\frac{\partial k \langle u_j \rangle}{\partial x_j} = \frac{\partial}{\partial x_j}\left[\left(\nu + \frac{\nu_t}{\sigma_k} \right) \frac{\partial k}{\partial x_j} \right] + \nu_t \langle S \rangle^2 - \varepsilon, \quad \langle S \rangle = \sqrt{2 \langle S_{ij} \rangle \langle S_{ij} \rangle} \quad , \tag{2.22}$$

$$\frac{\partial \varepsilon \langle u_j \rangle}{\partial x_j} = \frac{\partial}{\partial x_j}\left[\left(\nu + \frac{\nu_t}{\sigma_\varepsilon} \right) \frac{\partial \varepsilon}{\partial x_j} \right] + C_{\varepsilon 1} \nu_t \langle S \rangle^2 \frac{\varepsilon}{k} - C_{\varepsilon 2} \frac{\varepsilon^2}{k} \quad . \tag{2.23}$$

From k and ε the turbulent viscosity is computed as

$$\nu_t = C_\mu \frac{k^2}{\varepsilon} \tag{2.24}$$

A typical set of constants appearing in (2.22) – (2.24) is (σ_k, σ_ε, $C_{\varepsilon 1}$, $C_{\varepsilon 2}$, C_μ) = (*1.0, 1.3, 1.44, 1.92, 0.09*). The specific values of the constants in wind engineering applications will be further discussed in section 2.5 and 4.1.

The standard k-ε model is still the industry standard for a wide range of applications, despite its many well known limitations. However, one of these limitations leads to erroneous pressure distribution on buildings. The reason for this is an excessive production of k in stagnation flow regimes, known as stagnation point anomaly. The large values of k then lead to much too high pressures on the windward sides of the building and prevent separation at the frontal corners. Durbin and Petterson Reif (2001) give three reasons for the stagnation point anomaly:

- Deficient representation of Reynolds stress anisotropy.
- Quantitative overestimation of the production of k.
- Dissipation does not keep up with production.

To alleviate the stagnation point anomaly several ad hoc modifications of the standard k-ε model have been proposed. Kato and Launder (1993) redefined the production term in the k equation by including the modulus of the rotation rate tensor defined in (2.26). Tsuchiya et al. (1997) included this influence in the definition of C_μ. Both modifications reduce the production of k in stagnation regions and lead to improved predictions of the pressures on the windward side and the roof. However, the prediction of the velocity field is worse than with the standard model (Mochida et al., 2002; Wright and Easom, 1999).

An increased dissipation is obtained by using the ReNormalization Group (RNG) k-ε model of Yakhot and Orszag (1986). While the model itself is rigorously derived, the increased dissipation is due to an ad hoc modification. With the RNG model the pressure prediction is improved over the standard k-ε model and the velocity field is still in good agreement with experiments (Mochida et al., 2002; Wright and Easom, 2003). The RNG k-ε model therefore is the most general two equation turbulence model that should be used in wind loading predictions on structures.

Yet another approach to reduce the stagnation point anomaly was followed by Durbin (1996) who realized that the large production of k in stagnation regions can even lead to normal turbulent stresses that are negative and therefore violate the physics. When a turbulence model yields unphysical results it does not fulfill what is known as realizability (Schumann, 1977). Durbin (1996) remedied that problem of the standard k-ε model by limiting the turbulent viscosity.

Also motivated by realizability Shih et al. (1995) introduced a Realizable k-ε (RKE) model. With this model the Reynolds stresses no longer depend linearly on the mean velocity gradients, cf. Equation (2.21). The inclusion of a non linear dependence of the Reynolds stresses on the velocity gradients is in general done with so called non linear two equation models which are an extension of the linear Boussinesq hypothesis (2.21), see Pope (1975), and can be derived rigorously from the transport equations of the Reynolds stresses, which are described in the next section. With this models the isotropic viscosity hypothesis of (2.21) is removed. Therefore they can better represent the anisotropy of the Reynolds stresses, especially of the normal Reynolds stresses. Their application to wind loading predictions is however limited. E.g., Wright and Easom (2003) used the quadratic non linear k-ε model of Craft et al. (1996) to predict the pressure distribution on the Silsoe cube. With the quadratic dependence of the Reynolds stresses on the velocity gradients the approximation is

$$\begin{aligned}\langle u_i'u_j'\rangle - \frac{2}{3}k\delta_{ij} &= -2\nu_t\langle S_{ij}\rangle + C_1\nu_t\frac{k}{\varepsilon}\left(\langle S_{ik}\rangle\langle S_{kj}\rangle - \frac{1}{3}\langle S_{kl}\rangle\langle S_{kl}\rangle\delta_{ij}\right) \\ &\quad + C_2\nu_t\frac{k}{\varepsilon}\left(\langle \Omega_{ik}\rangle\langle S_{kj}\rangle - \langle \Omega_{jk}\rangle\langle S_{ki}\rangle\right) \\ &\quad + C_3\nu_t\frac{k}{\varepsilon}\left(\langle \Omega_{ik}\rangle\langle \Omega_{kj}\rangle - \frac{1}{3}\langle \Omega_{kl}\rangle\langle \Omega_{kl}\rangle\delta_{ij}\right)\end{aligned}, \quad (2.25)$$

with the rotation rate tensor

$$\langle \Omega_{ij}\rangle = \frac{1}{2}\left(\frac{\partial\langle u_i\rangle}{\partial x_j} - \frac{\partial\langle u_j\rangle}{\partial x_i}\right) \quad . \quad (2.26)$$

The results of Wright and Easom (2003) are shown in Figure 6 for the 0° and 45° approach flow direction. While still deviating from the experimental results, especially on the roof and at the leeward side, the results agree well with the full differential stress models described in the following section.

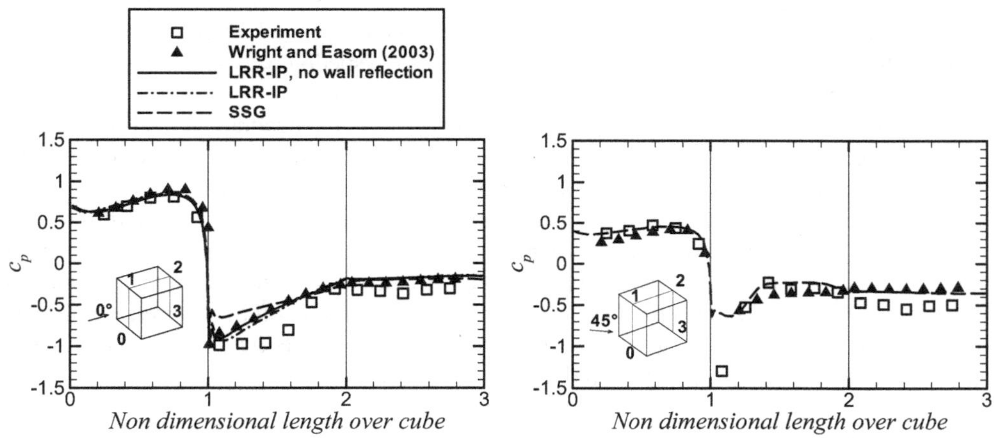

Figure 6. Pressure distribution over the center line of the Silsoe cube for 0° (left) and 45° (right) approach flow direction.

Differential stress models. The most natural way to take the anisotropy of the Reynolds stresses into account is to solve an additional transport equation for every of the Reynolds stresses.

$$\frac{\partial \langle u_i' u_j' \rangle \langle u_k \rangle}{\partial x_k} = \frac{\partial}{\partial x_k}\left(\nu \frac{\partial \langle u_i' u_j' \rangle}{\partial x_k} \right) - \langle u_i' u_j' \rangle \frac{\partial \langle u_j \rangle}{\partial x_k} - \langle u_j' u_k' \rangle \frac{\partial \langle u_i \rangle}{\partial x_k} + \Sigma_{ij} - \varepsilon_{ij} - \frac{\partial D_{ijk}}{\partial x_k} \tag{2.27}$$

Modeling is required for the last three terms in (2.27), the pressure strain correlation Σ_{ij}, the dissipation rate correlation tensor ε_{ij} and the diffusion correlation D_{ijk}. For the last term a gradient approximation is normally applied (Lien and Leschziner, 1994) and the dissipation rate correlation tensor is modeled by the scalar dissipation rate, $\varepsilon_{ij} \approx 2/3\varepsilon\delta_{ij}$. Thus seven additional transport equations have to be solved together with (2.18) and (2.19) in the Differential Stress Model (DSM) approach.

The model for the pressure strain correlation,

$$\Sigma_{ij} = \left\langle P' \left(\frac{\partial u_i'}{\partial x_j} + \frac{\partial u_j'}{\partial x_i} \right) \right\rangle \quad , \tag{2.28}$$

is regarded as the most important in DSM modeling. In the model of Launder et al. (1975), known as LRR-IP model, Σ_{ij} is split into three contributions, the slow pressure strain correlation, the fast pressure strain correlation and a wall reflection term. For the results presented in Figure 6 the models of Rotta (1951), Fu et al. (1987) and Gibson and Launder (1978) are used for the respective terms. The wall reflection model does however also lead to unrealistic turbulence values in stagnation regions (Craft et al., 1993; Murakami, 1998; Wright and Easom, 2003). Therefore its omission is often recommended. The influence of the wall reflection term is shown in Figure 6 for the 0° approach flow direction. Except for the suction peak at position 1 there are no differences in the results with and without the wall reflection term.

Besides the LRR-IP model with and without wall reflection the quadratic model for the pressure strain correlation of Speziale et al. (1991), known as SSG model, was used in the simulations. This model does not require the explicit inclusion of a wall reflection term. As can be seen from Figure 6 on the left, the results are comparable to the results of the LRR-IP model on the front and leeward side. On the roof however a lower suction is predicted. Also for the 45° approach flow case the SSG model cannot predict the large suction on the roof at position 1. Oliveira and Younis (2002) also used the SSG model for the prediction of the pressure distribution on a glass house. They obtained better agreement with the corresponding experimental results for the SSG model as compared to the standard k-ε model.

2.4 Unsteady RANS simulation and Detached Eddy Simulation (DES)

A third approach for defining the solution variables is unsteady RANS (URANS), also known as Very Large Eddy Simulation (VLES). In the URANS approach the resolved flow variables are defined with the aid of an ensemble average,

$$\langle \phi \rangle_N (x_i, t) = \lim_{N\to\infty} \frac{1}{N} \sum_{n=1}^{N} \phi_n (x_i, t) \quad . \tag{2.29}$$

This time dependent average conceptually results when performing N identical measurements of a flow variable and is the basis of every statistical description of turbulent flows (e.g., Pope, 2000). With regards to the turbulent energy spectrum shown in Figure 3 it is clear that this definition of the averaging does not introduce any cut off frequency or wave number and therefore leaves the spectrum unaltered.

The corresponding fluctuation is defined as

$$\phi_n'(x_i,t) = \phi_n(x_i,t) - \langle \phi \rangle_N (x_i,t) \tag{2.30}$$

Application of the averaging operator (2.29) to the basic equations (2.1) and (2.2) results in the URANS equations.

$$\frac{\partial \langle u_j \rangle_N}{\partial x_j} = 0 \tag{2.31}$$

$$\frac{\partial \langle u_i \rangle_N}{\partial t} + \frac{\partial \langle u_i \rangle_N \langle u_j \rangle_N}{\partial x_j} = -\frac{\partial \langle P \rangle_N}{\partial x_i} + \nu \frac{\partial}{\partial x_j}\left(\frac{\partial \langle u_i \rangle_N}{\partial x_j} + \frac{\partial \langle u_j \rangle_N}{\partial x_i} \right) - \frac{\partial \tau_{N,ij}}{\partial x_j} \tag{2.32}$$

Except for the unsteady term in the momentum equation (2.32), these equations are the same as the ones for the steady RANS approach, see equations (2.18) and (2.19). Also the closure problem in terms of the Reynolds stress tensor is formally the same.

$$\tau_{N,ij} = \langle u_i u_j \rangle_N - \langle u_i \rangle_N \langle u_j \rangle_N = \langle u_i' u_j' \rangle_N \tag{2.33}$$

This one to one correspondence between the URANS and the RANS equations results from the definition of the URANS solution variables as ensemble average. Sometimes the basic equations of the URANS approach are derived with the aid of a time filter,

$$\langle \phi \rangle_T (x_i,t) = \frac{1}{T} \int_{t-T/2}^{t+T/2} \phi(x_i,t')\, dt' \quad . \tag{2.34}$$

This is a moving average with the averaging interval T, which filters the high frequency components of the variable. The averaging operation is analogous to the box filter used in spatial filtering in the context of the LES approach, see Equation (2.6) in section 2.2. As the temporal filter width is in general constant no temporal commutation errors result when (2.34) is applied to the basic equations (2.1) and (2.2). But the definition of the Reynolds stress tensor in terms of the averaged velocities and their corresponding fluctuations changes due to the fact that (2.34) is no Reynolds operator fulfilling the Reynolds rules of averaging (see, e.g., Sagaut, 2002). Especially the following two inequalities,

$$\langle \langle u_i \rangle_T \rangle_T \neq \langle u_i \rangle_T \,, \qquad \langle u_i' \rangle_T \neq 0 \,, \tag{2.35}$$

where the fluctuation is defined in analogy to Equation (2.30), lead to the modified definition of the Reynolds stress tensor,

$$\tau_{T,ij} = \langle \langle u_i \rangle_T \langle u_j \rangle_T \rangle_T - \langle u_i \rangle_T \langle u_j \rangle_T + \langle \langle u_i \rangle_T u_j' \rangle_T + \langle u_i' \langle u_j \rangle_T \rangle_T + \langle u_i' u_j' \rangle_T \quad . \tag{2.36}$$

This definition only reduces to (2.33) when the inequalities in (2.35) become equalities, which is the case if the turbulent energy spectrum has a gap with zero energy at the cut off frequency π/T (Aldama, 1990). For meteorological data used in wind engineering a spectral gap is normally associated with the low but non zero turbulent kinetic energy content in a frequency range corresponding to periods from 10 minutes to about 5-10 hours, see, e.g., Dyrbye and Hansen (1997). Despite the differences in their definitions, (2.36) and (2.33) and are normally modeled by one of the turbulence models for the steady RANS approach, leading to closure of Equation (2.32).

A special version of URANS is the Detached Eddy Simulation (DES) approach which is intermediate between the LES and URANS approach (Spalart et al., 2006). Like in URANS a turbulence model originally proposed for the steady RANS equations is used. This is the Spalart-Allmaras turbulence model, a one equation model for the turbulent viscosity (Spalart and Allmaras, 1994). In the DES approach this turbulence model makes use of the local length scale defined by the grid or the distance from the next wall. If this distance is smaller than the local grid width then the RANS version of the turbulence model is used. Otherwise it behaves like a SGS model. Breuer et al. (2003) presented an analysis of the differences between the Spalart-Allmaras and the Smagorinsky SGS model and obtained comparable results with both models after minor modifications of the Spalart-Allmaras model.

2.5 Modeling at walls

Nearly all turbulence models presented so far for the LES, RANS and URANS approach need to be modified close to walls. The reason for this is that close to a wall the direct viscous effects, which normally are neglected in deriving the turbulence models, become important. Another problem is the increase of the velocity magnitude from zero at a wall to large values over a small wall normal distance, leading to large velocity gradients that are expensive to resolve in a computational simulation. Therefore simplified models are used at walls to overcome these problems. These so called wall functions are introduced in the following first for the RANS and URANS approaches and then for LES.

RANS, URANS and DES. Three modeling approaches can be discerned for RANS, URANS and DES at walls. The first approach, known as low-Reynolds number modeling, solves for the velocities down to the wall and incorporates the increasing importance of molecular viscosity by damping functions into the transport equations for the turbulence quantities (see, e.g., Patel et al., 1984). This approach normally (see Zhang et al., 1996, for an exemption) does not include the effect of wall roughness, which is important in wind engineering. In addition the resolution of the large velocity gradients requires a large number of grid points in wall normal direction, which makes this approach computationally very expensive.

The same computational cost is encountered with the two layer modeling approach for k-ε two equation turbulence models (Chen and Patel, 1988). Here again the velocity distribution is resolved down to the wall. A simplified model for the turbulence quantities is solved in the wall adjacent region and patched to the standard model at some distance from the wall. Durbin et al. (2001) extended this model to include wall roughness. Despite the fact that the two layer modeling approach leads to improved predictions over the wall function approach presented next (see, e.g., Rodi, 1991), it has not been used for loading predictions in wind engineering, presumably due to its high computational cost.

The most common approach to model the flow at the wall is known as the wall function approach. Here the viscosity dominated region close to the wall is bridged by wall functions, removing the necessity to resolve the large velocity gradients near the wall. This is achieved by placing the first computational node off the wall in the so called logarithmic region of the boundary layer. This logarithmic region can be seen in Figure 7, where the non dimensional velocity distribution U^+ in a boundary layer flow over a smooth wall is shown as function of the non dimensional wall distance z^+. These quantities are defined as

$$U^+ = \frac{\langle u \rangle (z)}{u_\tau}, \quad z^+ = \frac{u_\tau z}{\nu}, \quad u_\tau = \sqrt{\langle \tau_w \rangle / \rho} \quad , \tag{2.37}$$

where u_τ, often also written as u_*, is the friction velocity, determined from the mean wall shear stress.

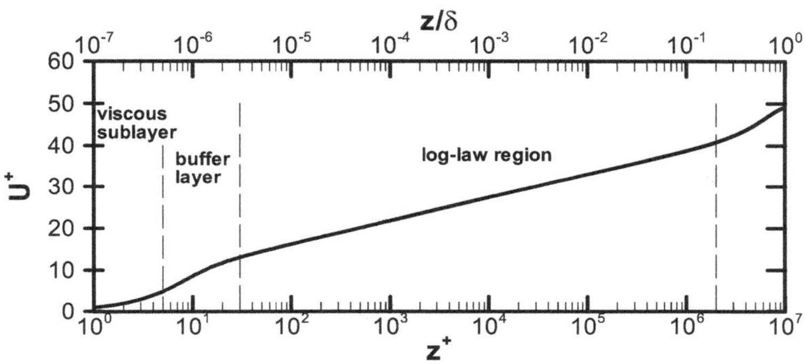

Figure 7. Non dimensional velocity distribution in a boundary layer.

The non dimensional velocity distribution can be divided into four regions. Close to the wall the viscous forces dominate in the viscous sublayer up to $z^+=5$, followed by the buffer layer. From $z^+=30 \div 70$ the log-law region starts which extends up to approximately $10 \div 30\%$ of the entire boundary layer thickness δ. In the final wake region the constant free stream velocity is approached.

In the logarithmic region the velocity is described by

$$U^+ = \frac{1}{\kappa} \ln\left(z^+\right) + B = \frac{1}{\kappa} \ln\left(E z^+\right), \quad E = \exp(\kappa B) \quad . \tag{2.38}$$

Here κ is the von Karman constant ($\kappa \approx 0.4$) and B $\approx$ 5. Within the wall function approach (2.38) is used at the first computational node off the wall to determine the wall shear stress. Therefore it has to be assured that this position is in the logarithmic region.

For flows over rough walls the log-law is still valid, the velocity however decreases due to the increased drag exerted by the roughness elements. The reduction of the velocity is taken into account by the inclusion of a measure of the roughness in the logarithmic velocity distribution. In meteorology and wind engineering the roughness is expressed in terms of the hydrodynamic roughness height z_0. The corresponding logarithmic velocity distribution is

$$U^+ = \frac{1}{\kappa}\ln\left(\frac{z-d}{z_0}\right) \quad \text{or} \quad U^+ = \frac{1}{\kappa}\ln\left(\frac{z+z_0-d}{z_0}\right) \quad , \tag{2.39}$$

where the second formulation is used to prevent a zero argument of the logarithm in case of a vanishing displacement height $d = 0$ and the wall being at $z = 0$.

Contrary to (2.39) the roughness is expressed in terms of the sandgrain roughness k_s or the equivalent sandgrain roughness in mechanical engineering. The velocity shift is normally included with an additional constant ΔB, depending on the non dimensional roughness height $k_s^+ = u_\tau k_s/\nu$, in the logarithmic distribution (2.38).

$$U^+ = \frac{1}{\kappa}\ln\left(Ez^+\right) - \Delta B\left[k_s^+\right] \tag{2.40}$$

Based on k_s^+ the influence of the roughness is divided into three regimes. For $k_s^+ \leq 2.25$ the roughness does not influence the velocity distribution and the wall is said to be hydrodynamically smooth. For $k_s^+ > 90$ the influence of the roughness is dominant and the regime is called fully rough. Between these values of k_s^+ the wall is transitionally rough.

Several empirical formulas describe the dependence of ΔB on k_s^+, see, e.g., Ligrani and Moffat (1986) and Cebeci and Bradshaw (1977). The latter provide the following formulation for the fully rough regime,

$$\Delta B = \frac{1}{\kappa}\ln\left(1 + C_r k_s^+\right), \tag{2.41}$$

which is also used in the commercial flow solver FLUENT (2005) with the constant $C_r \in [0,1]$ to be chosen by the user. Similar formulations are implemented in other commercial software (Blocken et al., 2006; Hargreaves and Wright, 2006).

When commercial general purpose CFD software is to be used for wind engineering applications one normally has to transform the hydrodynamic roughness length z_0 into the sandgrain roughness k_s. In the fully rough regime the following approximation is obtained when equating (2.39) and (2.40) for the case of vanishing displacement height $d = 0$.

$$k_s \approx \frac{E}{C_r} z_0 \tag{2.42}$$

For Equation (2.42) the first formulation in (2.39) has been used together with the approximation $\ln(1 + C_r\, k_s^+) \approx \ln(C_r\, k_s^+)$. Without this approximation and the second formulation of the velocity distribution in (2.39) the following equation is obtained for the determination of k_s, again for the case of vanishing displacement height $d = 0$.

$$k_s^+ = \frac{1}{C_r}\left(\frac{z_0 E z^+}{z + z_0} - 1\right) \tag{2.43}$$

This equation can be solved if for the wall normal distance the location of the first computational node off the wall is used, i.e. $z = z_P$. Thus equality of (2.39) and (2.40) is only required at this position. However, this is exactly where the wall function is used. The relation of k_s and z_p is shown in Figure 8 for $z_0 = 0.01$m, corresponding to the Silsoe case, and three values for C_r.

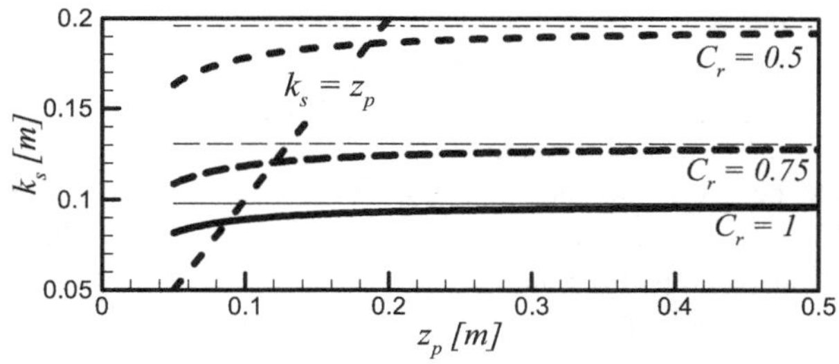

Figure 8. Sandgrain roughness as function of the first computational node off the wall for $z_0 = 0.01$.

The thick lines in Figure 8 are the solution of (2.43) for different C_r while the thin horizontal lines are the corresponding relations from the approximation (2.42). The thick dash double dot line shows the equality of k_s and z_P. Only values of z_P to the right side of this line are possible in most CFD codes, as the first computational node off the wall has to be at least k_s away from the wall (Blocken et al., 2006). The closer the first computational node shall be placed to this limit value the larger is the difference in k_s obtained from the two relations. The influence of the different values for k_s on the simulation of a boundary layer flow are shown in section 4.1.

For the turbulence quantities the following modifications are used at the first computational node in the wall function approach. For the k equation (2.22) the production is changed to contain only the wall normal velocity gradient of the wall parallel velocity component. k_P at the first computational node is used to compute the friction velocity, $u_\tau = k_P^{1/2}\, C_\mu^{1/4}$. The equation for the turbulent dissipation rate ε is not solved but ε itself computed at the first node under the assumption of an equilibrium between production and dissipation,

$$\varepsilon_P = \frac{u_\tau^3}{\kappa z_P} \quad \text{or} \quad \varepsilon_P = \frac{u_\tau^3}{\kappa\left(z_P + z_0\right)} \quad . \tag{2.44}$$

The two formulations correspond to the two velocity profiles in (2.39). The first of them is used in general in commercial CFD software.

LES. For LES also the two principally different modeling approaches at walls exist, which are described above for the RANS, URANS and DES approach. On one hand one can try to resolve the dynamic structures close to the wall within the viscous sublayer, corresponding to a DNS close to the wall. This requires very fine grids in wall normal and parallel direction. With this approach some SGS models have to be modified. While the dynamic approach can model the influence of wall on the SGS scales, e.g. the standard Smagorinsky model has to be modified. To include the reduction of the SGS stresses close to the wall a damping function is used, yielding

$$\nu_{sgs} = \left(C_s\Delta\right)^2\left\{1-\exp\left[\left(-z^+/25\right)^3\right]\right\}\left|\overline{S}\right| \tag{2.45}$$

instead of (2.13) (Sagaut, 2002).

Due to the very high computational cost of this first approach the wall function approach is also normally used in LES for wind engineering. The logarithmic velocity distribution is then assumed for the instantaneous resolved velocity component. Extensions of the smooth wall profile (2.38) for rough walls are described by Mason and Callen (1986) and Grötzbach (1977).

A completely different approach is also used in CWE, which models the flow through the roughness elements like the flow through a porous medium (see, e.g., Nakayama et al, 2005). The free flow area has to be prescribed and the drag force exerted by the roughness elements. The latter are then included as source terms in the momentum equation. This approach is known as canopy model, distributed drag force approach or discrete element roughness method. While for LES additional terms are only included in the momentum equations, the RANS and URANS versions of this approach also contain additional terms in the equations for the turbulence quantities (Maruyama, 1999).

3 Numerical solution with the Finite Volume method

The basic system of partial differential equations for each simulation approach described in section 2 cannot be solved analytically. The equations can also not directly be solved with a computer but first have to be transformed into algebraic equations by means of discretization in space and possibly in time. For this transformation several numerical approaches can be employed, of which only the Finite Volume method will be described in the following due to its widespread use especially in commercial general purpose CFD codes. Many of the presented aspects can however be directly transferred to other approaches like Finite Difference or Finite Element methods. The most important ones are the accuracy of different approximations expressed in terms of their order and the brief presentation of the iterative solution of the algebraic system of equations.

3.1 Integral form of the basic equations

As the name Finite Volume (FV) method implies, this approach for space discretization is based on the integral form of the basic equations. When integrating, e.g., the basic equations of the URANS approach (2.31) and (2.32) from section 2.4 over a volume V which is enclosed by the surface ∂V with unit normal vector $\boldsymbol{n}$, one obtains

$$\oint_{\partial V} n_j \langle u_j \rangle dA = 0 \quad , \tag{3.1}$$

$$\frac{\partial}{\partial t}\int_V \langle u_i \rangle dV + \oint_{\partial V} n_j \langle u_i \rangle \langle u_j \rangle dA = -\oint_{\partial V} n_i \langle P \rangle dA - \oint_{\partial V} n_j \tau_{ij} dA + \oint_{\partial V} n_j \left[\nu \left(\frac{\partial \langle u_i \rangle}{\partial x_j} + \frac{\partial \langle u_j \rangle}{\partial x_i} \right) \right] dA \tag{3.2}$$

for the integral continuity and momentum equation. Volume integrals have been converted into surface integrals by means of Gauss's theorem where possible and the index N, denoting the ensemble average (2.29), has been omitted for better readability.

The volumes for which (3.1) and (3.2) have to be evaluated are defined by the computational grid. In Figure 9 the most simple case of an equidistant Cartesian mesh is shown in 2D. For the presentation of the basic ingredients of the FV method this kind of mesh is sufficient.

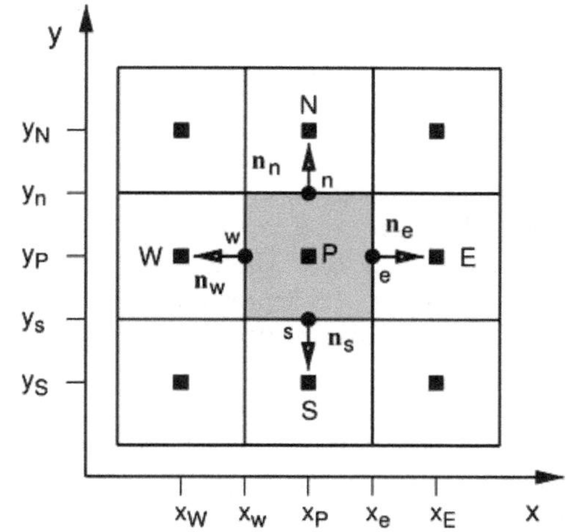

Figure 9. Cartesian computational grid and definition of the volume V_P using compass notation.

The quadrilateral cells in Figure 9 are enclosed by four distinct surfaces, which are lines in 2D. These surfaces are plane and characterized by their unit normal vectors. Labeling of the surfaces follows the so called compass notation, i.e. west, east, south and north surface. This common notation is also used for the geometric centers of the cells sharing surfaces with the volume V_P for which the integral equations will be discretized. However, to simplify the presentation the discretization will not be introduced for equations (3.1) and (3.2), but for the following generic scalar transport equation.

$$\underbrace{\frac{\partial}{\partial t}\int_{V_P}\phi\, dV}_{\text{unsteady term}} + \underbrace{\oint_{\partial V_P} n_j \phi u_j\, dA}_{\text{convective term}} - \underbrace{\oint_{\partial V_P} n_j\left(\Gamma\frac{\partial\phi}{\partial x_j}\right) dA}_{\text{diffusive term}} = \underbrace{\int_{V_P} S(\phi)\, dV}_{\text{source term}} \tag{3.3}$$

Here the scalar ϕ may represent any of the filtered or averaged quantities for which a transport equation was introduced in section 2. Equation (3.3) agrees best with the transport equations for the turbulent kinetic energy k, (2.22), and its dissipation rate ε, (2.23), presented in section 2.3 for the steady RANS approach. To recover, e.g., the continuity equation (3.1) one has to choose $\phi = 1$. To express the momentum equation (3.2) in the form of (3.3), the term containing the pressure and that part of the viscous and turbulent stresses that cannot be represented by the diffusion term in (3.3) are often collected in the source term. Due to their definition as surface integrals they should however also be approximated as surface integrals, which means that there is no one to one correspondence between (3.3) and (3.2). But the general results of the discretization can easily be transferred to equation (3.2).

Equation (3.3) can be further modified by introducing the volume average of ϕ in cell V_P,

$$^V\overline{\phi}_P \equiv \frac{1}{V_P}\int_{V_P}\phi\, dV \quad , \tag{3.4}$$

and by splitting the entire surface ∂V into the four distinct surfaces A_w, A_e, A_s and A_n.

$$\frac{\partial}{\partial t}\left({}^{V}\overline{\phi}_P V_P\right)+\sum_{l=w,e,s,n}\int_{A_l} n_j \phi u_j \, dA-\sum_{l=w,e,s,n}\int_{A_l} n_j\left(\Gamma\frac{\partial \phi}{\partial x_j}\right) dA={}^{V}\overline{[S(\phi)]}_P V_P \tag{3.5}$$

Equation (3.5), where the volume average of the source term is defined in analogy to (3.4), clearly shows that the flow variables that can be computed with the FV approach are the volume averages defined over each computational cell. These are spatial averages similar to the ones defining the resolved flow variables in the LES approach, see section 2.2. In fact, definition (3.4) instead of (2.3) or (2.8) was used by Schumann (1975) in his LES approach, known as volume-balance procedure, to define the spatially resolved flow variables. With this approach the spatial filter and the volume of the computational cell are identical, leading to a concise definition of the implicit LES approach. Due to the appearance of surface integrals in (3.5) the volume-balance procedure uses two different spatial averages which constitutes its main difference from the LES approach presented in section 2.2.

In the following the most common approximations for each of the four terms constituting the transport equation (3.5) will be introduced, based on Ferziger and Perić (2002).

3.2 Approximation of convective/advective fluxes

The convective term in Equation (3.5), which is also known as advective term, e.g., in meteorology, consists of four surface integrals. As the general procedure of approximating these integrals is the same for all surfaces only the *e* surface integral will be regarded in the following. Using the fact that the normal unit vector is constant on the surface the integral is reformulated as

$$n_{j,e}\int_{A_e} \phi u_j \, dA = n_{j,e}\,{}^{e}\overline{\phi u_j}\,A_e \quad . \tag{3.6}$$

in terms of the surface average, ${}^{e}\overline{\phi u_j}$, which is defined in analogy to the volume average (3.4). Knowing the geometrical quantities the approximation of the surface integral is now reduced to approximating the surface average by numerical integration. The most common approximation is the one point Gaussian quadrature, also known as midpoint rule. With the midpoint rule the surface average is approximated by the value in the geometric center of the surface.

The error of this approximation can be shown with the aid of a Taylor series expansion ($\Delta y = y_n - y_s$, see Figure 9).

$${}^{e}\overline{\phi u_j}=\left(\phi u_j\right)_e+\frac{(\Delta y)^2}{24}\left[\frac{\partial^2\left(\phi u_j\right)}{\partial y^2}\right]_e+\frac{(\Delta y)^4}{1920}\left[\frac{\partial^4\left(\phi u_j\right)}{\partial y^4}\right]_e+O\left[(\Delta y)^6\right] \tag{3.7}$$

The midpoint rule only retains the first term on the right hand side of (3.7). The method is therefore said to be second order accurate as the lowest exponent of the grid width in the neglected terms, which constitute the truncation error, is two. The order of a numerical approximation therefore mainly contains information on the dependence of the error on the grid width as the magnitude of the partial derivatives is normally unknown.

With the midpoint rule (3.6) is approximated as

$$n_{j,e} \int_{A_e} \phi u_j \, dA = n_{j,e} \, {}^{e}\overline{\phi u_j} \, A_e \approx n_{j,e} \left(\phi u_j\right)_e A_e = n_{j,e} \phi_e u_{j,e} A_e \quad . \tag{3.8}$$

The last step in the approximation of the convective flux is the approximation of the flow variables in the geometric center of the surface by interpolation. The interpolation is performed with the values of the variables in the geometric centers of the surrounding cells. These values themselves are approximated within second order in the grid width by the volume averages (3.4) over the cells, i.e.

$$\phi_P \approx {}^{V}\overline{\phi}_P \, . \tag{3.9}$$

With the cell center values a first order approximation of the surface center values is the upwind method,

$$\phi_e \approx \begin{cases} \phi_P & \text{for} \quad n_{j,e} u_{j,e} > 0 \\ \phi_E & \text{for} \quad n_{j,e} u_{j,e} < 0 \end{cases} \quad , \tag{3.10}$$

The first order upwind method is however viewed as too inaccurate to be used in CFD applications as it introduces a large amount of what is known as numerical diffusion into the solution. This can be understood when looking at the leading error term in the truncation error of this approximation. For flow in positive x-direction the Taylor series expansion is ($\Delta x = x_e - x_P$)

$$\phi_e = \phi_P + \frac{\Delta x}{2} \left(\frac{\partial \phi}{\partial x} \right)_e + O\left[(\Delta x)^2 \right] \, . \tag{3.11}$$

The first term in the truncation error contains the first derivative of the flow variable and therefore behaves like a diffusion term, cf. Equation (3.3). The numerical diffusion coefficient of this error is $\Gamma_{num} = n_{j,e} \, u_{j,e} \, \Delta x/2$, which is normally much larger than the molecular diffusion coefficient Γ for the high Reynolds number flows encountered in wind engineering.

When the first order upwind method (3.11) is applied to the convective term in the momentum equation (3.2) it also introduces additional numerical dissipation into the numerical solution. This numerical dissipation is especially problematic in unsteady simulations with either the LES, URANS or DES approach, where it substantially reduces the fluctuations in the solution.

A second order interpolation for the surface center values is linear interpolation,

$$\phi_e \approx \frac{x_e - x_P}{x_E - x_P} \phi_E + \frac{x_E - x_e}{x_E - x_P} \phi_P \quad . \tag{3.12}$$

This approximation does not introduce numerical diffusion. It may however result in oscillatory solutions if the mesh is too coarse in regions of high gradients. By reducing the mesh width these oscillations can be removed.

Another interpolation method often used in CWE is the QUICK (Quadratic Upwind Interpolation for Convective Kinematics) scheme introduced by Leonard (1979). Here a parabola is fit through three volume centers. Two values are taken from the downwind side, which is determined by the velocity direction in e like with the first order upwind method (3.10). The method is third order on equidistant grids and second order on non equidistant grids. The influence of the interpolation method on the pressure distribution at the centerline of the Silsoe cube is shown in Figure 10 for the 0° approach flow direction using the RNG k-ε turbulence model (cf. section

2.3). The first order upwind method (UPW1), the QUICK scheme and the second order upwind method (UPW2) have been used with FLUENT (2005). The second order upwind method also uses information from the downwind cell to compute the value in the cell center. The extrapolation is however not constant like with the first order upwind method (3.10), but uses the gradient of the variables in the cell center (cf. section 3.3) to obtain a second order extrapolation.

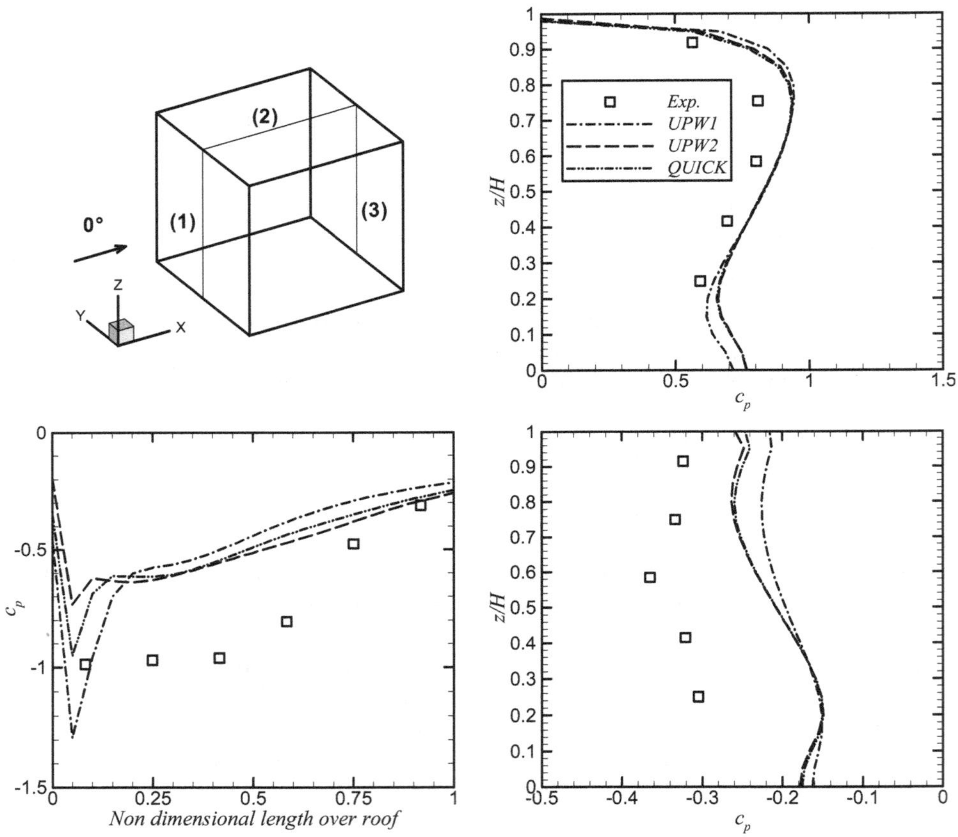

Figure [illegible] Influence of the approximation for the convective term on the pressure distribution.

The largest differences are obtained at the windward corner of the roof, where the QUICK scheme shows the best agreement with the experiments. At the windward and leeward side the results of the QUICK and the second order upwind method are nearly identical and do also not differ much from the first order upwind scheme.

For LES in general high order schemes are required. However, the numerical diffusion of some approximations is also used as the only SGS model in the aforementioned MILES approach to dissipate energy at the grid scale (Sagaut, 2002; Grinstein and Fureby, 2004).

3.3 Approximation of diffusive fluxes

The diffusive term in Equation (3.5) is also a surface flux. The numerical integration is again performed by the midpoint rule. For the e surface this results in

$$n_{j,e}\int_{A_e}\left(\Gamma\frac{\partial\phi}{\partial x_j}\right)dA = n_{j,e}\;{}^{e}\overline{\Gamma\frac{\partial\phi}{\partial x_j}}\,A_e \approx n_{j,e}\left(\Gamma\frac{\partial\phi}{\partial x_j}\right)_e A_e \quad , \tag{3.13}$$

leaving the approximation of the first derivative in the surface center. One possible approximation results when making again use of the assumption of a linear profile between the centers of cell P and E, as was already done with the linear interpolation described in the previous section. The derivative in the surface center is then approximated as

$$\left(\frac{\partial\phi}{\partial x_j}\right)_e \approx \frac{\phi_E-\phi_P}{x_E-x_P} \quad . \tag{3.14}$$

This approximation is second order on equidistant grids and first order on non equidistant grids.

Another possibility for the approximation of the derivatives in the surface centers is to first approximate them in the cell centers and then interpolate these approximations to the surface centers. Especially when the first derivatives in the cell centers are also used for the linear interpolation in the convective terms if the line connecting the cell centers used in the interpolation does not pass through the surface center, see section 3.2, this approach is preferred. To that end the derivative in the cell center is first approximated by the volume average of the derivative in the corresponding cell.

$$\left(\frac{\partial\phi}{\partial x_j}\right)_P \approx {}^{V}\overline{\left(\frac{\partial\phi}{\partial x_j}\right)_P} = \frac{1}{V_P}\int_{V_P}\frac{\partial\phi}{\partial x_j}dV = \frac{1}{V_P}\sum_{l=e,w,s,n}\int_{A_l} n_j\phi\, dA \approx \frac{1}{V_P}\sum_{l=e,w,s,n} n_{j,l}\,\phi_l\,A_l \tag{3.15}$$

In (3.15) the Gauss theorem has again be used to convert the volume integral into a surface integral. Approximation of the derivative in P is thus requires again the values of ϕ in the surface centers, like in the approximation of the convective term. The same interpolation can therefore be employed. Interpolation is finally also used to obtain the surface center derivatives from the cell center derivatives. As for the convective term the order of the entire approximation is determined by the minimum order of the single approximations. Normally second order methods are used.

3.4 Approximation of source terms

The volume averaged source term in Equation (3.5) is usually approximated as

$$^{V}\overline{[S(\phi)]_P} \approx S(\phi_P) \quad , \tag{3.16}$$

i.e. it is evaluated with the cell center values. When the dependence of S on ϕ is non linear then a linearization is performed to improve the numerical solution of the equation.

3.5 Approximation of unsteady terms

Inserting all the spatial approximations presented in the previous sections into Equation (3.5) and moving all these approximations to the right hand side yields the following semi discrete equation,

$$\frac{\partial^V \overline{\phi}_P}{\partial t} = R\left[{}^V\overline{\phi}(t)\right] \quad , \tag{3.17}$$

where R contains all the space approximations in terms of the volume averages. This equation is discrete in space but still continuous in time. For the discretisation in time Equation (3.17) has to be evaluated at discrete times, e.g. at t_{n+1}, where n is the time step index.

$$\left(\frac{\partial^V \overline{\phi}_P}{\partial t}\right)_{t=t_{n+1}} = R\left[{}^V\overline{\phi}(t_{n+1})\right] \tag{3.18}$$

The left and right hand side of Equation (3.18) have to be evaluated at the same time. In (3.18) the time derivative has to be approximated at the new time t_{n+1}, where the solution is required, with the aid of the known solutions at $t \leq t_n$. Like for the approximation in space approximations with different orders in the time step $\Delta t = t_{n+1} - t_n$ are possible. First order approximations again introduce numerical diffusion and dissipation and should therefore not be used. A common second order approximation for the time derivative for Δt = const. is the three time level method.

$$\left(\frac{\partial^V \overline{\phi}_P}{\partial t}\right)_{t=t_{n+1}} \approx \frac{3\,{}^V\overline{\phi}_P(t_{n+1}) - 4\,{}^V\overline{\phi}_P(t_n) + {}^V\overline{\phi}_P(t_{n-1})}{2\Delta t} \tag{3.19}$$

By evaluating (3.17) at the new time t_{n+1} an implicit Equation for the unknown solution is obtained which has to be solved by an iterative method, as described in the next section. An explicit equation for ${}^V\overline{\phi}_P(t_{n+1})$ results when Equation (3.17) is evaluated at the known time t_n or integrated over Δt. In the latter approach Runge-Kutta methods are often used for the numerical time integration. For explicit approximations in time the time step size that can be used in the time integration is however limited due to stability reasons. As this limitations in general leads to very small time steps, implicit methods are preferred as the choice of the time step size can be solely based on physical arguments, like the resolution of frequencies inferred from the turbulent kinetic energy spectrum.

3.6 Iterative solution of the algebraic system of equations

Inserting all the approximations described in the previous section into Equation (3.5) the following algebraic equation for the unknown value of ${}^V\overline{\phi}$ in cell P at time t_{n+1} is obtained (see Figure 9).

$$a_P^{(n+1)}\,{}^V\overline{\phi}_P(t_{n+1}) + \sum_{l=W,E,S,N} a_l^{(n+1)}\,{}^V\overline{\phi}_l(t_{n+1}) = b_P^{(n+1)} \tag{3.20}$$

The coefficients a contain geometrical quantities, fluid properties and the velocity. $b_P^{(n+1)}$ is the right hand side which contains all known values like, e.g., the solutions at the previous time steps.

An algebraic equations (3.20) is obtained for every cell in the computational domain, leading to an algebraic system of equations that must be solved.

$$\boldsymbol{A}^{(n+1)}\boldsymbol{\Phi}^{(n+1)} = \boldsymbol{b}^{(n+1)} \tag{3.21}$$

Here $\boldsymbol{A}^{(n+1)}$ is the M x M coefficient matrix containing the coefficients a, where M is the number of cells in the computational domain. $\boldsymbol{\Phi}^{(n+1)}$ is the vector of the M unknown solution variables and $b^{(n+1)}$ is the right hand side vector. The time dependence $(n+1)$ will be omitted in the following for better readability.

When solving (3.21) with an iterative method one first specifies an initial guess for the solution at $m=0$, where m is the iteration number. After m iterations a solution $\boldsymbol{\Phi}^{(m)}$ is obtained that does not satisfy (3.21), yielding the definition of the residual vector $\boldsymbol{r}^{(m)}$.

$$\boldsymbol{A}\boldsymbol{\Phi}^{(m)} - \boldsymbol{b}^{(m)} = \boldsymbol{r}^{(m)} \tag{3.22}$$

As can be seen from Equation (3.22), the iterative solution method will converge to the exact solution of the algebraic system of equations (3.21) when the residual vector vanishes. The behavior of the residual vector in dependence of the iteration number is normally monitored by its L_1 or L_2 norm, which are defined as

$$L_1 - \text{norm}: \quad \left\|\boldsymbol{r}^{(m)}\right\|_1 = \sum_{k=1}^{M}\left|r_k^{(m)}\right| \quad ; \quad L_2 - \text{norm}: \quad \left\|\boldsymbol{r}^{(m)}\right\|_2 = \sqrt{\sum_{k=1}^{M}\left|r_k^{(m)}\right|^2} \quad . \tag{3.23}$$

Convergence of the iterative solution is then usually judged by a normalized norm of the residual vector. If the actual norm of the residual vector is divided by its corresponding value after the first iteration,

$$r^{(m)} = \frac{\left\|\boldsymbol{r}^{(m)}\right\|}{\left\|\boldsymbol{r}^{(1)}\right\|} \tag{3.24}$$

then $r^{(m)}$ shows the reduction of the residual as a fraction of the initial residual norm. As the definition of the normalized residual is code dependent, the precise definition should always be given, if the iterative convergence of a solution is discussed. E.g., FLUENT (2005) uses for the continuity equation a residual that is normalized with the maximum residual norm of the first five iterations. For all other equations the normalized residual is defined as

$$r^{(m)} = \frac{\left\|\boldsymbol{r}^{(m)}\right\|}{\sum_{k=1}^{M}\left(a_P\, {}^V\overline{\phi}_P\right)_k} \tag{3.25}$$

The iterative solution of (3.21) is controlled by the user who defines a threshold value for the normalized residual. This threshold value is known as convergence criterion. If the normalized residuals for all the variables that are solved for have dropped below the corresponding convergence criteria, then the solution is stopped. Depending on the magnitude of the prescribed convergence criteria the computed solution still contains an error from the incomplete iterative solution, as the computed solution does not fulfill (3.21). The magnitude of this incomplete iterative solution error should be analyzed by comparing the results obtained with different values for the convergence criteria.

In Figure 11 the influence of the convergence criteria on the pressure distribution along the center line of the Silsoe cube is shown for the 0° approach flow direction, obtained with the standard k-ε model. For other turbulence models the results are similar.

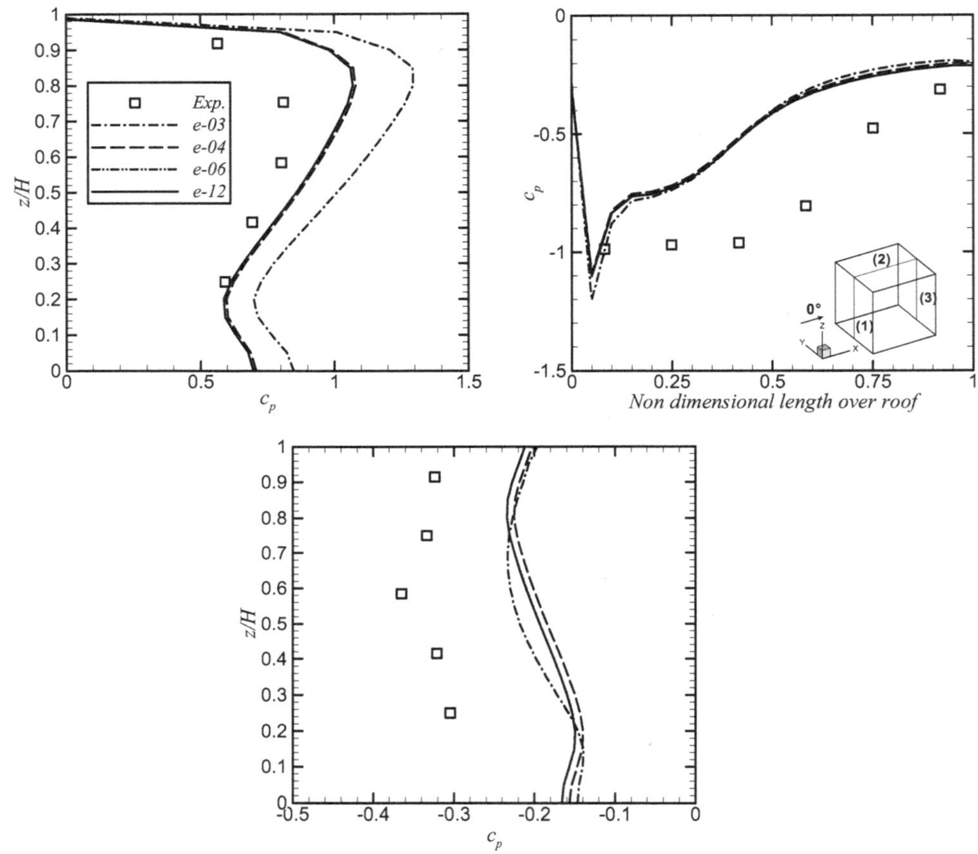

Figure 11. Influence of the convergence criteria on the pressure distribution along the front (up left), roof (up right) and leeward center line.

Clearly the default criteria of 10^{-3}, which are used in many commercial general purpose CFD codes as default values are insufficient to have a converged solution, especially at the windward face. With 10^{-4} the pressure coefficients differ only slightly from the results obtained with convergence criteria of 10^{-6} and 10^{-12}, which are indistinguishable. Therefore at least a convergence criterion of 10^{-4} should be used for the residuals defined in (3.24) and (3.25).

If the normalized residuals do not drop below the prescribed criteria or even do not converge at all but oscillate around large values, then the variables that determine the solution should be monitored in addition to the residuals. For wind loading this normally means monitoring the pressures at several positions on the structures or the entire load as a function of the iteration number. Judgement of the iterative convergence of the solution can then additionally be based on the behavior of the monitored variables. For steady state computations these variables should be constant or at least oscillate around a constant average value.

4 Boundary conditions and grids

In this section boundary conditions are presented. These have to be discussed together with the computational domain, especially its size. In addition the choice of the computational grid is briefly discussed in section 4.2. Important are the quality, the shape of the cells and the spatial variation of the cell size.

4.1 Boundary conditions and computational domain

The computational domain defines the region in which the flow field is computed. It first of all includes the structure for which the wind loads are to be determined. This structure should be represented geometrically as detailed as possible. Additionally all those buildings or topography that are assumed to have an influence on the structure of interest must be included in the computational domain. The further away these buildings are from the structure of interest the less important is their detailed geometry.

After having defined the built area, the distance of the boundaries of the computational domain from the built area and especially the structure of interest must be chosen. The distances are mainly determined by the type of boundary condition that will be applied at the corresponding boundary. For the common box shaped computational domain, like the ones shown in Figure 2, the fluid normally enters through one side, the inflow plane, and leaves through the outflow plane. Flow through the lateral sides is prevented by applying symmetry boundary conditions there. If the simulation results are to be compared with experiments from a boundary layer wind tunnel then wall boundary conditions should be used there, corresponding to the experimental boundary conditions. The cross section of the computational domain should ideally equal the wind tunnel cross section.

The symmetry boundary condition is also often used at the top side of the domain, but the application of fixed values for the velocities and turbulence quantities is to be preferred as will be shown in the following when discussing the inflow boundary condition. For the latter boundary condition there will also be no flow through the top boundary. Therefore the distance of the top boundary from the structure of interest should be five times its height to prevent an artificial acceleration of the flow over the structure. For the same reason the lateral sides of the computational domain should also be far enough away from the structure of interest. For single buildings in an otherwise empty domain a maximum blockage of 3% is recommended, where the blockage is defined as the ratio of the structures frontal area perpendicular to the flow and the cross section of the computational domain (Baetke et al., 1990). This requirement together with the recommended height can be used to determine the distance of the lateral boundaries.

The extent of the computational domain in flow direction and the boundary conditions applied at the inflow and outflow plane will be discussed separately for the steady RANS and the unsteady approaches, like LES, URANS or DES in the following.

Steady RANS. The most common outflow boundary condition is to either specify a constant pressure at this plane or to force a vanishing of the gradients of all flow variables over this plane, which is normally called outflow boundary conditions. As the normal gradients of the velocity components are also forced to vanish with the constant pressure boundary condition, both kinds of outflow conditions are ideally placed in regions where the flow field only changes moderately in flow direction. For a single structure in the domain a distance of the outflow plane from the structure of ten to fifteen times the structure's height is normally recommended.

The inflow plane should be located outside the influence region of the structure of interest. A distance of five times the structure's height is normally recommended. At the inflow plane normally equilibrium boundary layer profiles are prescribed for the velocity and turbulence quantities. Richards and Hoxey (1993) derived the corresponding profiles for the standard k-ε model under the assumptions of a one dimensional flow with constant shear stress, vanishing viscosity and vanishing normal velocity component.

$$\langle u \rangle (z) = \frac{u_\tau}{\kappa} ln\left(\frac{z + z_0}{z_0}\right) \tag{4.1}$$

$$k(z) = u_\tau^2 / \sqrt{C_\mu} \tag{4.2}$$

$$\varepsilon(z) = u_\tau^3 / [\kappa(z + z_0)] \tag{4.3}$$

When these profiles are used with the implementation of the rough wall function approach described in section 2.5 a substantial change of the profiles even inside an empty computational domain has been observed by several authors, see Blocken et al. (2006) and Hargreaves and Wright, (2006) for the most recent discussion of this topic. Several requirements have to be met to keep the changes as small as possible. First of all the profiles must be generated with the value for κ that is also used in the code. Then the constant σ_ε normally has to be changed according to

$$\sigma_\varepsilon = \frac{\kappa^2}{\sqrt{C_\mu}(C_{\varepsilon 2} - C_{\varepsilon 1})} \quad , \tag{4.4}$$

which is an additional condition under which (4.1) – (4.3) were obtained. The next requirement concerns the roughness on the floor of the computational domain. In case that wall functions based on z_0 are used, the z_0 specified in the boundary condition at the floor must be the same as the one used in (4.1) and (4.3). If wall functions based on the equivalent sandgrain roughness are used, then k_s has to computed from z_0 of the inflow boundary conditions as discussed in section 2.5. k_s then determines the minimum height of the computational cells at the floor, which must be twice as high as k_s to yield physically meaningful results. As k_s is normally much larger than z_0 this requirement can lead to very high cells at the floor. Blocken et al. (2006) discuss possible remedies for this problem.

Finally the boundary condition the top should reflect the assumption of a constant shear stress boundary layer. Therefore either this constant shear stress should be applied at the top boundary or constant values for the velocity components and turbulence quantities, corresponding to the profile values at that height. The often used symmetry boundary condition corresponds to a zero shear stress, incompatible with (4.1) – (4.3). The influence of this boundary condition on the profiles close to the floor of the computational domain are however small for typical distances between the inflow plane and the building. In Figure 12 the relative errors in the profiles for the velocity, turbulent kinetic energy and its dissipation are shown at a distance of 6.5H from the inflow plane. The relative error in, e.g., k is defined as $\Delta k = [k(x/H{=}6.5) - k_{in}]/k_{in}$, where k_{in} is the inflow profile computed from (4.2).

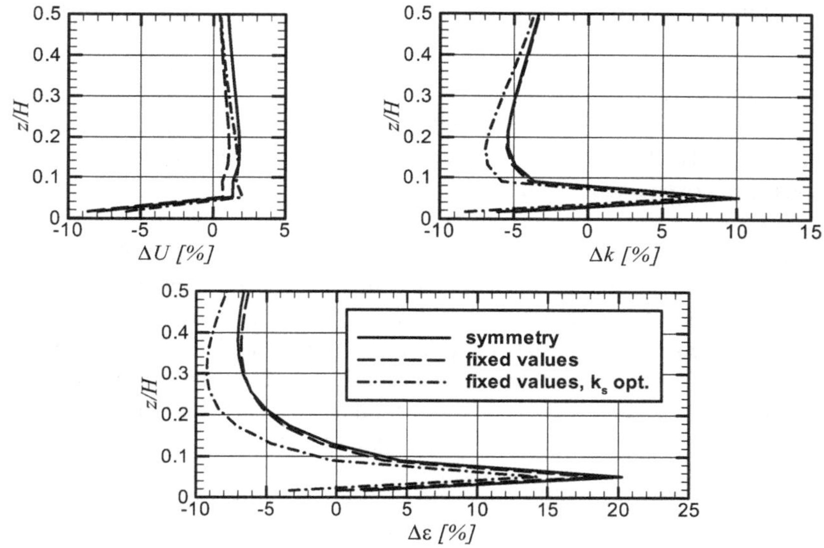

Figure 12. Relative errors of the computed profiles at 6.5H from the inlet in an empty domain.

The simulations are performed with the standard k-ε model in a 2D domain with the same mesh layout and other parameter settings (approximations in space, convergence criteria) as in the 3D case with 0° approach flow. The constant $\sigma_\varepsilon = 1.217$ has been used, which follows from (4.4) with $\kappa = 0.4187$ and the standard values for the other constants, cf. section 2.3. The use of the symmetry boundary condition at the top mainly changes the profiles close to the top boundary at least for a distance of 6.5H. Then the relative errors in all profiles are negligible down to a height of app. 2H. Below 2H the profiles again change substantially due to the wall boundary conditions. In Figure 12 the errors in the profiles are shown up to half the cube height to enhance the visibility of the differences. First of all it can be noted that the top boundary condition only has a small influence on the velocity profile with slightly larger errors for the symmetry condition. The turbulence quantities are unaffected. The error in the velocity is small except for the first cell at the wall, where the velocity is reduced by app. 10%. For k and ε the largest errors are obtained in the second cell off the wall. This can be attributed to a peak in the computed k in the second cell off the wall, which was also shown by Hargreaves and Wright (2006). The reason for this peak is at the moment not well understood.

The third parameter varied in the simulations is the sandgrain roughness at the floor. The optimized value for $k_s = 0.08903m$ is obtained from Equation (2.43) presented in section 2.5. For the other computations $k_s = 0.09793m$ was calculated from Equation (2.42). The optimized value of the roughness reduces the maximum errors in all profiles.

The presented results show the importance of checking the horizontal development of the prescribed inflow profiles. Ideally a simulation in an empty domain with the same grid layout and the same physical and numerical parameters should accompany a wind loading simulation. From the above results usage of a symmetry boundary condition at the top can be justified despite its known deficiencies.

Yet another problem in the prescription of inflow profiles is the magnitude of the turbulent kinetic energy computed from (4.2) with the standard coefficient $C_\mu = 0.09$. This value leads to a turbulent kinetic energy which is much lower than in full scale and wind tunnel experiments (see, e.g., Bottema, 1997). The lower value of C_μ is confirmed by Durbin and Petterson Reif (2001) for boundary layer flows over smooth wall.

Unsteady boundary conditions. For unsteady simulation approaches the inflow boundary conditions must also be time dependent. In the context of LES several approaches to generate time dependent velocities have been described. Kondo et al. (1998) generated correlated velocity fluctuations from measured spectra. Nozawa and Tamura (2002) modeled the flow over roughness elements in a domain with periodic boundary conditions with taking the increasing boundary layer height into account. Another approach is to explicitly model the approach flow like in wind tunnels by using a step, vortex generators and roughness elements placed on the floor (see, e.g., Nakayama et al., 2005).

Concerning the boundary conditions at walls similar requirements as for the steady RANS approach exist for the unsteady simulation approaches.

Finally, at the outflow plane the so called convective outflow boundary condition should be used (see, e.g., Fröhlich and Rodi, 2002).

4.2 Grids

The grids used in the computational domain do not only determine the spatial resolution of the solution but also have a substantial influence on the accuracy of the solution and the iterative convergence. The latter two aspects are mainly influenced by the type of grid that is used and by its quality. Ideally the grid is equidistant. Therefore, grid stretching/compression should be small in regions of high gradients to keep the truncation error small. The expansion ratio between two consecutive cells should be below 1.3 in these regions. Scaperdas and Gilham (2004), as well as Bartzis et al. (2004) even recommend a maximum of 1.2 for the expansion ratio. However, higher order numerical schemes might allow larger changes as the absolute value of the truncation error is smaller than with lower order schemes. In the context of Finite Volume methods another criterion for grid quality is the angle between the normal vector of a cell surface and the line connecting the midpoints of the neighbouring cells (Ferziger and Perić, 2002). Ideally these should be parallel. With regard to the shape of the computational cells, hexahedra are preferable to tetrahedra, as the former are known to introduce smaller truncation errors and display better iterative convergence (Hirsch et al., 2002). On walls the grid lines should be perpendicular to the wall (Casey and Wintergerste, 2000). Therefore if a tetrahedral grid is to be used, prismatic cells should be used at the wall with tetrahedral cells away from the wall.

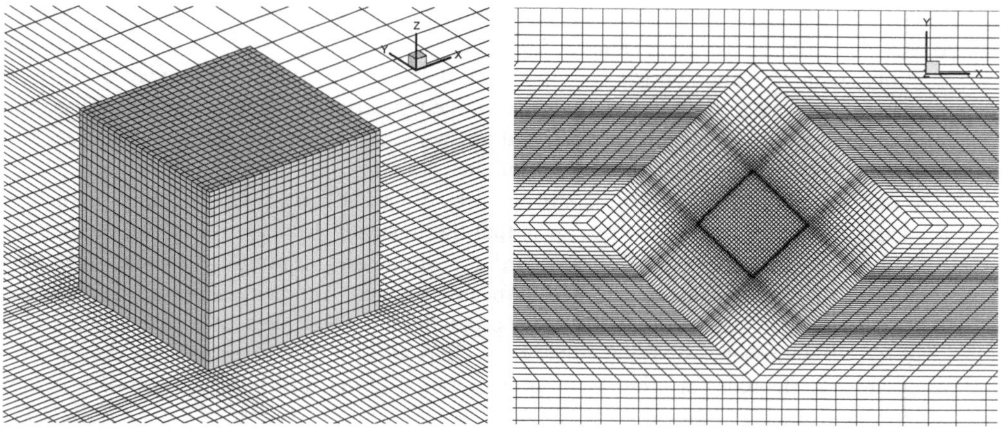

Figure 13. Surface mesh around the cube for the 0° (left) and 45° (right) approach flow for the coarse grids.

In Figure 13 the surface mesh close to the cube is shown for the 0° (left) and 45° (right) approach flow direction. The resolution is roughly identical for the both directions. In the 45° case a quadratic region around the cube has been used to guarantee that the grid lines are normal to the cube surface. All together six meshes have been used in the simulations, three systematically refined meshes for each case. For the 0° approach flow direction the fine mesh has 1 510 160, the medium mesh 753 632 and the coarse mesh 365 751 cells. For the 45° case the corresponding size are 1 857 024, 931 952 and 457 542. With the results on these meshes the dependence of the pressure distribution on the spatial resolution were determined quantitatively by visual comparison. The differences between the pressures obtained on the medium and fine mesh are small.

For a quantitative determination of the influence of the resolution on the results Richardson extrapolation should be used (Roache, 1997; Roy, 2005). With the results obtained on different grids it is then possible to estimate the numerical error due to the spatial discretization. In general three solutions corresponding to three different grids are required which makes the procedure expensive. However, or quality assurance of CFD codes and simulation results in the context of validation and verification Richardson extrapolation is required (Oberkampf et al., 2004).

When a global systematic grid refinement is not possible due to resource limitations then at least a local grid refinement around the structure should be used. Most commercial CFD codes offer the possibility to locally refine the grid in dependence of the gradients or curvature of flow variables.

5 Conclusions

In this chapter a basic introduction to the prediction of wind loads on structures by means of CWE has been given. The different simulation approaches, ranging from LES over URANS and DES to the predominating steady RANS, were introduced together with their most common turbulence models. The transformation of the resulting systems of non linear partial differential equations into systems of non linear algebraic equations was sketched for the Finite Volume method together with the iterative solution process of the latter systems. Finally common boundary conditions, their influence on the computational domain size and recommended grid layouts were described. Novices in the field of CFD for wind engineering applications should have found a first overview of the many approximations that are necessary in this discipline to, finally, obtain a computational result of the pressure distribution on a structure.

While the steady RANS approach is still the prevailing method in wind loading predictions, LES is increasingly used. The main reason is that the unsteady LES approach can predict time series of pressures with corresponding standard deviations and minimum and maximum pressures. The steady RANS approach is limited to mean pressures. Another reason for the preference of LES is that it in most cases also yields better predictions of mean pressures and velocities than the steady RANS approach. For statements about URANS and DES there are not yet sufficient applications of the methods for wind loading predictions. In the future these methods will however be more and more combined with the LES approach in the context of hybrid LES/RANS simulations. The still increasing availability of affordable computing power supports the increasing use of time dependent simulation approaches, which are also required in the emerging field of simulating the interaction of the fluid and the structure.

References

Aldama, A. A. (1990). *Filtering Techniques for Turbulent Flow Simulation*. Berlin: Springer Verlag.

Baetke, F., Werner, H. and Wengle, H. (1990). Numerical Simulation of Turbulent Flow over Surface Mounted Obstacles with Sharp Edges and Corners. *J. Wind Eng. Ind. Aerodyn.* 35:129-147.

Bardina, J., Ferziger, J. H., and Reynolds, W. C. (1983). *Improved turbulence models based on large eddy simulation of homogeneous, incompressible, turbulent flow*. Report TF-19, Thermosciences Division, Dept. Mechanical Engineering, Stanford University.

Bartzis, J.G., Vlachogiannis, D. and Sfetsos, A. (2004). Thematic area 5: Best practice advice for environmental flows. *The QNET-CFD Network Newsletter* 2:34-39.

Bitsuamlak, G., Stathopoulos, T., and Bedard, C. (2004). Numerical evaluation of wind flow over complex terrain: Review. *Journal of Aerospace Engineering* 17: 135-145.

Blocken, B., Stathopoulos, T. and Carmeliet, J. (2006). CFD simulation of the atmospheric boundary layer – wall function problems. *Atmospheric Environment*, accepted for publication.

Bottema, M. (1997). Turbulence closure model "constants" and the problems of "inactive" atmospheric turbulence. *J. Wind Eng. Ind. Aerodyn.* 67&68: 897-908.

Breuer, M., Jovičić, N., Mazaev, K. (2003). Comparison of DES, RANS and LES for the separated flow around a flat plate at high incidence. *Int. J. Numer. Meth. Fluids* 41:357-388.

Casey, M. and Wintergerste, T., eds. (2000). *ERCOFTAC SIG "Quality and Trust in Industrial CFD": Best Practice Guidelines*. ERCOFTAC.

Cebeci, T., and Bradshaw, P. (1977). *Momentum Transfer in Boundary Layers*. New York: Hemisphere.

Chen, C.-J., and Patel, V. C. (1988). Near-wall turbulence models for complex flows including separation. *AIAA Journal* 26: 641-648.

Craft, T.J., Graham, L.J.W. and Launder, B.E. (1993). Impinging jet studies for turbulence models assessment – II. An examination of the performance of four turbulence models. *Int. J. Heat Mass Transfer* 36: 2685-2697.

Craft, T.J., Launder, B.E. and Suga, K. (1996). Development and application of a cubic eddy-viscosity model of turbulence. *Int. J. Heat and Fluid Flow* 17: 108-115.

Durbin, P. A. (1996). On the k-ε stagnation point anomaly. *Int. J. Heat and Fluid Flow* 17: 89-90.

Durbin, P. A., and Petterson Reif, B. A. (2001). *Statistical Theory and Modeling for Turbulent Flows*. Chichester: Wiley.

Durbin, P. A., Medic, G., Seo, J.-M., Eaton, J. K., and Song, S. (2001). Rough wall modification of two-layer k-ε. *ASME Journal of Fluids Engineering* 123: 16-21.

Dyrbye, C., and Hansen, S. O. (1997). *Wind Loads on Structures*. Chichester: Wiley.

Ferziger, J. H., and Perić, M. (2002). *Computational Methods for Fluid Dynamics, 3rd Edition*. Berlin: Springer Verlag.

FLUENT (2005). *FLUENT V6.2 User's Guide*. Lebanon NH: Fluent Inc.

Franke, J., Hirsch, C., Jensen, A. G., Krüs, H. W., Schatzmann, M., Westbury, P. S., Miles, S. D., Wisse, J. A., and Wright, N. G. (2004). Recommendations on the Use of CFD in Wind Engineering. In van Beeck, J. P. A. J., ed., *Proceedings of the International Conference on Urban Wind Engineering and Building Aerodynamics: COST Action C14 - Impact of Wind and Storm on City Life and Built Environment*. Rhode-Saint-Genèse: von Karman Institute for Fluid Dynamics. C.1.1 - C.1.11.

Fröhlich, J., and Rodi, W. (2002). Introduction to Large Eddy Simulation of Turbulent Flows. In Launder, B. E., and Sandham, N. D., eds., *Closure Strategies for Turbulent and Transitional Flows*. Cambridge: Cambridge University Press. 267-298.

Fu, S., Launder, B. E., and Tselepidakis, D. P. (1987). *Accomodating the effects of high strain rates in modelling the pressure-strain correlation*. Thermofluids report TFD/87/5, Manchester: UMIST.

Germano, M., Piomelli, U., Moin, P., and Cabot, W. H. (1991). A dynamic subgrid-scale eddy viscosity model. *Phys. Fluids A* 3: 1760-1765.

Geurts, B. J. (2004). *Elements of direct and large-eddy simulation*. Philadelphia PA: Edwards.

Gibson, M. M., and Launder, B. E. (1978). Ground effects on pressure fluctuations in the atmospheric boundary layer. *J. Fluid Mech.* 86: 491-511.

Ghosal, S. and Moin, P. (1995). The basic equations for the large-eddy simulation of turbulent flows in complex geometry. *J. Comput. Phys.* 118: 24-37.

Grinstein, F. F. and Fureby, C. (2004). From canonical to complex flows: Recent progress on Monotonically Integrated LES. *Comput. Sci. and Eng.* 6: 37-49.

Grötzbach, G. (1977). Direct numerical simulation of secondary currents in turbulent channel flow. In Fiedler, H., ed., *Lecture Notes in Physics 76*. Berlin: Springer Verlag. 308-319.

Hargreaves, D.M. and Wright, N.G. (2006). On the use of the k-ε model in commercial CFD software to model the neutral atmospheric boundary layer. *J. Wind Eng. Ind. Aerodyn.* doi:10.1016/j.jweia. 2006. 08.002.

Hirsch C., Bouffioux, V. and Wilquem F. (2002). CFD simulation of the impact of new buildings on wind comfort in an urban area. In Augusti, G., Borri, C. and C. Sacré, C., eds., *Proceedings of the Workshop Impact of Wind and Storm on City life and Built Environment*. Nantes: CSTB. 164-171.

Holmes, J. D. (2001). *Wind Loading of Structures*. London: Spon Press.

Hoxey, R. P., Richards, P. J., and Short, J. L. (2002). A 6m cube in an atmospheric boundary layer flow. Part1. Full-scale and wind-tunnel results. *Wind and Structures* 5: 165-176.

Kato, M., and Launder, B.E. (1993). The modeling of turbulent flow around stationary and vibrating square cylinders. In *Proceedings of the Ninth Symposium on Turbulent Shear Flows*.

Kondo, K., Murakami, S. and Mochida, A. (1997). Generation of velocity fluctuations for inflow boundary condition of LES, *J. Wind Eng. Ind. Aerodyn.* 67&68: 51-64.

Launder, B. E. and Spalding, D. B. (1972). *Lectures in Mathematical Models of Turbulence*. London: Academic Press.

Leonard, B. P. (1979). A stable and accurate convection modelling procedure based on quadratic upstream interpolation. *Comp. Meth. Appl. Mech. Engineering* 19: 59-98.

Lien, F. S., and Leschziner, M. A. (1994). Assessment of turbulent transport models including non-linear RNG eddy-viscosity formulation and second-moment closure. *Computers & Fluids* 23: 983-1004.

Ligrani, P. M., and Moffat, R. J. (1986). Structure of transitionally rough and fully rough turbulent boundary layers. *J. Fluid Mech.* 162:69-98.

Maruyama, T. (1999). Surface and inlet boundary conditions for the simulation of turbulent boundary layer over complex rough surfaces. *J. Wind Eng. Ind. Aerodyn.* 81: 311-322.

Mason, P.J. and Callen, N.S. (1986). On the magnitude of the subgrid-scale eddy coefficient in large-eddy simulations of turbulent channel flow. *J. Fluid Mech.* 162: 439-462.

Mochida, A., Tominaga, Y., Murakami, S., Yoshie, R., Ishihara, T., and Ooka, R. (2002). Comparison of various k-ε models and DSM to flow around a high rise building - report of AIJ cooperative project for CFD prediction of wind environment. *Wind and Structures* 5: 227-244.

Murakami, S. (1998). Overview of turbulence models applied in CWE-1997. *J. Wind Eng. Ind. Aerodyn.* 74-76: 1-24.

Nakayama, H., Tamura, T. and Okuda, Y. (2005). LES study of fluctuating dispersion of hazardous gas in urban canopy. In Náprstek, J., and Pirner, M., eds., *Proceedings of the 4th European-African Conference on Wind Engineering*, Prague, Paper #168.

Nozawa, K. and Tamura, T. (2002). Large eddy simulation of the flow around a low-rise building immersed in a rough-wall turbulent boundary layer. *J. Wind Eng. Ind. Aerodyn.* 90: 1151-1162.

Nozawa, K., and Tamura, T. (2003). Numerical prediction of pressure on a high-rise building immersed in a turbulent boundary layer using LES. In *Proceedings of the Annual Meeting of JACWE*, 169-170.

Oberkampf, W. L., Trucano, T. G., and Hirsch, C. (2004). Verification, validation, and predictive capability in computational engineering and physics. *Appl. Mech. Rev.* 57: 345 - 384.

Oliveira, P. J., and Younis, B. A. (2002). On the prediction of turbulent flows around full-scale buildings. . *J. Wind Eng. Ind. Aerodyn.* 86: 203-220.

Patel, V. C., Rodi, W., and Scheurer, G. (1984). Turbulence modeling for near-wall and low Reynolds number flows: a review. *AIAA Journal* 23: 1308-1319.

Pope, S.B. (1975). A more general effective-viscosity hypothesis. *J. Fluid Mech.* 72: 331-340.

Pope, S. B. (2000). *Turbulent Flows*. Cambridge: Cambridge University Press.

Richards, P.J. and Hoxey, R.P. (1993). Appropriate boundary conditions for computational wind engineering models using the k-ε turbulence model. *J. Wind Eng. Ind. Aerodyn.* 46 & 47: 145-153.

Richards, P. J., Quinn, A. D., and Parker, S. (2002). A 6m cube in an atmospheric boundary layer flow. Part2. Computational solutions. *Wind and Structures* 5: 177-192.

Richards, P. J., Hoxey, R. P., Connell, B. D., and Lander, D. P. (2005). Wind tunnel modelling of the Silsoe cube. In Náprstek, J., and Pirner, M., eds., *Proceedings of the 4th European-African Conference on Wind Engineering*, Prague, Paper #313.

Roache, P. J. (1997). Quantification of Uncertainty in Computational Fluid Dynamics. *Annu. Rev. Fluid Mech.* 29: 123-160.

Rodi, W. (1991). Experience using two-layer models combining the k-ε model with a one-equation model near the wall. *AIAA Paper* 91-0609.

Rotta, J. (1951). Statistische Theorie nichthomogener Turbulenz. 1. Mitteilung. *Zeitschrift für Physik* 129: 547-572.

Roy, C. J. (2005). Review of code and solution verification procedures for computational simulation. *J. Comp. Phys.* 205: 131-156.

Sagaut, P. (2002). *Large Eddy Simulation for Incompressible Flows, 2nd edition*. Berlin: Springer Verlag.

Scaperdas, A. and Gilham, S. (2004). Thematic Area 4: Best practice advice for civil construction and HVAC. *The QNET-CFD Network Newsletter* 2: 28-33.

Schumann, U. (1975). Subgrid scale model for finite difference simulations of turbulent flows in plane channels and annuli. *J. Comput. Phys.* 18: 376-404.

Schumann, U. (1977). Realizability of Reynolds-stress turbulence models. *Phys. Fluids* 20: 721-725.

Shih, T.-H., Liou, W.W., Shabbir, A., Yang, Z. and Zhu, J. (1995). A New k-ε Eddy-Viscosity Model for High Reynolds Number Turbulent Flows - Model Development and Validation. *Computers & Fluids* 24: 227-238.

Simiu, E., and Scanlan, R. H. (1996). *Wind Effects on Structures, 3rd edition*. New York: Wiley.

Smagorinsky, J. (1963). General circulation experiments with the primitive equations. I: The basic experiment. *Month. Weath. Rev.* 91:99-165.

Spalart, P. R., and Allmaras, S. R. (1994). A one-equation turbulence model for aerodynamic flows. *La Rech. Aérospatiale* 1:5-21.

Spalart, P. R., Deck, S., Shur, M. L., Squires, K. D., Strelets, M. Kh., and Travin, A. (2006). A new version of detached-eddy simulation, resistant to ambiguous grid densities. *Theor. Comput. Fluid. Dyn.* 20: 181-195.

Speziale, C. G., Sarkar, S., and Gatski, T. B. (1991). Modeling the pressure-strain correlation of turbulence: an invariant dynamical systems approach. *J. Fluid. Mech.* 227: 245-272.

Stathopoulos, T. (1997). Computational wind engineering: Past achievements and future challenges. *J. Wind Eng. Ind. Aerodyn.* 67&68: 509-532.

Stathopoulos, T. (2002). The numerical wind tunnel for industrial aerodynamics: Real or virtual in the new millenium. *Wind and Structures* 5: 193-208.

Tamura, T., Nozawa, K., and Kondo, K. (2006). AIJ guide for numerical prediction of wind loads on buildings. In *Proceedings of the Fourth International Symposium on Computational Wind Engineering*, 161-164.

Tominaga, Y., Mochida, A., Shirasawa, T., Yoshie, R., Kataoka, H., Harimot, K. and Nozu, T. (2004). Cross Comparisons of CFD Results of Wind Environment at Pedestrian Level around a High-rise Building and within a Building Complex. *Journal of Asian Architecture and Building Engineering* 3.

Tsuchiya, M., Murakami, S., Mochida, A., Kondo, K., and Ishida, Y. (1997). Development of a new k-ε model for flow and pressure fields around bluff bodies. *J. Wind Eng. Ind. Aerodyn.* 67&68: 169-182.

VDI (2005). *Environmental meteorology – Prognostic micro-scale wind field models – Evaluation for flow around buildings and obstacles*. VDI Guideline 3783, Part 9. Berlin: Beuth Verlag.

Wright, N.G., and Easom, G.J. (1999). Comparsion of several computational turbulence models with full-scale measurements of flow around a building. *Wind and Structures* 2: 305-323.

Wright, N.G. and Easom, G.J. (2003). Non-linear k-ε turbulence model results for flow over a building at full scale. *Applied Mathematical Modelling* 27: 1013-1033.

Yakhot, V. and Orzsag, S.A. (1986). Renormalization group analysis of turbulence: I. Basic theory. *Journal of Scientific Computing* 1: 1-51.

Yoshie, R., Mochida, A., Tominaga, Y., Kataoka, H., Harimoto, K., Nozu, T., and Shirasawa, T. (2005). Cooperative project for CFD prediction of pedestrian wind environment in the Architectural Institute of Japan. In Náprstek, J., and Pirner, M., eds., *Proceedings of the 4th European-African Conference on Wind Engineering*, Prague, Paper #292.

Zhang, H., Faghri, M., and White, F. M. (1996). A new low-Reynolds-number k-ε model for turbulent flow over smooth and rough surfaces. *ASME Journal of Fluids Engineering* 118: 255-259.

Wind-induced random vibrations of structures

Prof. Dr.-Ing. R. Höffer, Dr. rer. nat. Mozes Galffy, Prof. em. Dr.-Ing. H.-J. Niemann

Faculty of Civil Engineering, Ruhr-University Bochum, Germany
Institute for Structural Engineering

Abstract. Modern wind codes indicate calculational procedures which allow engineers to deal with structural systems which are susceptical to conduct wind excited oscillations. In the codes approximative formulas are specified which relate the dynamic problem to rather abstract parameter functions. The complete theory behind is not visible in order to simplify the applicability of the procedures. This contribution derives important basic relations of the spectral method and explains simplified assumptions for its application in order to elucidate part of the theoretical background of computations after the new codes.

1 Introduction

1.1 Stochasticity and statistic parameters

Vibrations with non-regularly fluctuating amplitudes are considered to be *random* or *stochastic processes*. A stochastic process is a statistical collective consisting of individual realisations of random functions instead of discrete events. Basic features of stochastic and deterministic processes are compared in table 1. A deterministic function, e.g. defined by $x(t) = sin(\omega t)$, describes a well-defined value x which is assigned to the time t_i. The function is reproducible, and x is predictable. A random (e.g. time dependent) function has a random value at the time t_i. The function is not reproducible, and x is not predictable. The value of the function at time t, $x(t)$, is replaced by a probabilistic statement, e.g.

$$P_{x \leq a}(a,t) = \int_{-\infty}^{a} f_x(\xi,t)\, d\xi \tag{1.1}$$

$P_{x \leq a}(a,t)$ is the probability that a realisation with a value $x \leq a$ at the time t occurs, and $f_x(\xi,t)$ is the probability density function of the stochastic variable at level $x = \xi$ and at time t.

1.2 Statistical parameters – one variable

A structure is subjected to a single random excitation, e.g. due to wind, earthquake, waves. For the description of these loads, statistical estimators are applied. Important estimators are:

arithmetical mean: $$\overline{x} = \lim_{T\to\infty} \frac{1}{T} \int_{-T/2}^{T/2} x(t)\,dt \quad (1.2)$$

square mean: $$\overline{x^2} = \lim_{T\to\infty} \frac{1}{T} \int_{-T/2}^{T/2} x^2(t)\,dt \quad (1.3)$$

variance of scattering: $$\sigma_x^2 = \overline{\left(x-\overline{x}\right)^2} = \lim_{x\to\infty} \frac{1}{T} \int_{-T/2}^{T/2} \left[x(t)-\overline{x}\right]^2 dt \text{ or } \sigma_x^2 = \overline{x^2} - \overline{x}^2. \quad (1.4)$$

Statistical measures can practically only be obtained from limited collectives. Thus, they are only estimates of the expected, 'true' values of a collective with an infinite number of elements. In general, the probability density function is *non-stationary*, $f_x(\xi,t_1) \neq f_x(\xi,t_2)$, which means that the statistical properties in general are time-dependent. The essential conditions within in this section are:

1st assumption: Stochastic vibrations are supposed to be *stationary*, which means that $f_x(\xi,t) = f_x(\xi)$ is not dependent on the time t. Stationarity causes the joint probability density function between the collectives $X_1=\{x_1(t_1),x_2(t_1),\ldots,x_N(t_1)\}$ and $X_2=\{x_1(t_2),x_2(t_2),\ldots,x_N(t_2)\}$ to not depend on times t_1 and t_2 but on the time difference $\tau = t_2 - t_1$ only, and $\overline{x}_E(t_1) = \overline{x}_E(t_2) = \overline{x}_E(t)$.

2nd assumption: The process is ergodic; its statistical properties are completely reflected by each realisation which means that $\overline{x}_{T,1} = \overline{x}_{T,2} = \overline{x}_T$ and $\overline{x}_T = \overline{x}_E$. The simplest statistic measure is the (arithmetical) mean value $\overline{x}$. Averaging with respect to the ensemble at time t leads to (fig. 1):

$$\overline{x}_E(t) = \lim_{N\to\infty} \frac{1}{N} \sum_{i=1}^{N} x_i(t) \quad (1.5)$$

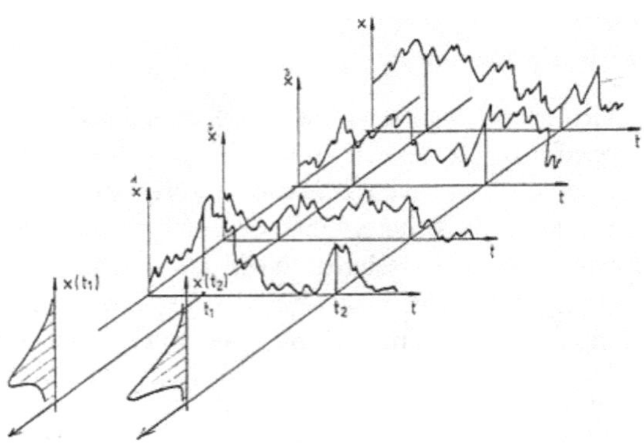

Figure 1: Realisations of random processes

Averaging over the time for the ith realisation results into:

$$\overline{x}_{T,i} = \lim_{T\to\infty} \frac{1}{T} \int_{-T/2}^{T/2} x_i(t)\,dt \quad (1.6)$$

The load is considered to be a Gaussian process with the probability density function

$$f(x)=\frac{1}{\sqrt{2\pi}\cdot\sigma_x}e^{-\frac{(x-\bar{x})^2}{2\sigma_x^2}} \quad \text{(normal or Gaussian distribution function)}. \tag{1.7}$$

The probability density function is completely described by the mean value $\bar{x}$ and the variance σ_x^2, where the latter value carries information about the probability of load amplitudes. Other representations of the scattering are:

- standard deviation: $\sigma=\sqrt{variance}$
- intensity: $I_x=\sigma/\bar{x}$

Statistical information about the properties of the process in time is required for a vibration analysis. In the time domain such information is provided by the autocorrelation function $R(\tau)$.
In the frequency – domain: spectral density $S(\omega)$, ω denotes the circular frequency. For Gaussian processes $R(\tau)$ allows for probability statements concerning the load amplitudes at two different time instants differing by τ can be made. More information and an overview about special features of the autocorrelation function are presented in the following chapters 1.3 and 1.4.

1.3 Autocorrelation

The autocorrelation function R is a statistical measure for the memory regarding the sequences of signals of a process under consideration. R is expressed as function of the argument τ, which is explained in fig. 2 as the time shift. A real stochastic process, e.g. the time history of velocity fluctuations in turbulent wind, is characterized by a quickly decaying autocorrelation function.

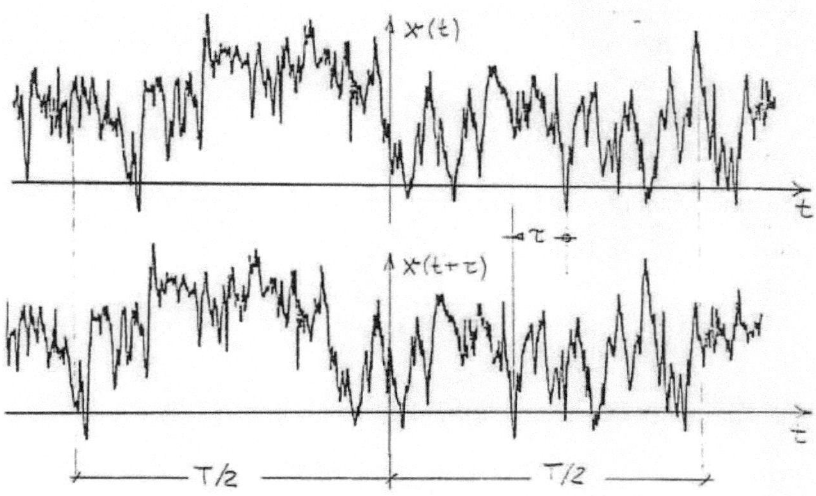

Figure 2: Autocorrelation

The autocorrelation function is defined by

$$R(\tau) = \lim_{T\to\infty} \frac{1}{T} \int_{-T/2}^{+T/2} x(t)\cdot x(t+\tau)d\tau \ .$$

The function can be characterized as the following properties:

1. $R(0) = \overline{x}^2$ - square mean
2. for $\overline{x} = 0$ (zero-mean process) is $R(0) = \sigma^2$- variance
3. $R(-\tau) = R(\tau)$ - the autocorrelation is an even function.

Fig. 3 shows examples of the behaviour of the autocorrelation function of broad band stochastic processes and narrow band stochastic processes.

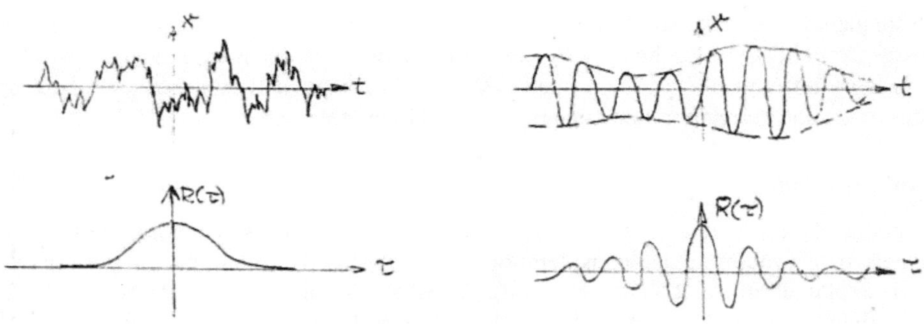

Figure 3: Broad band stochastic process (left) and narrow band stochastic process (right)

An autocorrelation function is a special case of a cross correlation function where two different processes are compared. Cross correlation functions describe the synchronism of stochastic signals.

2 Quantities and Units

The following basic quantities and units will be used in this section:

Basic Quantity		Dimension	Unit
Length	l,	[l]	m
Mass	m,	[m]	kg
Time	t,	[t]	s

The mechanical quantities refer to the basic units as follows:

Quantity		Dimension	Unit
Displacement	x, y, z	[l]	m
Velocity	$\dot{x}, \dot{y}, \dot{z}$	$[lt^{-1}]$	m/s
Acceleration	$\ddot{x}, \ddot{y}, \ddot{z}$	$[lt^{-2}]$	m/s^2

Force = mass · acceleration	[mlt^{-2}]	1kg · m/s² = 1 N
	or	103 kg . m/s2 = 1 kN
Work = force · displacement	[ml^2t^{-2}]	N · m
Density = mass /volume	[ml^{-3}]	kg/m³
	or	10^3 kg/m³ = t_0 /m³
Mass moment of inertia	[ml^2]	kg · m²
Impulse	[mlt^{-1}]	kg · m/s = N · s

3 Fourier representations

3.1 Fourier-coefficients

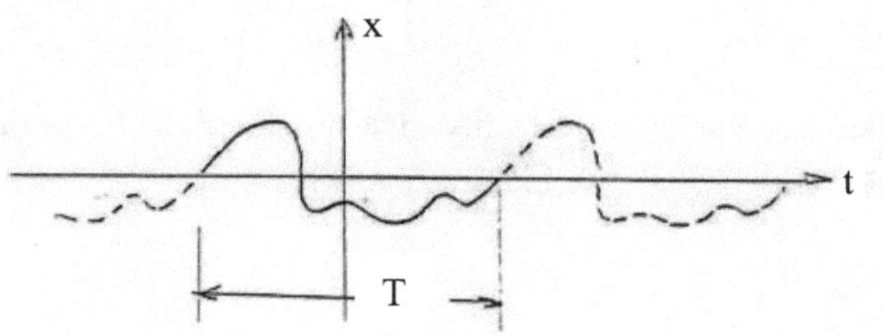

Figure 4: Periodic function

The periodic function $x(t)$ with period T can be represented as a Fourier-series:

$$x(t) = a_0 + a_1 \cos \omega_1 t + b_1 \sin \omega_1 t + a_2 \cos 2\omega_1 t + b_2 \sin 2\omega_1 t + ... + a_n \cos n\omega_1 t + b_n \sin n\omega_1 t + ...$$

where $\omega_1 = 2\pi/T$ is the circular frequency, and a_n and b_n are the Fourier-coefficients. This representation corresponds to the decomposition of the function in terms of the basis functions sin $n\omega_1 t$ and cos $n\omega_1 t$ with the real amplitudes a_n and b_n. The Fourier-coefficients can be calculated as

$$a_0 = \frac{1}{T} \int_{-T/2}^{+T/2} x(t) dt, \quad a_n = \frac{2}{T} \int_{-T/2}^{+T/2} x(t) \cos n\omega_1 t \, dt \tag{3.1a}$$

and

$$a_n = \frac{2}{T} \int_{-T/2}^{-T/2} x(t) \cos n\omega_1 t \, dt, \quad b_n = \frac{2}{T} \int_{-T/2}^{-T/2} x(t) \sin n\omega_1 t \, dt, \; for \; n \geq 1. \tag{3.1b}$$

The 1st Fourier-coefficient, a_0, is the mean value of the periodic function.
The mean square of the function is given by

$$\overline{x^2} = \frac{1}{T} \int_{-T/2}^{+T/2} (a_0 + a_1 \cos \omega_1 t + b_1 \sin \omega_1 t + ...)^2 dt.$$

The basis vectors are orthogonal:

$$\int_{-T/2}^{T/2} \cos n\omega_1 t \cos m\omega_1 t \, dt = \int_{-T/2}^{T/2} \sin n\omega_1 t \sin m\omega_1 t \, dt = 0 \quad for \quad n \neq m,$$

$$\int_{-T/2}^{T/2} \sin n\omega_1 t \sin m\omega_1 t \, dt = 0$$

$$\int_{-T/2}^{T/2} \sin n\omega_1 t \sin m\omega_1 t \; dt = 0 \;\; \text{for any } n,m.$$

The square of their norm is T for $n = 0$ and

$$\int_{-T/2}^{T/2} \cos^2 \omega_1 t \; dt = \int_{-T/2}^{T/2} \sin^2 n\omega_1 t \; dt = \frac{T}{2} \; for \; n \geq 1.$$

Consequently, the mean square of the function can be calculated as

$$\overline{x^2} = a_0^2 + \sum_{n=1}^{\infty} \frac{a_n^2 + b_n^2}{2} \; .$$

This equation is called the Parseval's equality. The variance of the periodic function can be calculated from $\sigma^2 = \overline{x^2} + \overline{x}^2 = \overline{x^2} - a_0^2$, which results in

$$\sigma^2 = \sum_{n=1}^{\infty} \frac{a_n^2 + b_n^2}{2} = \sum_{n=1}^{\infty} \Delta\sigma_n^2$$

Here, the term $\Delta\sigma_n^2$ represents the nth modal component of the variance.

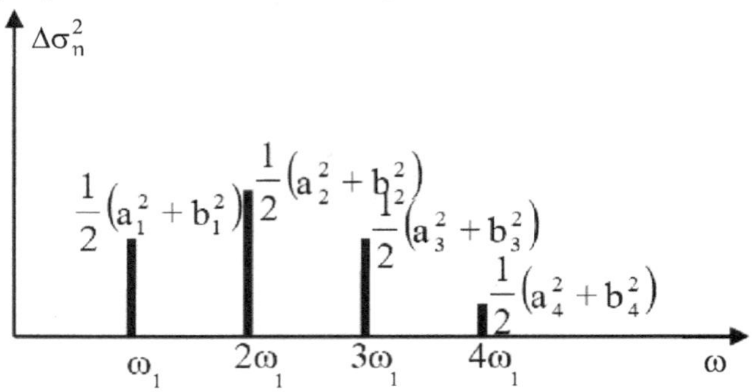

Figure 5: Modal decomposition of the variance

3.2 Fourier-integral

The harmonic components can also be written as complex exponential functions. Introducing the notation $\omega_n = n\omega_1 = n \cdot \dfrac{2\pi}{T}$ and using the identities

$\cos\omega_n t = \frac{1}{2}\left(e^{i\omega_n t} + e^{-i\omega_n t}\right)$, $\sin\omega_n t = -\frac{i}{2}\left(e^{i\omega_n t} - e^{-i\omega_n t}\right)$,

the periodic function is obtained as

$$x(t) = a_0 + \ldots + \frac{a_n - ib_n}{2} e^{i\omega_n t} + \frac{a_n + ib_n}{2} e^{-i\omega_n t} + \ldots$$

with $0 < n < \infty$. If the range of n is extended to negative numbers, $-\infty < n < \infty$, noticing that $\omega_{-n} = -\omega_n$, $a_{-n} = a_n$, $b_{-n} = -b_n$, the terms containing ω_n can be reduced to

$x(t) = \sum_{n=-\infty}^{\infty} \frac{a_n - ib_n}{2} \cdot e^{i\omega_n t} = \sum_{n=-\infty}^{\infty} c_n \cdot e^{i\omega_n t}$ with the complex modal amplitude

$$c_n = \frac{a_n - ib_n}{2}. \tag{3.2}$$

The variance of a centered process can be expressed in terms of the complex modal amplitudes:

$$\sigma^2 = \sum_{n=1}^{\infty} \frac{a_n^2 + b_n^2}{2} = 2\sum_{n=1}^{\infty} |c_n|^2 = \sum_{-\infty}^{\infty} |c_n|^2 = \sum_{n=1}^{\infty} \Delta\sigma_n^2 .$$

The complex amplitudes c_n can be obtained by substituting in (4.2) the real amplitudes a_n and b_n from (4.1):

$$c_n = \frac{1}{T} \int_{-T/2}^{T/2} x(t)\left(\cos\omega_n t - i \cdot \sin\omega_n t\right) dt = \frac{1}{T} \int_{-T/2}^{T/2} x(t) e^{-iw_n t} dt .$$

The difference between two consecutive frequencies is equal to the lowest circular frequency of the periodic process: $\Delta\omega = \omega_1 = 2\pi/T$.

The transition to continuous frequencies, and hereby to aperiodic processes, can be performed writing the process as

$$x(t) = \frac{1}{2\pi} \sum_{n=\infty}^{\infty} \left[\int_{-T/2}^{T/2} x(\tau) e^{-i\omega_n \tau} d\tau \right] e^{i\omega_n t} \Delta\omega.$$

If the function $x(t)$ is absolute integrable, i.e. if $\int_{\infty}^{-\infty} |x(t)| dt < \infty$,

then $T \to \infty$ can be applied. Then, the finite difference $\Delta\omega$ changes to the infinitezimal increment $d\omega$ and the circular frequencies transit from a discrete into a continuous set of values, $\omega_n \to \omega$. In consequence, the integral in the brackets,

$$F_T(\omega_n) = \int_{-T/2}^{T/2} x(t) \cdot e^{-i\omega_n t} dt$$

developes into the Fourier-integral of the direct transformation

$$F(\omega) = \int_{-\infty}^{\infty} x(t) \cdot e^{-i\omega t} dt \tag{3.3}$$

and the sum changes into the Fourier-integral of the inverse transformation,

$$x(t) = \frac{1}{2\pi} \int_{-\infty}^{\infty} F(\omega) e^{i\omega t} d\omega. \tag{3.4}$$

The complex amplitudes can be expressed in terms of the direct Fourier-transform,

$$c(\omega) = \lim_{T\to\infty} \frac{1}{T} F_T(\omega).$$

3.3 Spectral density

For discrete frequencies, the variance of the process is the sum of the contributions of the modal components $\Delta\sigma_n^2 = |c_n|^2 = \frac{1}{T^2}|F_T(\omega_n)|^2$. In the limit of infinitely large periods ($T \to \infty$), the finite differences $\Delta\sigma^2$ and $\Delta\omega$ change to the infinitezimal increments $d\sigma^2$ and $d\omega$. Taking into account that $\frac{1}{T} = \frac{\Delta\omega}{2\pi}$, the variance increment can be written as

$$d\sigma^2 = \frac{1}{4\pi}\left[\lim_{t\to\infty} \frac{2}{T}|F_T(\omega)|^2\right] d\omega.$$

The expression in the brackets is a function which describes the contribution of the circular frequency ω to the total variance σ^2. This real and all over positive function is called the spectral density function (SDF), or power spectral density, or simply spectrum of the function $x(t)$:

$$S(\omega) = \lim_{T\to\infty} \frac{2}{T}|F_T(\omega)|^2$$

The total variance can be obtained from an integral of the spectrum over all circular frequencies:

$$\sigma^2 = \frac{1}{4\pi}\int_{-\infty}^{+\infty} S(\omega)d\omega.$$

To obtain the commonly used form of this integral, one can make use of the fact that S is an even function of ω. Therefore, the domain can be limited to positive values $\omega \geq 0$, wich correspond to the case $n \geq 0$ with discrete circular frequencies. By replacing the circular frequency ω by the frequency f: $\omega = 2\pi f$ and using the identities

$$\int_{\infty}^{-\infty} S(\omega)d\omega = 2\int_{0}^{\infty} S(\omega)d\omega,\ d\omega = 2\pi df,\ d(\ln f) = \frac{df}{f}.$$

the commonly used form of the integral is obtained:

$$\int_0^\infty \frac{f\,S(f)}{\sigma^2} d(\ln f) = 1.$$

3.4 Examples

1) $x(t) = a_0 + a_1 \cos(2\pi f_1 t) + a_2 \cos(2\pi f_2 t)$

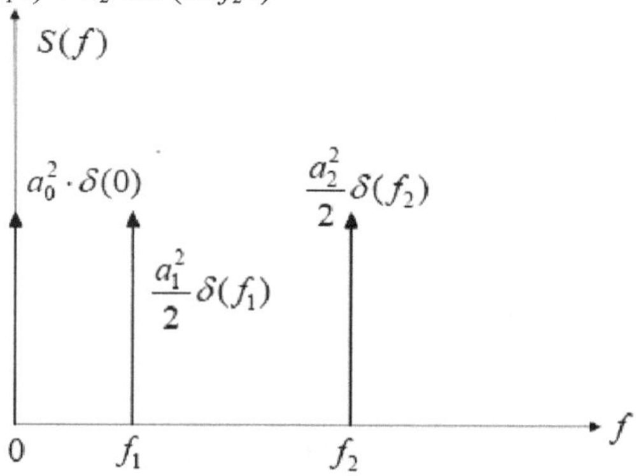

Figure 6: Spectral density for discrete frequencies

Because the function $x(t)$ only contains the discrete frequencies f_1 and f_2, the spectral density is infinite at these frequencies. The areas under the impulses correspond to the modal contributions $\Delta\sigma_1^2$ and $\Delta\sigma_2^2$. For a mean value $\alpha_0 \neq 0$, the SDF is also infinite at $f = 0$, the area under this impulse corresponding to $a_0^2 = \bar{x}^2$.

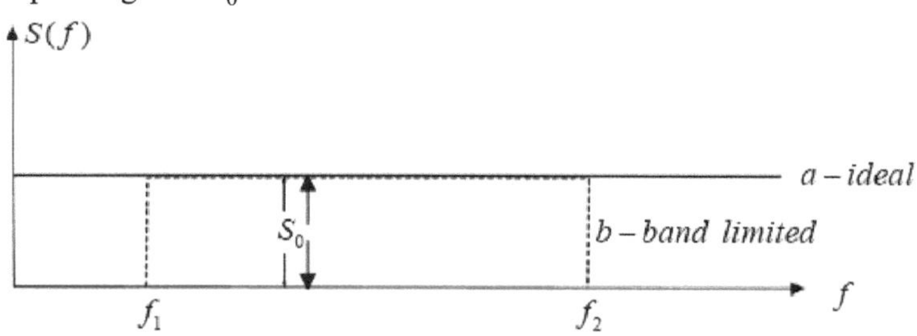

Figure 7: Spectrum of white noise

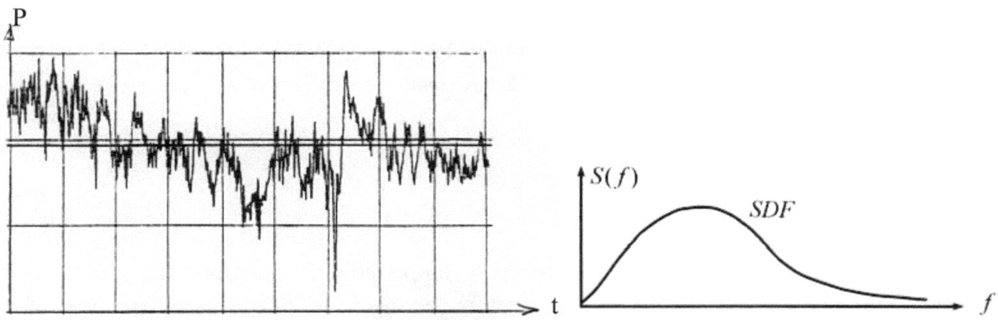

Figure 8: Time dependence (left) and spectrum (right) of a broad-band process (excitation)

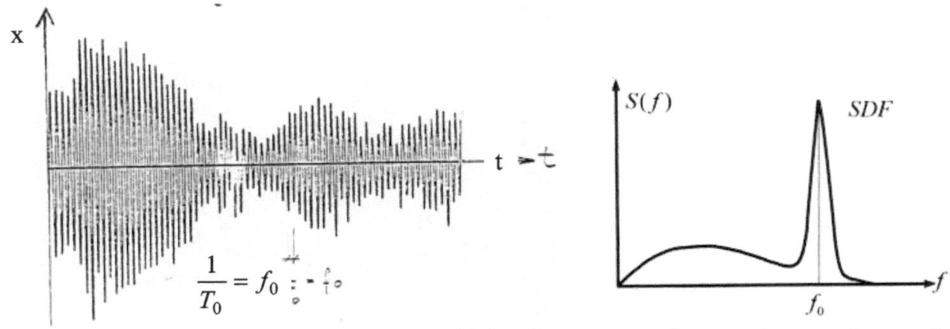

Figure 9: Time history (left) and spectrum (right) of a narrow-band process (structural reponse)

3.5 Autocorrelation and spectral density

In analogy to the time function $x(t)$, also the autocorrelation function can be split into frequency components by means of the Fourier-transformation. It can be shown that the spectral density $S(\omega)$ is the direct Fourier-transform of the autocorrelation function $R(\tau)$, and $R(\tau)$ is the inverse Fourier-transform of $S(\omega)$:

$$S(\omega) = 2\int_{-\infty}^{\infty} R(\tau)e^{-i\omega\tau}d\tau \quad \text{and} \quad R(\tau) = \frac{1}{4\pi}\int_{-\infty}^{\infty} S(\omega)e^{i\omega\tau}d\omega . \tag{3.5}$$

These equations are called the Wiener-Khinchine relations.

The value of the autocorrelation function at $\tau = 0$ is equal to the variance of the function $x(t)$,

$$R(0) = \frac{1}{4\pi}\int_{-\infty}^{+\infty} S(\omega)d\omega = \int_{0}^{\infty} S(f)\,df = \sigma^2 .$$

4 Transfer of a stochastic excitation I

The main goal of this chapter is to determine the steady state system response $x(t)$ to one or more stochastic excitation forces. This response is a scalar quantity, which can be a component of the displacement, velocity or acceleration vector of a selected nodal point, a component of a selected force or moment vector, a stress value, etc. As the system response can conveniently be described by the spectral density $S_x(\omega)$, this function will be evaluated for different oscillator systems in the following sections.

4.1 Single-degree-of-freedom oscillator - one force

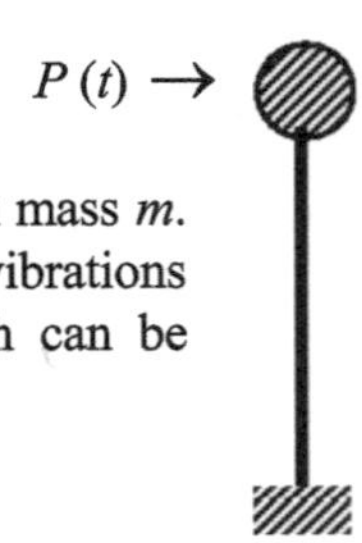

Consider a linear system consisting of an elastically supported and punctiform mass m. The damping ratio of the system is ξ, and the eigenfrequency of the undamped vibrations is $f_e = \omega_e/(2\pi)$. A load $P(t)$ with periode T is aplied to the system, which can be represented by its harmonic components over the frequencies $\omega_n = 2\pi n/T$:

$$P(t) = \sum_{n=\infty}^{+\infty} F_P(\omega_n) e^{i\omega_n t},$$

where $F_P(\omega_n)$ represent the complex Fourier amplitudes.
The system response in the steady state is

$$x(t) = \sum_{n=\infty}^{\infty} F_x(\omega n) e^{i\omega_n t} = \sum_{n=\infty}^{\infty} H(\omega n) F_P(\omega_n) e^{i\omega_n t},$$

where

$$F_x(\omega_n) = H(\omega_n)\, F_P(\omega_n) \tag{4.1}$$

is the complex Fourier amplitude of the nth modal component and

$$H(\omega_n) = \frac{1}{m\omega_e^2} \frac{1}{1-\beta_n^2 + 2i\xi\beta_n}$$

is the frequency-dependent mechanical transfer function. The nondimensional quantity $\beta_n = \omega_n / \omega_e$ is called the frequency ratio.

The mechanical transfer function has the absolute value

$$|H(\omega_n)| = \sqrt{H H^*} = \frac{1}{m\,\omega_e^2} \frac{1}{\sqrt{(1-\beta_n^2)^2 + 4\xi^2\beta_n^2}}.$$

The nth modal component contributes with

$$\Delta\sigma_{n,x}^2 = \left|F_x(\omega_n)^2\right| = F_x(\omega_n) F_x^*(\omega_n) = H(\omega_n) F_P(\omega_n) H^*(\omega_n) F_P^*(\omega_n) = |H(\omega_n)|^2 \Delta\sigma_{n,P}^2 \tag{4.2}$$

to the total variance of the displacement x. $\Delta\sigma_{n,P}^2$ denoting the variance of the nth modal component of the load.

4.1.1 Transition to stochastic excitations

The oscillations produced by stochastic excitations are also stochastic. In most of the practical cases, the system damping is low and, consequently, the mechanical transfer function exhibits a narrow peak at the eigenfrequency. Therefore, the system responds with a narrow-band vibration, even if the excitation is a broad-band process.

As stochastic processes are aperiodic, the excitation frequencies cover a continuous range. This can be mathematically described by setting $T \rightarrow \infty$, and replacing $\Delta\omega$ by $d\omega$ and $H(\omega_n)$ by $H(\omega)$. Doing so, the Fourier-coefficients $F_x(\omega_n)$ and $F_P(\omega_n)$ transit to the Fourier-Transforms $F_x(\omega)$ and $F_P(\omega)$. The quantities $4\pi\,\Delta\sigma^2_{n,x}/\Delta\omega$ and $4\pi\,\Delta\sigma^2_{n,P}/\Delta\omega$ transit to the spectral densities $S_x(\omega)$ and $S_P(\omega)$. Consequently, the equations (4.1) and (4.2) transit to

$$F_x(\omega) = H(\omega)\, F_P(\omega) \tag{4.3}$$

and

$$S_x(\omega) = |H(\omega)|^2\, S_P(\omega \tag{4.4}$$

4.1.2 Time response to a stochastic excitation

Applying the inverse Fourier-transformation (3.4) to the system response and the direct Fourier-transformation (3.3) to the load, and considering the transfer in the frequency domain (4.3), the general transfer equation in the time domain is obtained as

$$x(t) = \frac{1}{2}\int_{-\infty}^{\infty} H(\omega)e^{i\omega t}\, d\omega \int_{-\infty}^{\infty} P(\tau)e^{i\omega(t-\tau)}\, d\tau\,. \tag{4.5}$$

In this equation, the frequency response is given by $H(\omega)$, the time history by $e^{i\omega t}$ and the load amplitude by $\frac{d\omega}{2\pi}\int_{-\infty}^{\infty} P(\tau)e^{-i\omega\tau}d\tau\cdot$ The response in the time domain to a load with known time dependence $P(\tau)$ can be obtained from (5.5) and also from the Duhamel-integral

$$x(t) = \int_{0}^{t} P(\tau)\cdot h(t-\tau)d\tau\,,$$

$h(t)$ representing the impulse-response function, i.e. the system response to an excitation with an impulse of unity, $P(\tau) = I\,\delta(\tau)$, with $I = 1$ (δ is the Dirac-function). As both the load function P and the impulse response function h are 0 for negative arguments, the integration limits can be extended to infinity:

$$x(t) = \int_{-\infty}^{\infty} P(\tau)\cdot h(t-\tau)d\tau\cdot$$

From the comparision of the Duhamel-integral with the transfer equation,

$$\int_{-\infty} P(\tau)h(t-\tau)d\tau = \int_{-\infty} P(\tau)\left[\frac{1}{2\pi}\int_{-\infty} H(\omega)ei\omega(t-\tau)d\omega\right]d\tau.$$

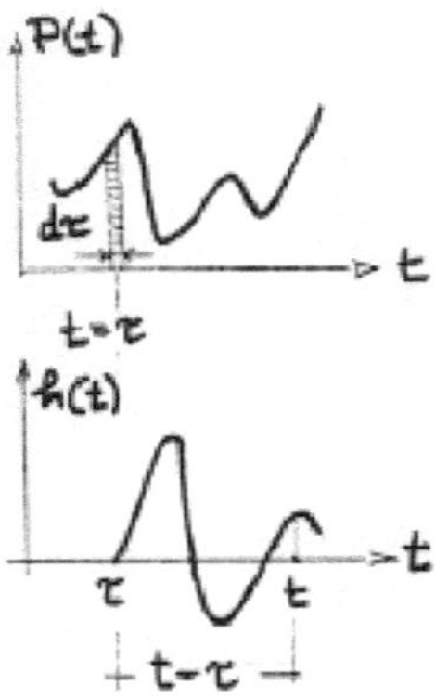

the impulse-response function is obtained as the inverse Fourier-tranform of the mechanical transfer function:

$$h(t) = \frac{1}{2\pi}\int_{-\infty}^{\infty} H(\omega)\cdot e^{i\omega t}d\omega\cdot$$

Accordingly, the mechanical transfer function is the direct Fourier-transform of the impulse-response function:

$$H(\omega) = \int_{-\infty}^{\infty} h(t)e^{-i\omega t}dt \tag{4.6}$$

4.1.3 Statistic parameters of the response

The static response of the system, i.e. the response to a constant load, is given by the arithmetical mean $\overline{x}$ of the time dependent response. This is obtained as

$$\overline{x} = \lim_{T\to\infty}\frac{1}{T}\int_{-T/2}^{T/2}\left[\int_{-\infty}^{\infty}P(\tau)h(t-\tau)d\tau\right]dt = \int_{-\infty}^{\infty}h(t-\tau)\left[\lim_{T\to\infty}\frac{1}{T}\int_{-T/2}^{T/2}P(\tau)d\tau\right]d(t-\tau),$$

or $\overline{x} = \overline{P}\int_{-\infty}^{\infty}h(t)dt,$

$\overline{P} = \lim_{T\to\infty}\frac{1}{T}\int_{-T/2}^{T/2}P(\tau)d\tau$ being the arithmetical mean of the load $P(\tau)$.

The quasistatic mechanical transfer function - the value for very low-frequency excitations - can be obtained from (4.6) for $\omega = 0$:

$$H(0) = \int_{-\infty}^{+\infty}h(t)\,dt\,.$$

Consequently, the mean response only depends on the mean load value and on the quasistatic mechanical transfer function:

$\overline{x} = \overline{P}H(0)$.

The dynamic response is described by the variance σ_x^2, which can be calculated in the frequency-domain, using the mechanical transfer function and the spectral density by

$$\sigma_x^2 = \int_0^{+\infty}\left|H(f)\right|^2\cdot S_P(f)df - \overline{x}^2,$$

or in the time-domain, using autocorrelation of the load and the impulse-response function,

$$\sigma_x^2 = \int_0^{\infty}\int_0^{\infty}R_P(\tau_1-\tau_2)\cdot h(\tau_1)\cdot h(\tau_2)d\tau_1 d\tau_2 - \overline{x}^2\cdot$$

If the load is a Gaussian process, then the response also shows a normal distribution, i.e. the probability density is completely described by $\overline{x}$ and. σ_x. Therefore, the probability function of the response - e.g. displacement or stress - can be specified.

4.2 Single-degree-of-freedom oscillator -- two forces

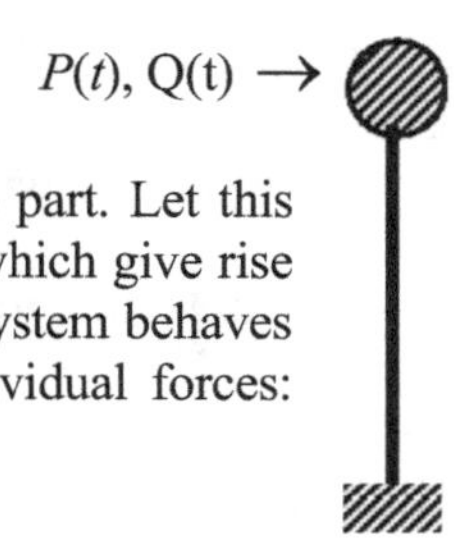

Consider the same single-degree-of-freedom oscillator,as in the previous part. Let this system be simultaneouslyexcited by two stochastic forces $P(t)$ and $Q(t)$, which give rise to the system responses $x_P(t)$ and $x_Q(t)$, respectively. As the considered system behaves linearly, the resulting response is the sum of the responses to the individual forces: $x(t) = x_P(t) + x_Q(t)$.

4.2.1 Spectral density function of the system response and cross-spectral density function

The spectral density function of the system response x is

$$S_x(\omega) = \lim_{T\to\infty} \frac{''}{T} \left|F_{x,T}(\omega)\right|^2,$$

where

$$F_{x,T}(\omega) = \int_{-T/2}^{T/2} x(t)\, e^{-i\omega t}\, dt$$

is the Fourier-transform of the system response $x(t)$. The calculation of $S_x(\omega)$ results in the task of evaluating the quantity $|F_{x,T}(\omega)|^2$.

As all quantities F_P, F_x, H and S are functions of ω, the argument will be omitted in the following calculation. From the Fourier-transforms of the response (4.3), results for finite T:

$F_{xP,T} = H\, F_{P,T}$ and $F_{xQ,T} = H\, F_{Q,T}$.

As the complex conjugates of these expressions are

$F^*_{xP,T} = H^* F^*_{P,T}$ and $F^*_{xQ,T} = H^* F^*_{Q,T}$,

the searched quantity is obtained as

$|F_{x,T}|^2 = [F^*_{xP,T} + F^*_{xQ,T}]\,[F_{xP,T} + F_{xQ,T}] = |H|^2\,[|F_{P,T}|^2 + F^*_{P,T} F_{Q,T} + F^*_{Q,T} F_{P,T} + |F_{Q,T}|^2]$.

The terms $|F_{xP,T}|^2$ and $|F_{xQ,T}|^2$ correspond to the spectra S_P and S_Q, and the mixed products correspond to the cross-spectral density functions, or simply cross-spectra

$$S_{PQ} = \lim_{T\to\infty} \frac{2}{T} F^*_{P,T} F_{Q,T} \qquad S_{QP} = \lim_{T\to\infty} \frac{2}{T} F^*_{Q,T} F_{P,T}$$

With these notations, the response spectrum is given by

$$S_x(\omega) = \left|H(\omega)\right|^2 \cdot \left[S_P(\omega) + S_{PQ}(\omega) + S_{QP}(\omega) + S_Q(\omega)\right]$$

The cross spectral density function can be expressed in terms of the cross-correlation function

$$R_{PQ}(\tau) = \lim_{T\to\infty} \frac{1}{T} \int_{-T/2}^{T/2} P(t) Q(t+\tau)\, dt$$

and vice versa. The expressions are analogous to the Wiener-Khinchine relations (3.5),

$$S_{PQ}(\omega) = 2 \int_{-\infty}^{\infty} R_{PQ}(\tau) e^{-i\omega\tau} dt \qquad R_{PQ}(\tau) = \frac{1}{4\pi} \int_{-\infty}^{\infty} S_{PQ}(\omega) e^{i\omega t} d\omega,$$

For $Q = P$, the cross-correlation is identical to the autocorrelation, $R_{PQ} = R_P$, and the cross spectral density function is the spectral density function, $S_{PQ} = S_P$. The cross spectral density function can be also written in the form

$$S_{PQ}(\omega) = 2 \int_{-\infty}^{\infty} R_{PQ}(\tau) \cos\omega\tau\, dt + 2i \int_{-\infty}^{\infty} R_{PQ}(\tau) \sin\omega\tau\, dt$$

where R_{PQ} is a real function. For $Q = P$, $R_{PQ}(\tau) = R_P(\tau)$ is an even function of τ and the second integral vanishes. Consequently, the spectral density function $S_P(\omega)$ is real. For $Q \neq P$, $R_{PQ}(\tau)$ is not an even function of τ and therefore the cross spectral density function $S_{PQ}(\omega)$ is complex.

Adopting

$$R_{QP}(-\tau) = \lim_{T\to\infty} \frac{1}{T} \int_{-T/2}^{T/2} Q(t)P(t-\tau)dt$$

and setting $t' = t - \tau$, $t = t' + \tau$, $dt = dt'$ one obtains:

$$R_{QP}(-\tau) = \lim_{T\to\infty} \frac{1}{T} \int_{-T/2}^{T/2} Q(t'+\tau)P(t')dt' = R_{PQ}(\tau)$$

and analogously,

$$R_{QP}(\tau) = R_{PQ}(-\tau)$$

Consequently, $S^*_{PQ}(\omega) = S_{QP}(\omega)$ and the response spectrum is real -- as required:

$$S_X(\omega) = |H(\omega)|^2 \cdot \left[S_P(\omega) + S^*_{PQ}(\omega) + S_{QP}(\omega) + S_Q(\omega)\right]$$
$$= |H(\omega)|^2 \cdot \left\{S_P(\omega) + 2\,\mathrm{Re}\left[S_{PQ}(\omega)\right] + S_Q(\omega)\right\}$$

The real part of the cross-spectrum, $\mathrm{Re}[S_{PQ}(\omega)] = \mathrm{Co}_{PQ}(\omega)$, which contains the in-phase components of the time-functions $P(t)$ and $Q(t)$, is called the coincidence-spectrum. The imaginary part, $\mathrm{Im}[S_{PQ}(\omega)] = \mathrm{Qu}_{PQ}(\omega)$ is called quadrature-spectrum, it contains the out-of-phase components of the time-functions (components shifted by a phase of $\pi/2$).

The square of the absolute value of the cross spectral density function is the sum of the squares of the coincidence and of the quadrature-spectrum:
$|S_{PQ}(\omega)|^2 = \mathrm{Co}^2_{PQ}(\omega) + \mathrm{Qu}^2_{PQ}(\omega)$.

The coherence function $\gamma^2_{PQ}(\omega)$ of the excitation processes $P(t)$ and $Q(t)$ is defined by

$$\gamma^2_{PQ}(\omega) = \frac{|S_{PQ}(\omega)|^2}{S_P(\omega)S_Q(\omega)}$$

It can be shown that $\gamma_{PQ}(\omega) \le 1$.

The phase shift $\varphi(\omega)$ between the processes can be obtained from

$$\tan\varphi(\omega) = -\frac{Qu_{PQ}(\omega)}{Co_{PQ}(\omega)} \tag{4.7}$$

4.2.2 Superposition of stochastic processes

The spectral density function of the response can also be determined from the direct superposition of the loads and in the time domain: $W(t) = P(t) + Q(t)$.
The autocorrelation of the process $W(t)$ can be evaluated as

$$R_W(\tau) = \lim_{T\to\infty} \int_{-T/2}^{T/2} [P(t)+Q(t)][P(t+\tau)+Q(t+\tau)]dt$$

$$= \lim_{T\to\infty} \int_{-T/2}^{T/2} [P(t)P(t+\tau)+P(t)Q(t+\tau)+Q(t)P(t+\tau)+Q(t)Q8t+\tau)]dt$$

$$= R_P(\tau)+R_{PQ}(\tau)+R_{QP}(\tau)+R_Q(\tau)$$

The covariance of two zero-mean processes $P(t)$ and $Q(t)$ with $\overline{P}=\overline{Q}=0$ is defined as the cross-correlation at $\tau = 0$, $\sigma_{PQ}^2 = R_{PQ}(0)$. Using this notation, the variance of the process $W(t)$ can be written as

$\sigma_W^2 = \sigma_P^2 + \sigma_{PQ}^2 + \sigma_{QP}^2 + \sigma_Q^2$.

If the processes $P(t)$ and $Q(t)$ are statistically independent, then $\sigma_{PQ} = \sigma_{QP} = 0$ and the variance is obtained as

$\sigma_W^2 = \sigma_P^2 + \sigma_Q^2$.

The measure of the statistical dependence of the processes $P(t)$ and $Q(t)$ is the correlation coefficient

$$\rho_{PQ}(\tau) = \frac{R_{QP}(\tau) - \overline{PQ}}{\sigma_P \sigma_Q}$$

It can be shown that $-1 \leq \rho_{PQ}(\tau) \leq 1$. For fully correlated processes $P(t)$ and $Q(t)$, the correlation coefficient at $\tau = 0$ is $\rho_{PQ}(0) = \sigma^2_{PQ}/(\sigma_P\ \sigma_Q) = \pm 1$. For statistically completely independent processes is $\rho_{PQ}(\tau) = 0$ for all τ.

4.2.3 Effect of *N* load processes

Consider a single-degree-of-freedom system, which is excited by N load processes $P_1(t)$, $P_2(t)$, ..., $P_N(t)$. Let the spectral density functions of the processes be denoted by $S_1(\omega)$, $S_2(\omega)$, ..., $S_N(\omega)$ and their cross-spectral density functions by $S_{12}(\omega)$, $S_{13}(\omega)$, etc. . The spectra and the cross-spectra can be specified in a cross-spectral matrix:

$$\underline{S}(\omega) = \begin{bmatrix} S_1(\omega) & S_{12}(\omega) & S_{13}(\omega) & \cdots & S_{1N}(\omega) \\ S_{21}(\omega) & S_2(\omega) & S_{23}(\omega) & \cdots & S_{2N}(\omega) \\ S_{31}(\omega) & S_{32}(\omega) & S_3(\omega) & \cdots & S_{3N}(\omega) \\ \vdots & \vdots & \vdots & \vdots & \vdots \\ S_{N1}(\omega) & S_{N2}(\omega) & S_{N3}(\omega) & \cdots & S_{NN}(\omega) \end{bmatrix}$$

4.3 Oscillator with two degrees of freedom -- two forces

4.3.1 The spectral densiy function of the system response

Consider a two-degree-of-freedom system, exposed to the simultaneous action of two forces: $P_1(t)$ acting on the 1st degree of freedom and $P_2(t)$ acting on the 2nd degree of freedom. As in the previous sections, the task is to determine the spectral density function $S_x(\omega)$ of the response $x(t)$, $x(t)$ being an arbitrary scalar quantity as specified at the beginning of this chapter.
Let $x_1(t)$ and $x_2(t)$ denote the effect of the forces $P_1(t)$ and $P_2(t)$ on the system response, when the forces are applied separately. Because the system behaves linearly, the response to the simultaneously applied forces is $x(t) = x_1(t) + x_2(t)$.
As the loads are applied on different degrees of freedom, the transfer of the excitations is described by two different mechanical transfer functions, $H_1(\omega)$ and $H_2(\omega)$. Those connect the Fourier-transforms of the responses and of the loads, $F_{x1}(\omega)=H_1(\omega)\,F_{P1}(\omega)$, $F_{x2}(\omega)=H_2(\omega)\,F_{P2}(\omega)$. The mechanical transfer functions can be expressed in terms of the load and response amplitudes, $|F_{Pj}|$ and $|F_{xj}|$, and the phase-difference φ_j between response and load:

$$H_j(\omega) = \frac{F_{xj}(\omega)}{F_{Pj}(\omega)} = \frac{|F_{xj}(\omega)|\, e^{-i\varphi_j(\omega)}}{|F_{Pj}(\omega)|} \rightarrow j = 1,2\,.$$

The Fourier-transform of the response is

$$F_{x,T}(\omega) = \int_{-T/2}^{T/2} [x_1(t) + x_2(t)] e^{-i\omega t} dt = F_{x1,T}(\omega) + F_{x2,T}(\omega) = H_1(\omega) F_{P1,T}(\omega) + H_2(\omega) F_{P2,T}(\omega)\,,$$

and its complex conjugate is

$$F^*_{x,T}(\omega) = F^*_{x1,T}(\omega) + F^*_{x2,T}(\omega) = H^*_1(\omega) F^*_{P1,T}(\omega) + H^*_2(\omega) F^*_{P2,T}(\omega).$$

Consequently, the response spectrum

$$S_x(\omega) = \lim_{T\to\infty} \frac{2}{T} F^*_{x,T}(\omega) F_{x,T}(\omega)$$

is obtained as

$$S_x(\omega) = |H_1(\omega)|^2 S_{P1}(\omega) + H^*_1(\omega) H_2(\omega) S_{P12}(\omega) + H^*_2(\omega) H_1(\omega) S_{P21}(\omega) + |H_2(\omega)|^2 S_{P2}(\omega),$$

where

$$S_{P1}(\omega) = \lim_{T\to\infty} \frac{2}{T} \left|F^*_{Pj,T}(\omega)\right|^2, \quad j = 1,2$$

are the spectra of the loads $P_j(t)$, and

$$S_{P1}(\omega) = \lim_{T\to\infty} \frac{2}{T} F^*_{Pj,T}(\omega) F_{Pk,T}(\omega), \quad j,k = 1,2, \quad j \neq k$$

are the cross-spectral density functions. Similarly to the single-degree-of-freedom case, the spectra are real, while the cross-spectra are complex, and

$$S_{P21}(\omega) = S^*_{P12}(\omega).$$

Because the cross-products are complex conjugates,

$$H^*_1(\omega) H_2(\omega) S_{P12}(\omega) = \left(H_1(\omega) H^*_2(\omega) S^*_{P12}(\omega)\right)^*$$

all imaginary parts are cancelled by the summation. Consequently, the response spectrum is real:

$$S_x(\omega) = |H_1(\omega)|^2 S_{P1}(\omega) + 2\,\mathrm{Re}\left[H_1^*(\omega)H_2(\omega)S_{P12}(\omega)\right] + |H_2(\omega)|^2 S_{P2}(\omega)$$

The cross products can also be written as:

$$H_1^*(\omega)H_2(\omega)S_{P12}(\omega) = |H_1(\omega)||H_2(\omega)||S_{P12}|e^{i[-\varphi_1(\omega)+\varphi_2(\omega)+\varphi_{12}(\omega)]}$$

$$H_1(\omega)H_2^*(\omega)S_{P12}^*(\omega) = |H_1(\omega)||H_2(\omega)||S_{P12}|e^{i[-\varphi_1(\omega)-\varphi_2(\omega)-\varphi_{12}(\omega)]}$$

where $\varphi_1(\omega)$ and $\varphi_2(\omega)$ are the phases of the mechanical transfer functions, and $\varphi_{12}(\omega)$ is the phase difference between the excitations $P_1(t)$ and $P_2(t)$, see equation (4.7). As can be seen from the above equations, the phases of the load processes have an influence on Re $[S_{P12}(\omega)]$ and consequently on the response spectrum $S_x(\omega)$. For completely uncorrelated processes $P_1(t)$ and $P_2(t)$, $S_{P12}(\omega) = S_{P21}(\omega) = 0$ for all ω, i.e. no cross-products occur. In general, the cross-products are in the same order of magnitude with the terms $|H_j(\omega)|^2 S_{Pj}(\omega)$ and cannot be neglected. For small phase shifts $\varphi_1(\omega) - \varphi_2(\omega) - \varphi_{12}(\omega)$, the real part of the cross-products can be replaced by their absolute value.

4.3.2 Matrix-notation and covariance matrix

As all quantities H and S are functions of ω, in the following the argument will be omitted. With the simplified notations $S_{P1} = S_{11}$, $S_{P12} = S_{12}$, $S_{P21} = S_{21}$ and $S_{P2} = S_{22}$, the mechanical transfer matrix $\underline{H}$, its adjoint $\underline{H}^+$ and the spectrum matrix $\underline{S}_P$ can be written as

$$\mathrm{H} = \begin{bmatrix} H_1 \\ H_2 \end{bmatrix}, \quad \underline{H}^+ = \begin{bmatrix} H_1^* & H_2^* \end{bmatrix}, \quad \mathrm{S_P} = \begin{bmatrix} S_{11} & S_{12} \\ S_{21} & S_{22} \end{bmatrix}$$

and the response spectrum is then obtained as $S_x = \underline{H}^+ \underline{S}_P \underline{H}$. Adopting the notation H_{kl} for the mechanical transfer function from the load $P_l(t)$ acting on the lth degree of freedom, to the kth displacement $x_k(t)$, the transfer matrix for a two-degree-of-freedom system under the influence of two loads can be written as

$$\underline{H} = \begin{bmatrix} H_{11} & H_{12} \\ H_{21} & H_{22} \end{bmatrix}.$$

It can be shown that the complete matrix of the displacement spectra is given by

$$S_{xx} = \begin{bmatrix} S_{x_1x_1} & S_{x_1x_2} \\ S_{x_2x_1} & S_{x_2x_2} \end{bmatrix} = \underline{H}^+ \underline{S}_P \underline{H} = \begin{bmatrix} H_{11}^* & H_{21}^* \\ H_{12}^* & H_{22}^* \end{bmatrix} \begin{bmatrix} S_{11} & S_{12} \\ S_{21} & S_{22} \end{bmatrix} \begin{bmatrix} H_{11} & H_{12} \\ H_{21} & H_{22} \end{bmatrix},$$

where the first term is obtained as

$$S_{x_1x_1} = |H_{11}|^2 S_{11} + H_{11}H_{12}^*S_{21} + H_{11}^*H_{12}S_{12} + |H_{12}|^2 S_{22}.$$

The other terms can be calculated analogously. The diagonal terms in the spectral matrix S_{xx} are the real spectral densities of the displacements x_1 and x_2, and the off-diagonal terms are the complex cross-spectral densities. The covariance-matrix of the displacements can be constructed from the integrals of the frequency-dependent spectral density functions.

$$\sigma_{x_j}^2 = \int_0^\infty S_{x_jx_j}(f)df, \quad \sigma_{x_jx_k}^2 = \frac{1}{2}\int_{-\infty}^{\infty} S_{x_jx_k}(f)df \;\; (j \neq k), \quad \sigma_x^2 = \begin{bmatrix} \sigma_{x_1}^2 & \sigma_{x_1x_2}^2 \\ \sigma_{x_2x_1}^2 & \sigma_{x_2}^2 \end{bmatrix}.$$

4.4 Mechanical transfer function for a two-degree-of-freedom oscillator

4.4.1 Modal analysis

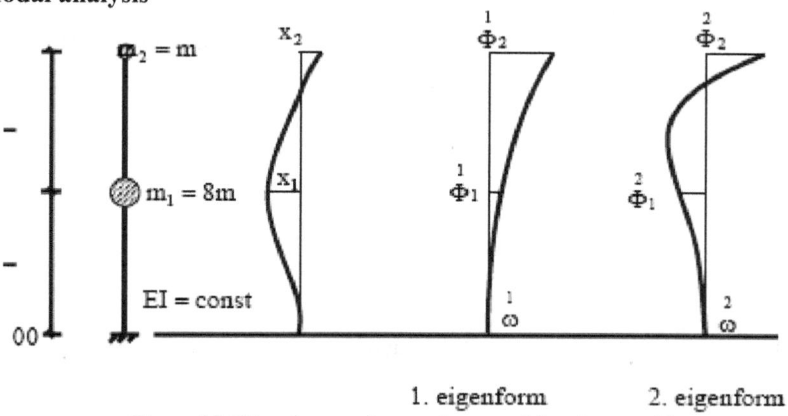

Figure 10: Eigenforms of a two-degree-of-freedom oscillator

Consider the two-degree-of-freedom oscillator shown in the figure, where the two nodal points can only oscillate in the direction x. The equations of motion for the two degrees of freedom, constructed with the mass-matrix $\underline{M}$, the damping matrix $\underline{C}$ and the stiffness matrix $\underline{K}$,

$$\underline{M}\,\underline{\ddot{x}} + \underline{C}\,\underline{\dot{x}} + \underline{K}\,\underline{x} = \underline{P}$$

are coupled. The eigenfrequencies $\overset{\upsilon}{\omega}$ and the normalized eigenforms (modal vectors) $\overset{\upsilon}{\underline{\Phi}}$ can be obtained by solving the eigenvalue problem for the undamped oscillator,

$$\underline{K}\,\Phi = \omega^2\,\underline{M}\,\Phi.$$

The time-dependent displacements of the nodal points 1 und 2 can be transformed into a basis consisting of the modal vectors $\overset{1}{\underline{\Phi}}$ and $\overset{2}{\underline{\Phi}}$,

$$x_1(t) = \overset{1}{\Phi}_1\, q_1(t) + \overset{2}{\Phi}_1\, q_2(t)$$

$$x_2(t) = \overset{1}{\Phi}_2\, q_1(t) + \overset{2}{\Phi}_2\, q_2(t)$$

or, using a matrix-notation and omitting the argument t for simplicity,

$$\underline{x} = \underline{\Phi}\,\underline{q}. \tag{4.8}$$

In this equation,

$\underline{x} = \begin{bmatrix} x_1 \\ x_2 \end{bmatrix}$ represents the vector of deformations,

$\underline{q} = \begin{bmatrix} q_1 \\ q_2 \end{bmatrix}$ represents the vector of generalized coordinates and

$\underline{\Phi} = \begin{bmatrix} \overset{1}{\Phi}_1 & \overset{2}{\Phi}_1 \\ \overset{1}{\Phi}_2 & \overset{2}{\Phi}_2 \end{bmatrix}$ represents the matrix of the normalized eigenvectors = modal matrix .

The modal matrix contains the normalized eigenforms in its columns. Applying the transformation (4.8), the matrices $\underline{M}$ and $\underline{K}$ become diagonal. If the damping matrix fulfills the Rayleigh-condition, $\underline{C} = \alpha\underline{M} + \beta\underline{K}$, then also this matrix is diagonalized by the same transformation, and the ecuations of motion written in generalized coordinates q_υ become decoupled:

$$\ddot{q}_\upsilon + 2\overset{\nu}{\xi}\overset{\nu}{\omega} q_\nu(t) + \overset{\nu^2}{\omega} q_\nu(t) = \frac{\overset{\nu}{P}}{\overset{\nu}{m}}, \quad \nu = 1,2.$$

Hereby,

$\overset{\nu}{\underline{\Phi}}{}^T \underline{M}\,\overset{\nu}{\underline{\Phi}} = \overset{\nu}{m}$ is the modal mass,

$\overset{\nu}{\omega}{}^2 = \frac{\overset{\nu}{k}}{\overset{\nu}{m}}$ is the square of the circular eigenfrequency,

$\overset{\nu}{k} = \overset{\nu}{\underline{\Phi}}{}^T \underline{K}\,\overset{\nu}{\underline{\Phi}}$ is the modal stiffness,

$2\overset{\nu}{\xi}\overset{\nu}{m}\overset{\nu}{\omega} = \overset{\nu}{\underline{\Phi}}{}^T \underline{C}\,\overset{\nu}{\underline{\Phi}}$ is the modal damping and

$\overset{\nu}{P} = \overset{\nu}{\underline{\Phi}}{}^T \underline{P}$ is the modal load.

Analogously to the two-degree-of-freedom system, an oscillator with n degrees of freedom is represented as n single-degree-of-freedom oscillators described by n modal equations of motion.

4.4.2 Solution of the modal equations of motion for harmonical excitations in the steady state and mechanical transfer function

Consider the two-degree-of-freedom oscillator shown in the figure, which is excited by a harmonic load at the node point 1.

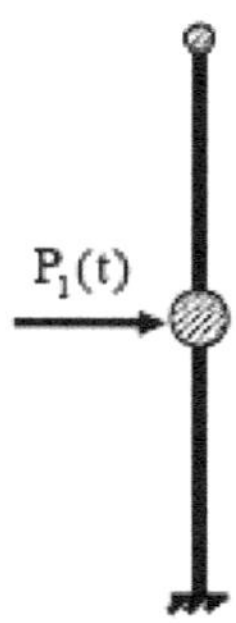

$$\underline{P} = \begin{bmatrix} P_1 \\ 0 \end{bmatrix}; \qquad P_1(t) = \hat{P}_1 e^{i\omega t}$$

$$\overset{\nu}{P} = \overset{\nu}{\Phi}_1 \hat{P}_1 e^{i\omega t}$$

Figure 11: Two-degree-of-freedom oscillator loaded at point 1

The complex load amplitude $\hat{P}_1 = |\hat{P}_1| e^{i\varphi_P}$ is specified by the real amplitude $|\hat{P}_1|$ and by the phase φ_P.

The response in generalized coordinates is similar to the single-degree-of-freedom oscillator:

$$\overset{\nu}{q}(t)=\overset{\nu}{\Phi}_1\,\overset{\nu}{H}(\omega)\hat{P}_1e^{i\omega t}=\frac{\overset{\nu}{\Phi}_1\,\hat{P}_1e^{i\omega t}}{\overset{\nu}{m}\left(\overset{\nu}{\omega}{}^2-\omega^2+2i\overset{\nu}{\xi}\overset{\nu}{\omega}\omega\right)}=\frac{\overset{\nu}{\omega}{}^2-\omega^2-2i\overset{\nu}{\xi}\overset{\nu}{\omega}\omega}{\overset{\nu}{m}\left[\left(\overset{\nu}{\omega}{}^2-\omega^2+4\overset{\nu}{\xi}{}^2\overset{\nu}{\omega}{}^2\omega^2\right)\right]}\overset{\nu}{\Phi}_1\,\hat{P}_1e^{i\omega t},$$

where

$$\overset{\nu}{H}=\frac{1}{\overset{\nu}{m}\left(\overset{\nu}{\omega}{}^2-\omega^2+2i\overset{\nu}{\xi}\overset{\nu}{\omega}\omega\right)}$$

is the mechanical transfer function for the νth mode. Its magnitude, written as a function of the frequencies f is

$$\left|\overset{\nu}{H}(f)\right|=\frac{1}{4\pi^2f_\nu^2\,\overset{\nu}{m}\sqrt{\left(1-f^2/f_\nu^2\right)^2+4\overset{\nu}{\xi}{}^2\,f^2/f_\nu^2}}.$$

For the static or quasi-static case $f \ll f_\nu$, the modal transfer function can be approximated by

$$\overset{\nu}{H}(0)=\frac{1}{4\pi^2f_\nu^2\,\overset{\nu}{m}}=\frac{1}{\overset{\nu}{k}}$$

i.e. the quasistatic modal mechanical transfer function is the reciprocal of the modal stiffness. The phase between the generalized coordinate $q_\nu(t)$ and the load $P_1(t)$ can be obtained from

$$\tan\varphi_\upsilon=\frac{2\overset{\nu}{\xi}\overset{\nu}{\omega}\omega}{\overset{\nu}{\omega}{}^2-\omega^2}$$

The displacement of the node 2, $x_2(t)$, is obtained as

$$x_2(t)=\overset{1}{\Phi}_2\,q_1(t)+\overset{2}{\Phi}_2\,q_2(t)=\left[\frac{\overset{1}{\Phi}_1\overset{1}{\Phi}_2}{\overset{1}{m}\left(\overset{1}{\omega}{}^2-\omega^2+2i\overset{1}{\xi}\overset{1}{\omega}\omega\right)}+\frac{\overset{2}{\Phi}_1\overset{2}{\Phi}_2}{\overset{2}{m}\left(\overset{2}{\omega}{}^2-\omega^2+2i\overset{2}{\xi}\overset{2}{\omega}\omega\right)}\right]\hat{P}_1e^{1\omega t}$$

Due to $x_2(t)=H_{21}(\omega)\hat{P}_1e^{i\omega t}$ the mechanical transfer function from node 1 to node 2 is

$$\begin{aligned}H_{21}(\omega)&=\frac{\overset{1}{\Phi}_1\overset{1}{\Phi}_2}{\overset{1}{m}\left(\overset{1}{\omega}{}^2-\omega^2+2i\overset{1}{\xi}\overset{1}{\omega}\omega\right)}+\frac{\overset{2}{\Phi}_1\overset{2}{\Phi}_2}{\overset{2}{m}\left(\overset{2}{\omega}{}^2-\omega^2+2i\overset{2}{\xi}\overset{2}{\omega}\omega\right)}\\&=\overset{1}{\Phi}_1\overset{1}{\Phi}_2\overset{1}{H}(\omega)+\overset{2}{\Phi}_1\overset{2}{\Phi}_2\overset{2}{H}(\omega)\end{aligned}\tag{4.9}$$

4.5 The oscillator with *n* degrees of freedom

4.5.1 Mechanical transfer function

Generalizing equation (5.9) to oscillators with *n* degrees of freedom, the mechanical transfer function expressing the displacement or rotation of the *k*th degree of freedom, caused by a harmonic excitation (force or moment) applied on the *l*th degree of freedom is obtained as

$$H_{kl}(\omega) = \sum_{\nu=1}^{n} \overset{\nu}{\Phi}_l \overset{\nu}{\Phi}_k \overset{\nu}{H}(\omega) \sum_{\nu=1}^{n} \frac{\overset{\nu}{\Phi}_l \overset{\nu}{\Phi}_k}{\overset{\nu}{m}\left(\overset{\nu}{\omega}^2 - \omega^2 + 2i\overset{\nu}{\xi}\overset{\nu}{\omega}\omega\right)}$$

Remind that $\overset{\nu}{\Phi}_l$ is the *l*th component of the *ν*th mode shape and $\overset{\nu}{H}(\omega)$ is the mechanical transfer function for the *ν*th mode. It is straightforward to calculate the mechanical transfer function for arbitrary scalar response quantities *Y*(*t*), which can be a component of the velocity or acceleration vector of a selected nodal point, a component of a selected force or moment vector, a stress value, etc. . Let $\overset{\nu}{\Phi}_Y$ denote the contribution of the *ν*th eigenform to the response *Y*. Analogously to the displacement $x_k(t)$, the mechanical transfer function from the load acting on the *l*th degree of freedom to the response *Y*(*t*) is obtained as

$$H_{Y,l}(\omega) = \sum_{\nu=1}^{n} \overset{\nu}{\Phi}_l \overset{\nu}{\Phi}_Y \overset{\nu}{H}(\omega) = \sum_{\nu=1}^{n} \frac{\overset{\nu}{\Phi}_l \overset{\nu}{\Phi}_Y}{\overset{\nu}{m}\left(\overset{\nu}{\omega}^2 - \omega^2 + 2i\overset{\nu}{\xi}\overset{\nu}{\omega}\omega\right)}.$$

4.5.2 Spectral density function of an arbitrary scalar response

For simplicity, the mechanical transfer function $H_Y(z,P_k,\omega)$ linking a load P_k, applied on the *k*th degree of freedom, and the response *Y* at position *z*, will be denoted as H_k.

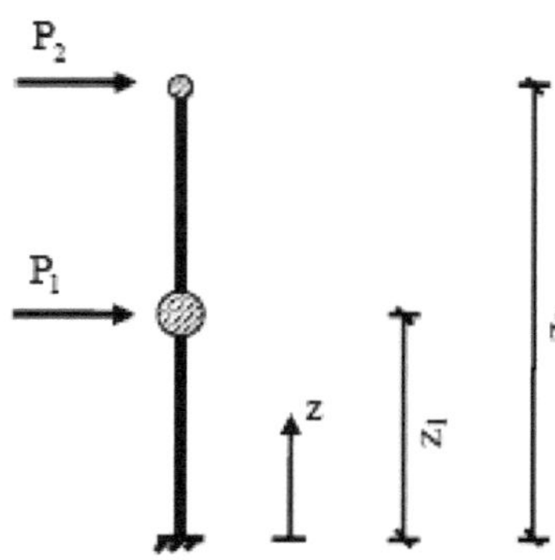

Figure 12: Two-degree-of-freedom oscillator loaded at point 1

For a two-degree-of-freedom system, the mechanical transfer function from the kth degree of freedom to the scalar quantity Y is

$$H_k = \sum_{\nu=1}^{2} \overset{\nu}{\Phi}_k \overset{\nu}{\Phi}_Y \overset{\nu}{H},$$

and the spectral density function of the response is given by

$$S_Y = \mathrm{Re}\left(H_1^* H_1 S_{11} + H_1^* H_2 S_{12} + H_2^* H_1 S_{21} + H_2^* H_2 S_{22}\right)$$

Generalizing for n degrees of freedom,

$$H_k = \sum_{\nu=1}^{n} \overset{\nu}{\Phi}_k \overset{\nu}{\Phi}_Y \overset{\nu}{H}, \text{ and } S_Y = \mathrm{Re}\left(\sum_{1}^{n}\sum_{1}^{n} H_k^* H_l S_{kl}\right)$$

is obtained. Expressing H_k and H_l in terms of the mode shapes and of the modal transfer functions, the spectral density function of the response Y becomes

$$S_Y = \mathrm{Re}\left(\sum_{1}^{n}\sum_{1}^{n} \overset{\nu}{\Phi}_Y \overset{\mu}{\Phi}_Y \overset{\nu}{H}{}^* \overset{\mu}{H} \sum_{1}^{n}\sum_{1}^{n} \overset{\nu}{\Phi}_k \overset{\mu}{\Phi}_l S_{kl}\right).$$

The complex modal transfer function H_ν can be split into a real and an imaginary part:

$$\overset{\nu}{H} = \left|\overset{\nu}{H}\right|\left(\cos\varphi_\nu - i \sin\varphi_\nu\right) = \left|\overset{\nu}{H}\right| \cos\varphi_\nu \left(1 - i \tan\varphi_\nu\right).$$

The phase difference $\overset{\nu}{\varphi}$ has been introduced with a negative sign because it represents the phase shift by which the response lags behind the excitation. Amplifying by the complex conjugate of the denominator, the modal transfer function becomes

$$\overset{\nu}{H} = \frac{\overset{\nu}{\omega}{}^2 - \omega^2 + 2i \overset{\nu}{\xi} \overset{\nu}{\omega} \omega}{\overset{\nu}{m}\left[\left(\overset{\nu}{\omega}{}^2 - \omega^2\right)^2 + 4 \overset{\nu}{\xi}{}^2 \overset{\nu}{\omega}{}^2 \omega^2\right]}.$$

Using the frequencies f instead of the circular frequencies ω, the real part of the modal transfer function can be written as

$$Re\left(\overset{\nu}{H}\right) = \frac{1 - f^2/f_\nu^2}{4\pi^2 f_\nu^2 \overset{\nu}{m}\left[\left(1 - f^2/f_\nu^2\right)^2 + 4 \overset{\nu}{\xi}{}^2 f^2/f_\nu^2\right]},$$

and the phase shift is obtained from

$$\tan\varphi_\nu = \frac{2\overset{\nu}{\xi} f/f_\nu}{1 - f^2/f_\nu^2}$$

The excitation spectrum consists of a real part called coincidence-spectrum and of an imaginary part called quadrature-spectrum, $S_{kl} = \mathrm{Co}_{kl} - i\,\mathrm{Qu}_{kl}$. Finally, the spectral density function of the response Y is obtained as:

$$S_Y = \sum_{\nu=1}^{n} \sum_{\mu=1}^{n} \overset{\nu}{\Phi}_Y \overset{\mu}{\Phi}_Y \left|\overset{\nu}{H}\right| \left|\overset{\mu}{H}\right| \sum_{k=1}^{n} \sum_{l=1}^{n} \overset{\nu}{\Phi}_k \overset{\mu}{\Phi}_l \left[\cos(\varphi_\nu - \varphi_\mu) Co_{kl} + \sin(\varphi_\nu - \varphi_\mu) Qu_{kl}\right]$$

From this equation it can be seen that both the real and the imaginary parts of the excitation spectrum, Co_{kl} and Qu_{kl}, contribute to the response. The eigenforms are linked by the cross-products of the transfer functions

$$\left|\overset{\nu}{H}\right| \left|\overset{\mu}{H}\right|, \nu \neq \mu.$$

In the case of resonance, $f = f_\nu$, small damping and sufficently seperated eigenfrequencies, these cross-products are significantly smaller than the square of the transfer function of the resonant mode:

$$\left|\overset{\nu}{H}(f_\nu)\right| \left|\overset{\mu}{H}(f_\nu)\right| << \left|\overset{\nu}{H}(f_\nu)\right|^2 .$$

Apart from resonance the phase shift $\varphi_\nu - \varphi_\mu$ is 0 or π and the contribution of the quadrature-spectrum is zero. Consequently, with systems with low damping and sufficently seperated eigenfrequencies, the spectral density function of the response can be evaluated with a good approximation neglecting the cross-products of the transfer functions. Neglecting of the cross-products $\nu \neq \mu$, the spectral density function can be approximated as

$$S_Y(f) = \sum_1^n \overset{\nu}{\Phi}{}_Y^2 \left|\overset{\nu}{H}\right| \sum_1^n \sum_1^n \overset{\nu}{\Phi}_k \overset{\nu}{\Phi}_l Co_{kl}(f), \text{ with } \overset{\nu}{Co}(f) = \sum_1^n \sum_1^n \overset{\nu}{\Phi}_k \overset{\nu}{\Phi}_l Co_{kl}(f) = \underline{\overset{\nu}{\Phi}}^T \underline{Co(f)} \underline{\overset{\nu}{\Phi}}$$

is representing the excitation spectrum with modal weights, i.e. the modal spectral density function of the excitation.

5 Spectral density of the wind load

5.1 Auto-SDF

The wind force $P'(t) = c_F(f) \cdot A \cdot q'(t)$ in a turbulent flow depends on

- the area A and aerodynamic shape of application area of the load,
- the actual dynamic pressure,
- the frequency f of the fluctuations of the velocities,

$c_F(f)$ is the frequency dependant force coefficient. The normalization of the equation with reference to the mean load results into

$$\frac{P'(t)}{\overline{P}} = \frac{c_F(f)}{c_F(0)} \cdot \frac{q'(t)}{\overline{q}}, \text{ or } \frac{P'(t)}{\overline{P}} = 2 \cdot \chi_F \cdot \frac{u'}{\overline{u}} \qquad (5.1)$$

if the definition of the aerodynamic transfer function

$$\chi_F = \frac{c_F(f)}{c_F(0)}$$

and the approximation

$$\frac{q'}{\overline{q}} = 2 \cdot \frac{u'}{\overline{u}}$$

are used. The transformation in the frequency domain results into the following SDF:

$$\frac{S_P}{\overline{p}^2} = 4 \cdot \chi_F^2 \cdot \frac{S_u}{\overline{u}^2}, \text{ or } \frac{S_P \cdot f}{\sigma_P^2} = \left(\frac{2 \cdot I_u}{I_P}\right)^2 \cdot \chi_F^2 \cdot \frac{S_u \cdot f}{\sigma_P^2} \tag{5.2}$$

if the the following normalizations are used:

$I_u = \sigma_u / \overline{u}$, $I_P = \sigma_p / \overline{P}$.

The aerodynamic transfer function χ_F has the same meaning for the dynamic excitation as the aerodynamic coefficient has for the static wind load. In general, it has to be determined by experiments. χ_F depends on the position of the considered point and on the ratio L/λ.

$\lambda = \overline{u}/f$ is the wave length, and L_A is a characterizing quantity of the loaded area A, $L_A = \sqrt{A}$. For $f \cdot \sqrt{A}/\overline{u} \rightarrow 0$ is $\chi \rightarrow 1$ and for $f \cdot \sqrt{A}/\overline{u} \rightarrow \infty$ is $\chi \rightarrow 0$. The assumption $\chi \rightarrow 1$ for the caculation of design loads is on the safe side. The following estimation was empirically determined for sharp-edged bodies:

$$\chi_F = \frac{1}{1 + \left(2 \cdot \frac{f \cdot \sqrt{A}}{\overline{u}}\right)^{4/3}} \tag{5.3}$$

Applying χ_F=1 as an estimation on the safe side it follows that $2 \cdot I_u / I_p = 1$, and this results into:

$$\frac{S_P \cdot f}{\sigma_P^2} = \frac{S_u \cdot f}{\sigma_u^2} \quad . \tag{5.4}$$

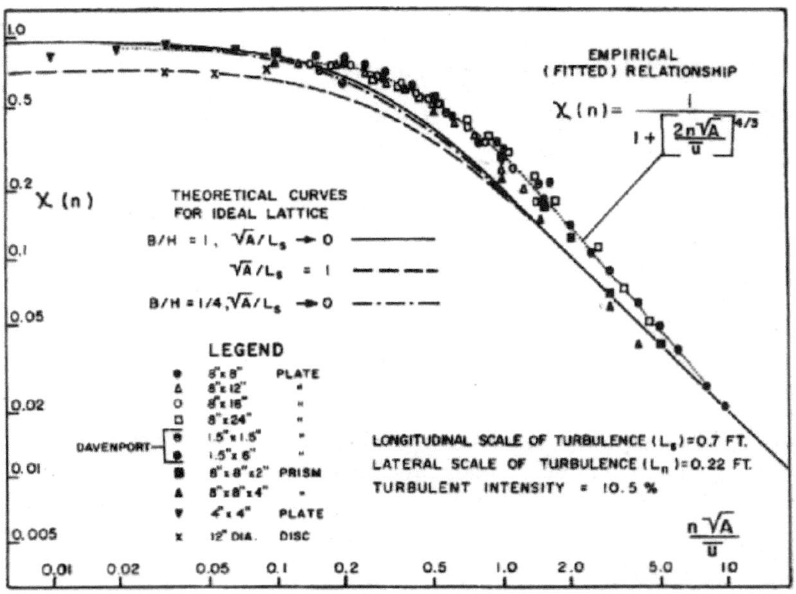

Figure 13: Empiric relation from Vickery (1966)
based on experiments conducted by Vickery and Davenport

5.2 Cross – SDF:

The cross – SDF of two wind forces P_1 and P_2 is complex. It can be seen in analogy to the auto – SDF and the cross – SDF of the wind velocities u_1 and u_2.

$$\frac{S_{P12}(if)}{\bar{P}_1 \cdot \bar{P}_2} = 2 \cdot \chi_{F1}(f) \cdot 2 \cdot \chi_{F2}(f) \cdot \frac{S_{u12}(if)}{\bar{u}_1 \cdot \bar{u}_2}$$

The square contribution of a cross – SDF is real:

$$|S_{12}|^2 = S_{12} \cdot S_{12}^* = Co_{12}^2 \cdot Qu_{12}^2$$

so that $S_{12}(if) = |S_{12}| \cdot e^{i \cdot \phi_{12}}$,

where the phase is a function of the frequency:

$$\phi_{12} = \phi_{12}(f) \cdot$$

Standardisation of the modulus yields the coherence:

$$\gamma_{12}(f) = \frac{|S_{12}|}{\sqrt{S_1 \cdot S_2}}$$

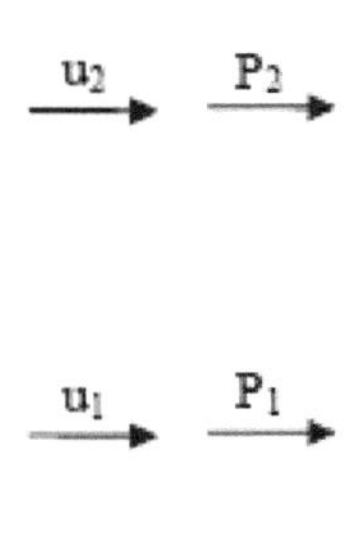

Figure 14: Transfer from wind fluctuations to force fluctuations

The cross – SDF of the velocity is expressed by modulus and phase:

$$S_{u12} = \gamma_{u12} \cdot \sqrt{S_{u1} \cdot S_{u2}} \cdot e^{i \cdot \phi_{12}}$$

Neglecting the height dependency of S_u:, $S_{u1} = S_{u2} = S_u$, $S_{u12} = \gamma_{u12} \cdot e^{i \cdot \phi} \cdot S_u$ and applying the approximation: $\chi_{F1} = \chi_{F2} = 1$ and $\sigma_{u1} = \sigma_{u2} = \sigma_u$ it follows that:

$$\frac{S_{P12} \cdot f}{\sigma_{P1} \cdot \sigma_{P2}} = \gamma_{u12} \cdot e^{i \cdot \phi_{12}} \cdot \frac{S_u \cdot f}{\sigma_u^2}$$

5.3 Empirical approach for coherence and phase:

The coherence and the phase between point 1 and 2 (fig. 14) can be approximated by

$$\gamma_{u_j u_k} = e^{-c_z \cdot f \cdot |\Delta z| / \bar{u}_m},$$

$$\phi_{12}(f) = -c_z' \cdot \Delta z \cdot \frac{f}{\bar{u}_m} \quad .$$

$\Delta z = z_j - z_k$:
vertical distance between points of observation

$$\bar{u}_m = \frac{1}{2}(\bar{u}_j + \bar{u}_k)$$

mean value of velocities

c_z: empirical constant depending on terrain, e.g. open terrain $c_z = 7$. The sign of the phase can be positive or negative which depends on whether points above or below are related to the point of reference. Hereby the groundswell due to an arriving gust and the tailing off after of the gust body can be taken in consideration. The empirical constant c_z' depends on the terrain. A value of $c_z' = 2{,}6$ can be assumed for open terrain.

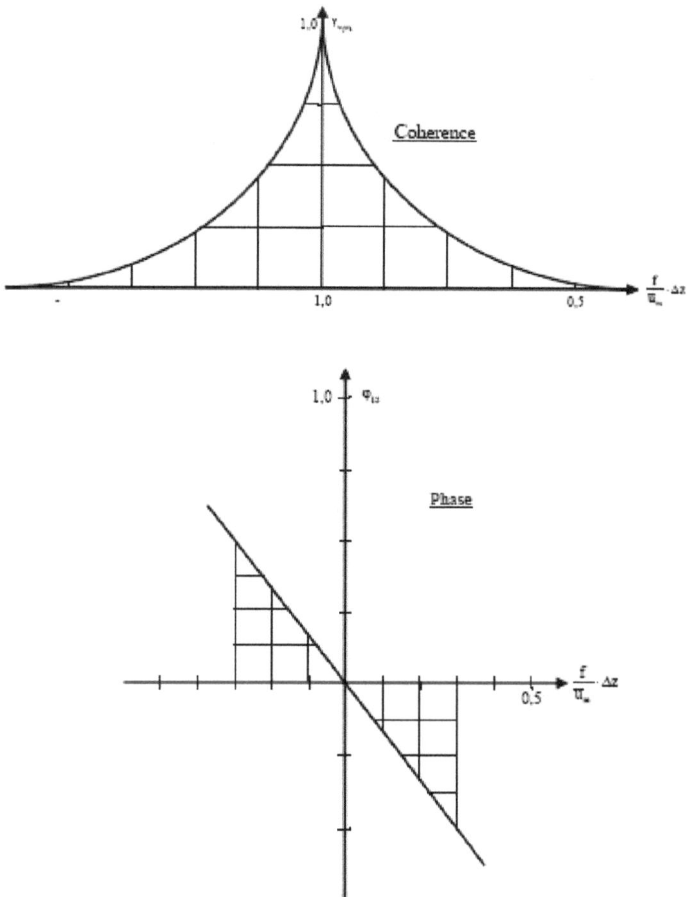

Figure 15 Semi-empirical coherence function and phase function of velocity fluctuations in turbulent wind

6 Application of the spectral method on a cantilever

6.1 System and fundamental mode shapes

In this chapter an example for the application of the spectral method on a cantilever structure under wind loading is given. Wind forces are acting on two points, the considered response quantity is the base bending moment Y. Three eigenforms are evaluated (fig. 16).

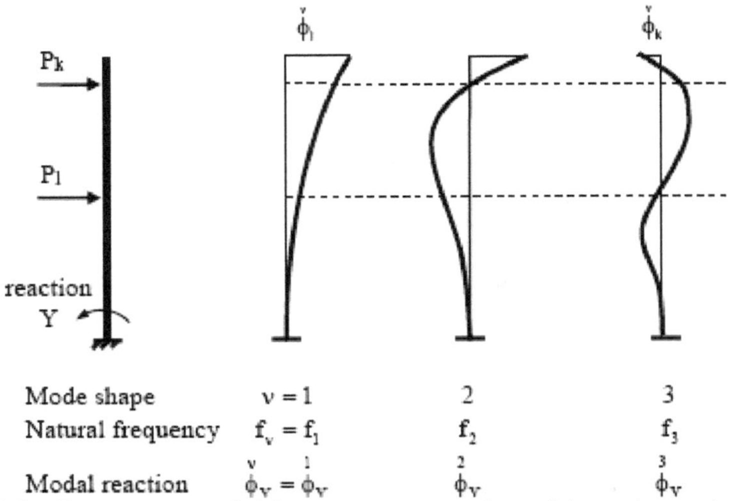

Figure 16 Cantilever structure, wind forces, response quantity, and three relevant eigenforms

6.2 Fundamental equation

The fundamental equation is: $S_{\underline{y}}(f) = \mathrm{Re}\{\underline{H}^*_y(if) \cdot \underline{S}^*_p(if) \cdot \underline{H}^T_y(if)\}$

6.2.1 Real part

The spectral density function of the excitation is defined by:

$$\underline{S}_p(if) = \underline{Co}(f) - i \cdot \underline{Qu}(f) \tag{6.1}$$

and contains the quadratic and symmetrical coincidence-spectrum and the quadratic and non-symmetrical quadrature-spectrum.

$$\underline{Co} = \begin{bmatrix} Co_{11} & & Co_{kl} \\ & Co_{22} & \\ Co_{kl} & & Co_{33} \end{bmatrix} \qquad \underline{Qu} = \begin{bmatrix} Qu_{11} & & Qu_{kl} \\ & Qu_{22} & \\ -Qu_{kl} & & Qu_{33} \end{bmatrix} \tag{6.2}$$

In modal analysis the mechanical transfer function for the reaction y due to load k is $H_{yk}(if)$

$$H_{yk} = \sum_{\nu} \overset{\nu}{H}_{yk} \quad \underline{H}^*_y = \sum_{\nu} \overset{\nu}{\underline{H}^*} \ ; \ \underline{H}^T_y = \sum_{\nu} \overset{\kappa}{\underline{H}^T_y}$$

From the previous considerations follows

$$\begin{aligned} S_{\underline{y}}(f) &= \mathrm{Re}\left\{ \overset{1}{\underline{H}^*_y} \cdot \underline{S}_p \cdot \overset{1}{\underline{H}^T_y} + \overset{1}{\underline{H}^*_y} \cdot \underline{S}_p \cdot \overset{2}{\underline{H}^T_y} + \ldots \overset{2}{\underline{H}^*_y} \cdot \underline{S}_p \cdot \overset{1}{\underline{H}^T_y} + \overset{2}{\underline{H}^*_y} \cdot \underline{S}_p \cdot \overset{2}{\underline{H}^T_y} \ldots \right\} \\ &= \mathrm{Re}\left\{ \sum_{\nu} \sum_{\kappa} \overset{\nu}{\underline{H}^*_y} \cdot \underline{S}_p \cdot \overset{\kappa}{\underline{H}^T_y} \right\} \end{aligned} \tag{6.3}$$

If all mixed terms ($\nu \neq \kappa$) are neglected (which usually is the case) the cartesian product between the different mode shapes is

$$S_{\underline{y}}(f) = \mathrm{Re}\left\{ \sum_{\nu} \overset{\nu}{\underline{H}}{}_y^{*} \cdot \underline{S}_p \cdot \overset{\nu}{\underline{H}}{}_y^{T} \right\} \tag{6.4}$$

The response of the ν-th mode shape is in general (fig. 17)
$H_k(if) = |H_k|(\cos(\varphi_k) + i \cdot \sin(\varphi_k))$.

In one mode shape the phase φ_k between response y and the excitation k is equal for all load points,
$\overset{\nu}{\varphi}_k = \overset{\nu}{\varphi}_1 = \overset{\nu}{\varphi}$.

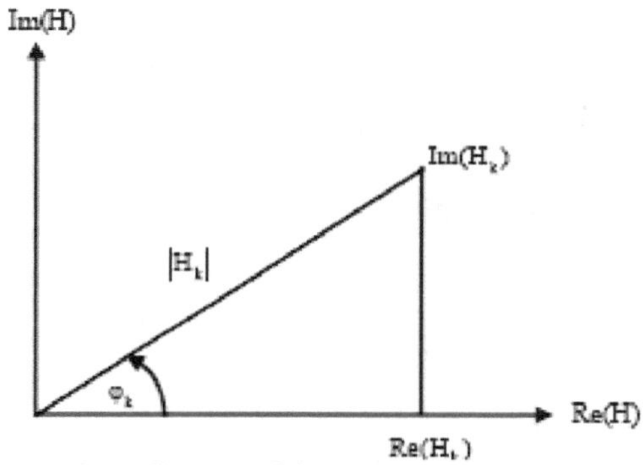

Figure 17: Phase diagram of the vector of the transfer function

The transfer matrix is

$$\overset{\nu}{\underline{H}}_y = \left|\overset{\nu}{\underline{H}}_y\right| \cdot \left(\cos\left(\overset{\nu}{\varphi}\right) + i \cdot \sin\left(\overset{\nu}{\varphi}\right) \right). \tag{6.5}$$

The resulting spectral density function of the response is

$$\overset{\nu}{S}_y(f) = \mathrm{Re}\left\{ \sum_{\nu} \overset{\nu}{\underline{H}}{}_y^{*} \cdot \underline{S}_p \cdot \overset{\kappa}{\underline{H}}{}_y^{T} \right\}$$

$$= \mathrm{Re}\left\{ \left|\overset{\nu}{\underline{H}}_y\right| \cdot \left(\underline{Co} - i \cdot \underline{Qu}\right) \cdot \left|\overset{\nu}{\underline{H}}_y\right|^T \right\} \cdot \underbrace{\left(\cos\left(\overset{\nu}{\varphi}\right) - i \cdot \sin\left(\overset{\nu}{\varphi}\right) \right) \cdot \left(\cos\left(\overset{\nu}{\varphi}\right) + i \cdot \sin\left(\overset{\nu}{\varphi}\right) \right)}_{=1}.$$

Herefrom, the basic equation of the modal system analysis is

$$\overset{\nu}{S}_y(f) = \sum_{\nu} \left|\overset{\nu}{\underline{H}}_y\right| \cdot \underline{Co} \cdot \left|\overset{\nu}{\underline{H}}_y\right|^T \tag{6.6}$$

6.3 Evaluation of the basic equation

For the most cases of wind action on cantilevered structures it is adequate to restrict the calculations to the natural frequency, i.e. the first mode.

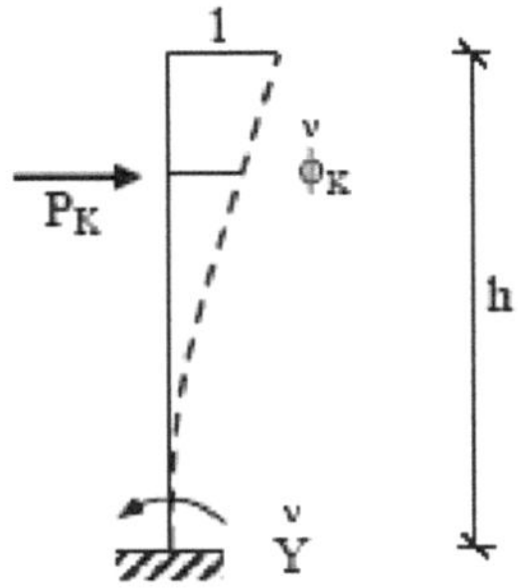

$$\hat{\check{y}} = \check{H}_{yk} \cdot \hat{P}_k = \check{\phi}_y \cdot \check{\phi}_k \cdot \check{H} \cdot \hat{P}_k$$

$$\left|\check{H}_{yk}\right| = \check{\phi}_y \cdot \check{\phi}_k \cdot \left|\check{H}\right|$$

with $\check{\phi}_k$ - normalized oscillation amplitude of ν-th mode shape

$$\left[\check{\phi}_k\right] = 1$$

$\check{\phi}_y$ corresponding value of the reaction y

Figure 18: 1st eigenform of the cantilever and basic equations

The response quantity is calculated with

$$\left|\check{H}\right| = \frac{\check{V}}{\check{m} \cdot \check{\omega}^2} \text{ and the magnification } \check{V} = \frac{1}{\left[\left(1-\check{\eta}^2\right)^2 + \left(\frac{\delta}{\pi}\right)^2 \check{\eta}^2\right]^{\frac{1}{2}}}$$

where $\check{\eta} = \frac{\omega}{\check{\omega}}$ is the reduced frequency calculated form the natural frequency . $\check{\omega} = 2\pi f_\nu$. The modal mass is

$$\check{m} = \int_h \check{\phi}^2(z) \cdot \mu(z) \cdot dz\,,\ \left[\check{m}\right] = kg\,,\ \mu = \text{mass per unit length.}$$

The base bending moment Y is calculated from the modal value of the base bending moment. The idealized basic equation of the modal system analysis for the νth mode shape is:

$$\left|\underline{\check{H}}_y\right| \cdot \underline{Co} \cdot \left|\underline{\check{H}}_y\right|^T = \check{\phi}_\nu^2 \cdot \left|\check{H}\right|^2 \cdot \underline{\check{\phi}} \cdot \underline{Co} \cdot \underline{\check{\phi}}^T\,,\ \check{\phi}_y = \int_h \mu(z) \cdot \check{\omega}^2\, \check{\phi}(z) \cdot z \cdot dz\,,\ \check{Co} = \sum_k \sum_l \check{\phi}_k \cdot Co \cdot \check{\phi}_l$$

or equivalently $\check{Co} = \int_h \int_h \check{\phi}(z_k) \cdot \check{\phi}(z_l) \cdot Co(z_k, z_l) \cdot dz_k \cdot dz_l$). The response spectrum results into

$$S_y = \sum_\nu \frac{\check{\phi}_y^2 \cdot \check{V}^2}{\check{\omega} \cdot \check{m}^2} \check{Co}$$

6.4 Calculation of the resonant response

6.4.1 First approach: calculation with the real part of the cross-product

The coincidence spectrum of the excitation $Co(z_k, z_l)$ is the modulus of the complete spectral density. This leads to the model

$$S_p(z_k, z_l) = \gamma(\Delta z) \cdot \left(S_p(z_k) \cdot S_p(z_l)\right)^{1/2}$$

$\gamma = e^{-c_z \frac{|\Delta z|}{\bar{u}_m} \cdot f}$ coherence

$l_c(f) = \dfrac{\bar{u}_m}{-c_z \cdot f}$ correlation length, from which follows that $\gamma = e^{-\frac{|\Delta z|}{l_c}}$

$$\overset{\vee}{Co} = \int_{z_k=0}^{h} \int_{z_l=0}^{h} \overset{\vee}{\phi}(z_k) \cdot [S_p(z_k)]^{1/2} \cdot \overset{\vee}{\phi}(z_l) \cdot [S_p(z_l)]^{1/2} \cdot e^{-\frac{|\Delta z|}{l_c}} dz_k dz_l$$

$$= 2 \cdot \int_{z_k=0}^{h} \overset{\vee}{\phi}(z_k) \cdot [S_p(z_k)]^{1/2} \cdot \left[\int_{z_k=z_l}^{h} e^{-\frac{|\Delta z|}{l_c}} \cdot \overset{\vee}{\phi}(z_l) \cdot [S_p(z_l)]^{1/2} \cdot dz_l \right] dz_k$$

It can be assumed close to the natural frequency that he correlation length $l_c << h$, so that $S_p(z_l) \cong const = S_p(z_k)$, $\overset{\vee}{\phi}(z_l) \cong const = \overset{\vee}{\phi}(z_k)$, and $\bar{u}_m \cong const = \bar{u}_m(z_k)$, which delivers $\overset{\vee}{Co} \cong 2 \cdot \int_{z_k=0}^{h} \overset{\vee}{\phi}^2(z_k) \cdot S_p(z_k) \left[\int_{\Delta z=0}^{h} e^{-\frac{|\Delta z|}{l_c}} \cdot d(\Delta z) \right] dz_k$. Evaluating $\int_{\Delta z=0}^{h} e^{-\frac{|\Delta z|}{l_c}} \cdot d(\Delta z) = \int_{\Delta z=0}^{\infty} e^{-\frac{|\Delta z|}{l_c}} \cdot d(\Delta z) = l_c$

leads to $\overset{\vee}{Co} = 2 \cdot \int_0^h l_c \cdot \overset{\vee}{\phi}^2(z) \cdot S_p(z)\, dz$ and the load spectrum $\dfrac{S_p}{\bar{p}^2} = (2 \cdot \chi_F \cdot I_u)^2 \cdot \dfrac{S_u}{\sigma_u^2}$

6.4.2 Second approach: S_u / σ^2_u const. in z

The mean wind load, the correlation length, and the turbulence intensity are here

$$\bar{p}^2(z) = \bar{p}_h^2 \cdot \left(\frac{z}{h}\right)^{4\alpha}, \quad l_c(f,z) = \frac{\bar{u}}{c_z \cdot f} = \frac{\bar{u}_h}{c_z \cdot f} \cdot \left(\frac{z}{h}\right)^{\alpha}, \quad I_u(z) = I_{uh} \cdot \left(\frac{z}{h}\right)^{-\alpha}.$$

The wind velocity spectrum for the higher frequency range is simplified as

$\dfrac{S_u \cdot f}{\sigma_u^2} = 0{,}115 \cdot \left(\dfrac{L_{ux}(h)}{h}\right)^{-\frac{2}{3}} \cdot \left(\dfrac{h \cdot f}{\bar{u}_h}\right)^{-\frac{2}{3}}$, with the reduced frequency defined by $\tilde{f} = L_{ux} \dfrac{f}{\bar{u}(z)} > 1$

from which follows

$$\frac{\overset{\vee}{Co} \cdot f}{\bar{p}_h^2 \cdot h^2} = \frac{2}{c_z} \cdot 0{,}115 \cdot \left(\frac{\bar{u}_h}{f \cdot h}\right)^{\frac{5}{3}} \cdot (2 \cdot \chi_F \cdot I_{uh})^2 \cdot \left(\frac{h}{L_{ux}(h)}\right)^{\frac{2}{3}} \cdot \int_0^h \left(\frac{z}{h}\right)^{3\alpha} \cdot \overset{\vee}{\Phi}^2 \cdot \frac{z}{h} \cdot d\left(\frac{z}{h}\right)$$

.

6.4.3 List of results

1. Mean value of the response quantity $\bar{Y} = \bar{F}_h \cdot \dfrac{h^2}{2} \cdot \dfrac{1}{(1+\alpha)}$

2. Variance of the resonant response $\overset{\vee}{S}_y = \dfrac{\overset{\vee}{\phi}_y^2}{\overset{\vee}{\varpi}^4 \cdot \overset{\vee}{m}^2} \cdot \overset{\vee}{V}^2(f) \cdot \overset{\vee}{Co}(f) \cong \dfrac{\overset{\vee}{\phi}_y^2}{\overset{\vee}{\varpi}^4 \cdot \overset{\vee}{m}^2} \cdot \overset{\vee}{V}^2(f) \cdot \overset{\vee}{Co}(f_v)$

$$\check{\sigma}_{yR}^2 = \int_0^\infty \check{S}_y(f)\cdot df = \frac{\check{\phi}_y^2}{\check{\varpi}^4\cdot\check{m}}\cdot \check{Co}(f)\cdot\int_0^\infty \check{V}^2(f)\cdot df$$

with: $\frac{\check{\phi}_y^2}{\check{\varpi}^4\cdot\check{m}^2} = \left(\frac{\int\mu\cdot\check{\phi}(z)\cdot z\cdot dz}{\int\mu\cdot\check{\phi}^2\cdot dz}\right)^2$ $\qquad \frac{\check{\sigma}_{yR}^2}{\overline{Y}^2} = \frac{\pi^2}{\delta}\cdot(1+\alpha)\cdot\left(\frac{\int\mu\cdot\check{\phi}\cdot\frac{z}{h}\cdot d\left(\frac{z}{h}\right)}{\int\mu\cdot\check{\phi}^2\cdot d\left(\frac{z}{h}\right)}\right)^2\cdot\frac{\check{Co}(f_v)}{\overline{p}^2\cdot h^2}$

3. Simplified evaluation with μ= const and the approximation $\overset{1}{\phi}(z)\cong\frac{z}{h}$ for the 1st mode shape:

$$\frac{\sigma_{yR}^2}{\overline{Y}^2} = \frac{\pi^2}{\delta}\cdot(1+\alpha)\cdot\frac{0{,}077}{c_z}\left(\frac{u_h}{f_v\cdot h}\right)^{\frac{5}{3}}\cdot\left(2\cdot\chi_F(f_v)\cdot I_{uh}\right)^2\cdot\left(\frac{h}{L_{ux}(10)}\right)^{\frac{2}{3}}\cdot\left(\frac{h}{10}\right)^{\frac{2\alpha}{3}}$$

7 Bibliography

Benaroya, H., Han, S.M. (2005). *Probability Models in Engineering and Science*, Taylor & Francis, Boca Ratoon

Davenport, A.G., (1962). The application of statistical concepts to the wind loading of structures. in *Proceedings of the Institute of Civil Engineering*, No.6480, Vol.19.

Simiu,E., Scanlan, R.H. (1996). Win*d Effects on Structures - Fundamentals and Applications to Design*. 3rd ed., chapter 5, John Wiley & Sons, New York.

Vickery, B.J. (1966). Fluctuating lift and drag on a long cylinder of square cross-section in a smooth and turbulent stream. *Journal of Fluid Mechanics*, Vol. 25, S.481-494, 1966.

Dynamic Approach to the Wind Loading of Structures: Alongwind, Crosswind and Torsional Response

Giovanni Solari and Federica Tubino

Department of Structural and Geotechnical Engineering, University of Genoa, Genoa, Italy

Abstract. This paper provides a general framework of the dynamic approach to the wind loading of structures, in particular of vertical structures such as buildings, towers and chimneys. The first part of the paper deals with the classical problems of the dynamic alongwind response and of the equivalent static forces, defining a chain of nondimensional coefficients, referred to as the gust factors, aimed at determining the maximum response for engineering applications to structural design. The second part of the paper generalises the above formulations and concepts to the crosswind and torsional response of structures due to gust buffeting; the focal problem of the vortex shedding is introduced but not developed in the due detail. The formulations and the applications discussed focuse on the analytical tools developed at the University of Genoa.

1 Introduction

Gust buffeting and wind-induced dynamic response of structures represent main topics of the research developed in wind engineering over the last 40 years. Such research was carried out by means of several tools including mathematical formulations, analytical methods, numerical solutions, wind-tunnel tests and full-scale experiments. A dominant aspect of wind engineering is the joint use of different tools in order to obtain robust results. This paper deals with the dynamic approach to the wind loading of structures, with special concern for its formulation. A wider prospect on this complex and articulated matter may be found in (Solari 1999).

The study of the alongwind response, in particular of vertical structures (e.g. buildings, towers and chimneys), began in the 60's thanks to the pioneering contributions of Davenport (1961, 1964, 1967). He took into account only the fundamental vibration mode and expressed the mean value of the maximum alongwind displacement, called for sake of simplicity maximum alongwind displacement, as the product of the mean static displacement by a non-dimensional constant coefficient, the Gust Response Factor (GRF). Davenport also defined the Equivalent Static Force (ESF) as the force that, statically applied on the structure, produces the maximum alongwind displacement. Exploiting structural linearity, the ESF is the product of the mean static force by the GRF.

Studies carried out in the 70's followed two distinct lines. On the one hand, Vellozzi and Cohen (1968), Vickery (1970) and Simiu (1976, 1980) perfected Davenport's method, especially with reference to wind and aerodynamic modelling; on the other hand, ESDU (1976) and ECCS

(1978) introduced procedures to determine the maximum values of any load effect (e.g. the bending moment and the shear force) by the influence function technique. All these criteria provided results by means of graphs obtained through numerical integrations.
Starting from Simiu's formulation, at the beginning of the 80's Solari (1982, 1983) obtained a Closed Form Solution (CFS) of the dynamic alongwind response of three standard models called point-like, vertical and horizontal. Continuing this research, Solari (1988) developed also the Equivalent Wind Spectrum Technique (EWST), a method to schematise wind as an equivalent velocity field identically coherent in space. Thanks to this criterion, a new CFS of the alongwind response was developed (Solari 1993), which represents a substantial advance with respect to the previous solution.
Research carried out in the 90's followed two distinct lines. The first was aimed at determining the maximum effects due to the alongwind response. The second extended the original methods from the alongwind response to the crosswind and torsional responses.
The first line derived from the remark that the ESF conceived by Davenport (1967) and used by most subsequent authors usually involves correct estimates of the maximum displacement, but may give place to significant errors for other load effects, above all the bending moment and the shear force. With the aim of avoiding such errors, Kasperski (1992) used the influnce function technique to develop the Load Response Correlation (LRC) method; among infinite possible ESFs, it selects the most probable load pattern whose application produces the maximum value of the quasi-static part of a selected load effect. Holmes (1994, 1996) applied this concept to derive a CFS for the maximum displacement, bending moment and shear force at any level of a free-standing lattice tower. Coherently with this solution, the ESF was found by separating the quasi-static and the resonant contributions: the former was obtained through the LRC method; the latter implies a load pattern that deforms the structure according to its fundamental vibration mode. Several authors applied analogous methods to different structural types (Davenport 1995, Zhou et al 1999, Zhou and Kareem 2001, Holmes 2002), some times arriving at a CFS.
The second line of research was aimed at determining the three-directional (3-D) wind-induced response of structures according to two main projects. The first, carried by Tamura et al (1996), derived analytical formulae of the maximum alongwind, crosswind and torsional responses of buildings with un-coupled modes of vibration, and the corresponding ESFs, by fitting the results of a wide campaign of wind-tunnel tests. The second was developed by Piccardo and Solari (1996) with reference to slender structures and structural elements. Alongwind, crosswind and torsional actions were schematised by the quasi-steady theory, as a linear combination of the longitudinal, lateral and vertical turbulence components; the crosswind forces and the torsional moments due to vortex shedding were superimposed assuming wake actions as independent of oncoming turbulence. Alongwind, crosswind and torsional responses, considered as un-coupled and only depending on the corresponding fundamental modes, were determined in closed form (Piccardo and Solari 1998a, 2000) by the Generalised Equivalent Spectrum Technique (GEST) (Piccardo and Solari 1998b), a method to extend the applicability range of the EWST from the alongwind response to the 3-D response. This formulation involves a 3-D GRF that generalises the GRF introduced by Davenport.
The 3-D Gust Effect Factor (GEF) technique (Piccardo and Solari 2002) may be interpreted as the junction point for several methods described above. It provides the maximum values and

the ESF for the most relevant wind load effects on cantilever slender vertical structures, through a single non-dimensional quantity, the 3-D GEF, whose expression was found in closed form. In such a framework, the GRF and the 3-D GRF techniques were derived as particular cases. A general criterion to classify the behaviour of structures under wind loading was established (Solari and Repetto 2002). The development of this new method created the bases to establish generalised expressions for the ESFs, and to classify their expressions into three families referred to as the Gust Factor (GF) technique, the Load Combination (LC) technique and the Global Loading (GL) technique (Repetto and Solari 2004).

A different strategy for determining the wind-induced dynamic response of structures involves a joint and rational use of analytical methods, numerical solutions and experimental results. Zhou et al (2003), Zhou and Kareem (2003) and Chen and Kareem (2005) determined the response of buildings with both un-coupled and coupled modes of vibration starting from wind-tunnel tests involving the use of high-frequency force-balances and aeroelastic balances. Katsumura et al (2004) determined a universal ESF that reproduces the maximum wind load effects on a complex structure, using wind tunnel pressure tests and the proper orthogonal decomposition.

This paper provides a general framework of the above matter, with special regard to the methods developed at the University of Genoa. Sections 2 and 3 illustrate, respectively, the dynamic alongwind response and the three-directional (alongwind, crosswind and torsional) response of structures, focusing on maximum wind load effects and the corresponding ESFs. Section 4 illustrates an application involving analytical and numerical solutions. Section 5 provides some synthetic conclusions.

2 Dynamic alongwind response

This section provides a general and homogeneous framework of the dynamic alongwind response of structures, by analysing the chain of five classical problems. Section 2.1 deals with the maximum wind velocity and local pressure, introducing the velocity gust factor and the local pressure gust factor. Section 2.2 analyses the maximum resultant force on a surface and the related equivalent pressure, introducing the resultant force or equivalent pressure gust factor. Section 2.3 determines the dynamic alongwind response of a single-degree-of-freedom structural system; accordingly, it defines an equivalent static force and an equivalent static pressure based on the gust response factor. Section 2.4 generalises the concepts developed in Section 2.3 to the dynamic alongwind response of a vertical structure. Section 2.5 generalises this formulation to any wind-excited load effect and the corresponding equivalent static force, introducing the gust effect factor. It is worth noting that each step forward of the above chain of problems may be dealt with as a generalisation of the previous one; on the other hand, each step backward of the same chain may be interpreted as a particular case of the subsequent one.

2.1. Wind velocity and local pressure

Let us consider, for sake of simplicity, a flat homogeneous terrain and a point M at height z over the ground, in the internal boundary layer (Solari and Piccardo 2000). Alongwind gust buffeting is traditionally analyzed ignoring the lateral and vertical turbulence components, and expressing the instantaneous wind velocity as:

$$u(M;t)=\bar{u}(z)+u'(M;t) \tag{1}$$

where t is the time, $\bar{u}$ is the mean wind velocity, u' is the zero-mean longitudinal turbulence component, treated here as a stationary Gaussian random process characterized, in the domain of frequency n, by its cross-power spectral density function:

$$S_{uu}(M,M';n)=\sqrt{S_u(z;n)S_u(z';n)}\mathrm{Coh}_{uu}(M,M';n) \tag{2}$$

where $S_u(z;n)$ is the power spectral density function (psdf) of $u'(M;t)$; $\mathrm{Coh}_{uu}(M,M';n)$ is the coherence function of $u'(M;t)$ and $u'(M';t)$.
The mean value of the maximum wind velocity over a given time interval T can be expressed as:

$$\bar{u}_{max}(z)=\bar{u}(z)+g_u(z)\sigma_u(z)=G_u(z)\bar{u}(z) \tag{3}$$

where σ_u and g_u are, respectively, the root mean square (rms) value and the peak factor of u'; G_u is referred to as the velocity gust factor and is given by the relationship:

$$G_u(z)=1+g_u(z)I_u(z) \tag{4}$$

where $I_u(z)=\sigma_u(z)/\bar{u}(z)$ is the longitudinal turbulence intensity. A closed form expression for G_u is reported by Solari (1993).
Let us consider the windward face of a fixed surface orthogonal to the direction of the oncoming flow. The instantaneous local pressure p exerted by wind in a point M of the surface at the height z is given by:

$$p(M;t)=\bar{p}(M)+p'(M;t) \tag{5}$$

$\bar{p}$ and p' being, respectively, the mean value and the zero-mean fluctuation of the local pressure. Using the strip and quasi-steady theory (Davenport 1961), and assuming that turbulence is small with respect to the mean wind velocity, i.e. $u'/\bar{u}\ll 1$, $\bar{p}$ and p' can be expressed as:

$$\bar{p}(M)=\frac{1}{2}\rho\bar{u}^2(z)c_p(M) \qquad p'(M;t)=\rho\bar{u}(z)u'(M;t)c_p(M) \tag{6}$$

where ρ is the air density and $c_p>0$ is the mean pressure coefficient in the point M. It is worth noting that this formulation does not apply in separation flow zones where $c_p<0$.
Based on Eq. (6), likewise the wind velocity u, also the local pressure p is a stationary Gaussian random process. Thus, the mean value of the maximum local pressure can be expressed as:

$$\bar{p}_{max}(M)=\bar{p}(M)+g_p(M)\sigma_p(M)=G_p(z)\bar{p}(M) \tag{7}$$

where σ_p is the rms value of p, $g_p(M)=g_u(z)$ is the peak factor of p', G_p is referred to as the local pressure gust factor and is given by the relationship:

$$G_p(z) = 1 + 2g_u(z) I_u(z) \tag{8}$$

The deep analogy between Eqs (4) and (8) is apparent.

2.2. Resultant force and equivalent pressure

Let us consider the windward face of a fixed surface of area A, orthogonal to the direction of the oncoming flow. The resultant force r acting on this surface is given by:

$$r(t) = \int_A p(M;t)\,\mathrm{d}A \tag{9}$$

where M is a generic point of the surface A. Replacing Eq. (5) into Eq. (9):

$$r(t) = \bar{r} + r'(t) \tag{10}$$

$$\bar{r} = \int_A \bar{p}(M)\,\mathrm{d}A \qquad r'(t) = \int_A p'(M;t)\,\mathrm{d}A \tag{11}$$

$\bar{r}$ and r' being, respectively, the mean vaue and the zero-mean fluctuation of the resultant wind force. In particular, expressing $\bar{p}$ and p' by means of Eq. (6), likewise the local pressure p, also the resultant force r is a stationary Gaussian random process.
The problem may be simplified by introducing the following hypotheses: 1) the mean pressure coefficient is constant over the surface A, i.e. $c_p(M) = c_p$; 2) the surface A is small enough to consider the mean wind velocity and the psdf of the longitudinal turbulence as constant on A, i.e. $\bar{u}(z) = \bar{u}(h_0) = \bar{u}$, $S_u(z;n) = S_u(h_0;n) = S_u(n)$, h_0 being the height over the ground of the centre of A. It follows that the mean value and the psdf of the resultant wind force r are given by:

$$\bar{r} = \frac{1}{2}\rho\bar{u}^2 c_p A \qquad S_r(n) = \rho^2\bar{u}^2 S_u(n)\chi_u(n) c_p^2 A^2 \tag{12}$$

where $\chi_u(n) \leq 1$ is a linear filter referred to as the aerodynamic admittance function (Vickery and Davenport 1967):

$$\chi_u(n) = \frac{1}{A^2}\int_A\int_A \mathrm{Coh}_{uu}(M, M';n)\,\mathrm{d}A\mathrm{d}A' \tag{13}$$

Based on this model, the mean value of the maximum resultant force can be expressed as:

$$\bar{r}_{max} = \bar{r} + g_r\sigma_r = G_r\bar{r} \tag{14}$$

where σ_r and g_r are, respectively, the rms value and the peak factor of r'; G_r is referred to as the resultant force gust factor and is given by:

$$G_r = 1 + g_r\frac{\sigma_r}{\bar{r}} = 1 + 2g_r I_u B \tag{15}$$

where $I_u = I_u(h_0)$; B is a non-dimensional quantity provided by the relationship:

$$B = \frac{1}{2I_u}\frac{\sigma_r}{\bar{r}} = \frac{1}{\sigma_u}\sqrt{\int_0^\infty S_u(n)\chi_u(n)\mathrm{d}n} \tag{16}$$

A closed form expression for G_r is reported by Solari (1993).
As a consequence of this formulation, the equivalent pressure is defined as the uniform pressure over the surface of area A that causes the mean value of the maximum resultant force. Thus:

$$p_{eq} = \frac{\bar{r}_{max}}{A} = G_r\bar{p} \tag{17}$$

where $\bar{p}$ is the mean local pressure provided by the first Eq. (6) for $z = h_0$ and $c_p(M) = c_p$. Due to this definition, G_r is called also the equivalent pressure gust factor.
Some general tendencies are noteworthy. For A tending to zero, $\chi_u(n) = 1$ for any n (Eq. (13)), $B = 1$ (Eq. (16)) and $g_r = g_u$ (Solari 1993); it follows that $G_r = G_p$ (Eqs. (8) and (15)) and $p_{eq} = \bar{p}_{max}$ (Eqs. (7) and (17)). On the other hand, for A tending to infinite, $\chi_u(n) = 0$ for any $n > 0$ (Eq. (13)) and $B = 0$ (Eq. (16)); it follows that $G_r = 1$ (Eq. (15)) and $p_{eq} = \bar{p}$ (Eq. (17)). Thus, due to the non-contemporaneity of the maximum local pressures, the equivalent pressure reduces on increasing the area of the bluff surface, and $\bar{p} \le p_{eq} \le \bar{p}_{max}$.

2.3. Dynamic response and equivalent static pressure for SDOF systems

Consider a Single-Degree-Of-Freedom (SDOF) dynamic system (Figure 1a). It is admitted that the wind pressure is applied on a rectangular surface of area A, whose centre is placed at the height h_0 over the ground (Figure 1b) and coincides with the barycentre of the structural mass. The resultant force exerted by wind on the bluff surface is characterized by the mean value and the psdf in Eq. (12). It is assumed here, for sake of simplicity, that the mean pressure coefficient c_p in Eq. (12) is now representative of the drag coefficient of the bluff surface, including the negative pressure on the leeward side of the surface. This hypothesis, frequently used in wind engineering, is not appropriate since the signature turbulence, i.e. the turbulence caused by flow separation in the rear of the surface, has a different frequency content from that associated with the oncoming turbulence.

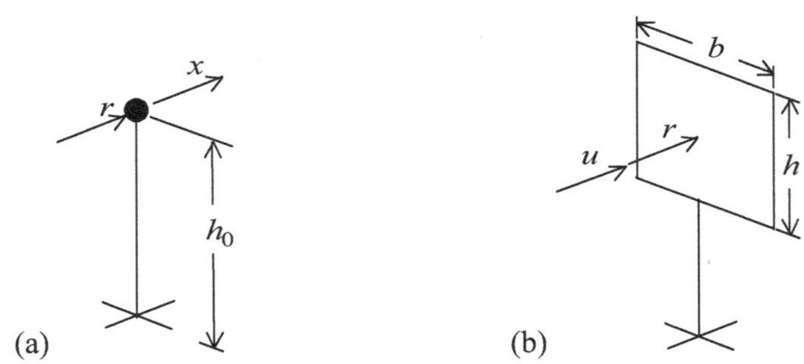

Figure 1. Wind-excited SDOF dynamic system.

The dynamic alongwind response can be expressed as:

$$x(t) = \overline{x} + x'(t) \tag{18}$$

where $\overline{x}$ and x' are, respectively, the mean static value and the zero-mean dynamic fluctuation of the response in the alongwind direction. In particular, dealing with a linear structural system, likewise the resultant force r, also the displacement x is a stationary Gaussian random process, characterized by a mean value and a psdf of the fluctuation provided by the relationships:

$$\overline{x} = \frac{\overline{r}}{m(2\pi n_0)^2} \qquad S_x(n) = \left|H(n)\right|^2 S_r(n) \tag{19}$$

In Eq. (19), m is the mass, n_0 is the fundamental frequency, $k = m(2\pi n_0)^2$ is the stiffness; $H(n)$ is the complex frequency response function defined as:

$$H(n) = \frac{1}{m(2\pi n_0)^2} \frac{1}{1 - \dfrac{n^2}{n_0^2} + 2i\xi\dfrac{n}{n_0}} \tag{20}$$

where i is the imaginary unit and ξ is the damping ratio.
Based on this model, the mean value of the maximum alongwind response can be expressed as:

$$\overline{x}_{max} = \overline{x} + g_x\sigma_x = \overline{x} + g_x\sqrt{\sigma_{Bx}^2 + \sigma_{Rx}^2} = G_x\overline{x} \tag{21}$$

where σ_x and g_x are, respectively, the rms value and the peak factor of x'; σ_{Bx} and σ_{Rx} are the rms values of the background or quasi-static part and of the resonant part of the fluctuating response, respectively; G_x is referred to as the Gust Response Factor (GRF) and is given by:

$$G_x = 1 + g_x\frac{\sigma_x}{\overline{x}} = 1 + 2g_xI_u\sqrt{B_x^2 + R_x^2} \tag{22}$$

where B_x and R_x are non-dimensional quantities provided by the relationships:

$$B_x = \frac{1}{2I_u}\frac{\sigma_{Bx}}{\overline{x}} \qquad R_x = \frac{1}{2I_u}\frac{\sigma_{Rx}}{\overline{x}} \tag{23}$$

$$\sigma_{Bx} = \frac{\sigma_r}{m(2\pi n_0)^2} \qquad \sigma_{Rx} = \frac{\sqrt{S_r(n_0)}}{m(2\pi n_0)^2}\sqrt{\frac{\pi n_0}{4\xi}} \tag{24}$$

being $B_x = B$ (Eq. (16)). A closed form expression for G_x is reported by Solari (1993).
The Equivalent Static Force (ESF) is the resultant force that statically applied to the structural system causes the mean value of the maximum alongwind response:

$$r_{es} = k\overline{x}_{max} \tag{25}$$

Accordingly, the equivalent static pressure is the pressure that uniformly applied on the surface of area A produces the ESF and the mean value of the maximum alongwind response. Thus:

$$p_{es} = \frac{r_{es}}{A} = G_x \bar{p} \tag{26}$$

Some general tendencies are noteworthy. On increasing the stiffness and the damping ratio, R_x tends to zero (Eq. (23)) and g_x tends to g_r (Solari 1993); it follows that G_x tends to G_r (Eqs. (15) and (22)) and p_{es} tends to p_{eq} (Eqs. (17) and (26)). On the other hand, on decreasing the stiffness and the damping ratio, R_x tends to become much greater than B_x; it follows that the GRF tends to the expression $G_x = 1 + 2g_x I_u R_x$ (Eq. (22)). In any case, $p_{eq} \le p_{es}$; instead, no general rule can be established to compare p_{es} with $\bar{p}_{max}$.

Eqs. (22), (16) and (23) show that the GRF depends on the size of the wind-exposed surface and on the structural properties: it is maximum for small, flexible and low damped structures, it is mimimum for large, stiff and highly damped structures. In any case $G_x \ge 1$.

As an alternative to Eq. (26), the equivalent static pressure can be expressed as:

$$p_{es} = C_d \bar{p}_{max} \tag{27}$$

where C_d is referred to as the dynamic coefficient:

$$C_d = \frac{G_x}{G_p} = \frac{1 + 2g_x I_u \sqrt{B_x^2 + R_x^2}}{1 + 2g_u I_u} \tag{28}$$

Differently from the GRF G_x, the dynamic coefficient C_d may be greater, equal or less than one. In these cases, respectively, the equivalent static pressure is greater, equal or less than the mean value of the maximum local pressure. The situation in which $G_x \ge G_p$ and $p_{es} \ge \bar{p}_{max}$ is typical of small, flexible and low damped structures; the opposite situation in which $G_x \le G_p$ and $p_{es} \le \bar{p}_{max}$ is typical of large, stiff and highly damped structures. It is worth noting that $B_x = 1$ and $R_x = 0$ is a sufficient but not necessary condition to obtain $G_x = G_p$ and $p_{es} = \bar{p}_{max}$.

Eq. (27) may be rewritten by expressing the dynamic coefficient C_d as the product of two partial coefficients, C_s and C_a, referred to as the size and the amplification coefficients, respectively:

$$p_{es} = C_s C_a \bar{p}_{max} \tag{29}$$

$$C_s = \frac{G_r}{G_p} = \frac{1 + 2g_r I_u B_x}{1 + 2g_u I_u} \qquad C_a = \frac{G_x}{G_r} = \frac{1 + 2g_x I_u \sqrt{B_x^2 + R_x^2}}{1 + 2g_r I_u B_x} \tag{30}$$

The size coefficient $C_s \le 1$ depends on the area A of the wind-exposed surface; it decreases on increasing A; in particular, $C_s = 1$ for $A = 0$, $C_s = 1/G_p$ for A tending to infinite. The amplification coefficient $C_a \ge 1$ depends on the dynamic properties of the structure and on its size; it reduces on increasing the fundamental frequency n_0, the damping ratio ξ and the area A of the wind-exposed surface; in particular, $C_a = 1$ for n_0 or ξ or A tending to infinite.

2.4. Dynamic response and equivalent static force for vertical structures

Let us consider a vertical structure schematised as a continuous cantilever beam (Figure 2a). The height, the width and the depth of the structure are denoted by h, b and d, respectively

(Figure 2b). Generalising Eqs. (18) and (21), the alongwind response and its mean maximum value can be expressed as:

$$x(z;t)=\bar{x}(z)+x'(z;t) \tag{31}$$

$$\bar{x}_{max}(z)=\bar{x}(z)+g_x(z)\sigma_x(z)=\bar{x}(z)+g_x(z)\sqrt{\sigma_{Bx}^2(z)+\sigma_{Rx}^2(z)}=G_x(z)\bar{x}(z) \tag{32}$$

where z is the height over the ground; $\bar{x}$ and x' are, respectively, the mean static value and the zero-mean dynamic fluctuation of the alongwind displacement; σ_x and g_x are, respectively, the rms value and the peak factor of x'; σ_{Bx} and σ_{Rx} are, respectively, the rms values of the quasi-static part and of the resonant part of the fluctuating response; G_x is the GRF defined as:

$$G_x(z)=1+g_x(z)\frac{\sigma_x(z)}{\bar{x}(z)}=1+2g_x(z)I_u\sqrt{B_x^2(z)+R_x^2(z)} \tag{33}$$

where $I_u=I_u(h_0)$, h_0 being a suitable reference height (h_0 = 0.6h in Solari 1982, 1993); B_x and R_x are non-dimensional quantities provided by the relationships:

$$B_x(z)=\frac{1}{2I_u}\frac{\sigma_{Bx}(z)}{\bar{x}(z)} \qquad R_x(z)=\frac{1}{2I_u}\frac{\sigma_{Rx}(z)}{\bar{x}(z)} \tag{34}$$

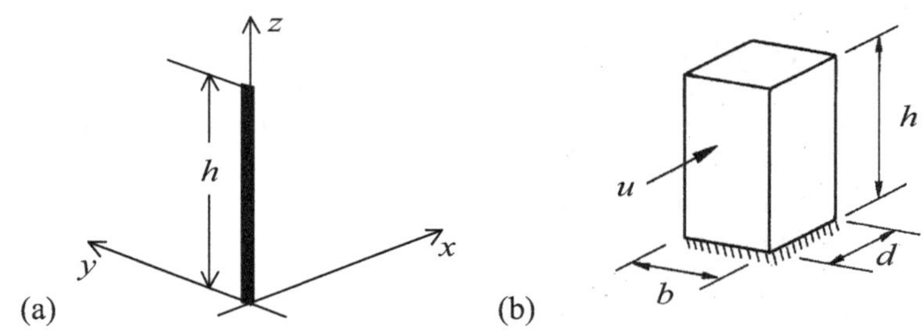

Figure 2. Vertical structure: (a) dynamic model; (b) structural geometry.

Assuming that the structure has classical normal modes of vibration, that natural frequencies are well separated and that damping ratios are small:

$$\bar{x}(z)=\sum_k\frac{\psi_{xk}(z)}{m_{xk}(2\pi n_{xk})^2}\int_0^h\bar{F}_x(\zeta)\psi_{xk}(\zeta)d\zeta \tag{35}$$

$$\sigma_{Bx}(z)=\sum_k\frac{\psi_{xk}(z)}{m_{xk}(2\pi n_{xk})^2}\sqrt{\int_0^\infty\left[\int_0^h\int_0^h S_{F_xF_x}(\zeta,\zeta';n)\psi_{xk}(\zeta)\psi_{xk}(\zeta')d\zeta d\zeta'\right]dn} \tag{36}$$

$$\sigma_{Rx}(z)=\sum_k\frac{\psi_{xk}(z)}{m_{xk}(2\pi n_{xk})^2}\sqrt{\int_0^h\int_0^h S_{F_xF_x}(\zeta,\zeta';n_{xk})\psi_{xk}(\zeta)\psi_{xk}(\zeta')d\zeta d\zeta'}\sqrt{\frac{\pi n_{xk}}{4\xi_{xk}}} \tag{37}$$

where $\overline{F}_x(z)$ and $S_{F_xF_x}(z,z';n)$ are, respectively, the mean static force and the cross-psdf of the fluctuating force per unit height; n_{xk}, m_{xk}, ξ_{xk}, $\psi_{xk}(z)$ are, respectively, the natural frequency, the modal mass, the modal damping ratio and the modal shape of the k-th mode of vibration in the alongwind direction.

Adopting the wind loading model described in the previous sections:

$$\overline{F}_x(z) = \frac{1}{2}\rho\overline{u}^2(z)c_d b \tag{38}$$

$$S_{F_xF_x}(z,z';n) = \rho^2\overline{u}(z)\overline{u}(z')\sqrt{S_u(z;n)S_u(z';n)}\chi_u(z,z';n)\gamma_u(n)c_d^2b^2 \tag{39}$$

where c_d is the drag coefficient; $\chi_u \leq 1$ and $\gamma_u \leq 1$ are linear filters referred to as aerodynamic admittance functions:

$$\chi_u(z,z';n) = \frac{1}{b^2}\int_0^b\int_0^b \mathrm{Coh}_{uu}(M,M';n)\,\mathrm{d}y\mathrm{d}y' \tag{40}$$

$$\gamma_u(n) = \frac{1}{c_d^2}\left[c_w^2 + 2c_wc_l\kappa_u(n) + c_l^2\right] \tag{41}$$

c_w and c_l are the mean pressure coefficients of the windward and leeward faces of the structure, respectively, being $c_d = c_w + c_l$; $\kappa_u(n)$ is a function that takes into account the coherence of the pressures on the windward and leeward structural faces (Dalgliesh 1971, Holmes 1975, Simiu 1980). Assuming such pressures as identically coherent, $\kappa_u = 1$ and $\gamma_u = 1$. Moreover, assuming the structure as slender line-like, i.e. $b \ll h$, $\chi_u(z,z';n) = \mathrm{Coh}_{uu}(z,z';n)$.

It is worth noting that the harmonic content of the longitudinal turbulence component decreases on increasing the frequency. Thus, provided that the second mode of vibration is well separated from the first one, only the first mode contributes to the dynamic response. It follows that the summations in Eqs. (35)-(37) may be limited to only one term with $k = 1$. In this case, likewise for a SDOF system (Section 2.3), also for a vertical structure B_x, R_x, g_x and G_x are constant quantities. A closed form expression for G_x is reported by Solari (1993).

This remark greatly simplifies the definition of the ESF, i.e. the force that statically applied to the structure gives rise to the mean value of the maximum displacement $\overline{x}_{max}(z)$. Generalising Eqs. (25) and (26), it is given by:

$$F_{x,es}(z) = G_x\overline{F}_x(z) \tag{42}$$

Thus, exploiting the structural linearity and the hypothesis that the response is dominated by only one natural mode, the ESF involves a load pattern proportional to the mean static force through the GRF.

2.5. Wind-excited load effects for vertical structures

Let us consider again the alongwind response of the vertical structure depicted in Figure 2. A generic load effect e_x at the height r over the ground can be expressed as:

$$e_x(r;t) = \overline{e}_x(r) + e'_x(r;t) \tag{43}$$

where $\bar{e}_x$ and e'_x are, respectively, the mean static part and the zero-mean fluctuation of e_x. Generalising Eq. (32), the mean value of the maximum load effect e_x can be expressed as:

$$\bar{e}_{x,max}(r) = \bar{e}_x(r) + g_x^e(r)\sigma_x^e(r) = \bar{e}_x(r) + g_x^e(r)\sqrt{\left[\sigma_{Bx}^e(r)\right]^2 + \left[\sigma_{Rx}^e(r)\right]^2} = G_x^e(r)\bar{e}_x(r) \quad (44)$$

where σ_x^e and g_x^e are, respectively, the rms value and the peak factor of e'_x; σ_{Bx}^e and σ_{Rx}^e are the rms values of the quasi-static part and of the resonant part of e_x, respectively; G_x^e is referred to as the Gust Effect Factor (GEF):

$$G_x^e(r) = 1 + g_x^e(r)\frac{\sigma_x^e(r)}{\bar{e}_x(r)} = 1 + 2g_x^e(r)I_u\sqrt{\left[B_x^e(r)\right]^2 + \left[R_x^e(r)\right]^2} \quad (45)$$

where:

$$B_x^e(r) = \frac{1}{2I_u}\frac{\sigma_{Bx}^e(r)}{\bar{e}_x(r)} \qquad R_x^e(r) = \frac{1}{2I_u}\frac{\sigma_{Rx}^e(r)}{\bar{e}_x(r)} \quad (46)$$

It is admitted here that the static and the quasi-static parts of the load effect depend on all the modes of vibration; instead, the resonant part of the load effect is related to only the fundamental mode. Thus, using the influence function technique, these quantities can be expressed as:

$$\bar{e}_x(r) = \int_0^h \bar{F}_x(z)\eta_x^e(r;z)\mathrm{d}z \quad (47)$$

$$\sigma_{Bx}^e(r) = \sqrt{\int_0^\infty \left[\int_0^h\int_0^h S_{F_xF_x}(z,z';n)\eta_x^e(r;z)\eta_x^e(r;z')\mathrm{d}z\mathrm{d}z'\right]dn} \quad (48)$$

$$\sigma_{Rx}^e(r) = \frac{m_x^e(r)}{m_{x1}}\sqrt{\int_0^h\int_0^h S_{F_xF_x}(z,z';n_{x1})\psi_{x1}(z)\psi_{x1}(z')\mathrm{d}z\mathrm{d}z'}\sqrt{\frac{\pi n_{x1}}{4\xi_{x1}}} \quad (49)$$

$$m_x^e(r) = \int_0^h \mu(z)\psi_{x1}(z)\eta_x^e(r;z)\mathrm{d}z \quad (50)$$

where μ is the structural mass per unit length; $\eta_x^e(r,z)$ is the influence function of $e_x(r)$, i.e. the value of e_x at the height r due to a unit static force applied at height z. Focusing attention on the displacement (*e=d*), on the bending moment (*e=b*) and on the shear force (*e=s*), the related influence functions result:

$$\eta_x^d(r;z) = \frac{\psi_{x1}(r)\psi_{x1}(z)}{m_{x1}(2\pi n_{x1})^2} \qquad \eta_x^b(r;z) = (z-r)H(z-r) \qquad \eta_x^s(r;z) = H(z-r) \quad (51)$$

$H(\bullet)$ being the Heaviside's function. The first Eq. (51) implies that the displacement depends on only the first mode of vibration.

The ESF is defined as the force that, statically applied to the structure, produces a specified load effect. Unfortunately, this definition is not unique as in the case of a SDOF structural system (Section 2.3) or of a vertical structure whose response depends on only one mode of

vibration (Section 2.4). Among the infinite possible ways of defining such force, three criteria deserve a special concern: the Gust Effect Factor (GEF) technique, the Load Combination (LC) technique and the Global Loading (GL) technique (Repetto and Solari 2004).
Using the GEF technique, the ESF is defined as:

$$F^e_{x,es}(r,z) = G^e_x(r)\bar{F}_x(z) \tag{52}$$

According to Eq. (52), independently of the load effect e_x, any ESF is proportional to the mean force. Thus, engineering calculations may be carried out applying on the structure only one static load, $\bar{F}_x(z)$. Static load effects $\bar{e}_x(r)$ are then evaluated by equilibrium relationships. The mean values of the maximum load effects are finally obtained by scaling such effects by the appropriate GEF.
It is worth noting that, for e=d, the above formulation coincides with that developed in Section 2.4. In particular, since $B^d_x(r) = B_x$, $R^d_x(r) = R_x$, $g^d_x(r) = g_x$, the GEF for the displacement does not depend on r and it coincides with the GRF, i.e. $G^d_x(r) = G_x$. Thus, the GEF technique represents the generalisation of the GRF technique to any wind load effect; on the other hand, the GRF technique may be regarded as a particular case of the GEF technique.
The LC technique assignes the ESF as a combination of three distinct load patterns associated, respectively, with the static, the quasi-static and the resonant parts of the response. It is assumed here, conventionally, that the quasi-static and the resonant parts of the ESF give rise to σ^e_{Bx} and σ^e_{Rx}, respectively. Thus, the maximum load effect $\bar{e}_{x,max}$ can be evaluated through Eq. (44).
The static and the resonant parts of the ESF are conveniently defined as independent of the load effect. This aim is fulfilled, respectively, by the mean static force $\bar{F}_x(z)$ itself (Eq. (38)), and by the inertial load that produces the rms value of the resonant response, i.e.:

$$F_{Rx,es}(z) = \mu(z)(2\pi n_{x1})^2 \sigma_{Rx}(z) = 2\mu(z)(2\pi n_{x1})^2 \bar{x}(z) I_u R_x \tag{53}$$

The quasi-static part of the ESF is more complex to define since no load pattern provides, simultaneously, the full scenario of the quasi-static effects; thus, it depends on the considered load effect and on its height, and such dependence is not univocal. Among the infinite possible ways of defining such force, three criteria deserve a special concern: the Load-Response-Correlation (LRC) method (Kasperski 1992), the Gust Effect Factor (GEF) technique and the Global Loading (GL) technique (Repetto and Solari 2004).
Based on the GEF technique, the quasi-static part of the ESF is given by:

$$F^e_{Bx,es}(r;z)=2\bar{F}_x(z) I_u B^e_x(r) \tag{54}$$

Using the LRC method, the quasi-static part of the ESF is defined as the most probable load pattern for each specified load effect; it results:

$$F^e_{Bx,es}(r;z)=2\bar{F}_x(z) I_u B^e_x(r)\Delta^e_x(r,z) \tag{55}$$

where Δ^e_x is a non-dimensional coefficient referred to as the LRC factor:

$$\Delta_x^e(r;z)=\frac{\bar{e}_x(r)}{\bar{F}_x(z)\left[\sigma_{Bx}^e(r)\right]^2}\int_0^\infty\left[\int_0^h S_{F_xF_x}(z,z';n)\eta_x^e(r;z)\,\mathrm{d}z'\right]\mathrm{d}n \tag{56}$$

Comparing Eqs. (54) and (56) it is apparent that Δ_x^e transforms the quasi-static part of the ESF provided by the GEF technique into the most probable load pattern.
The GL technique provides a unique loading condition able to furnish a correct scenario of n specified quasi-static load effects. This requirement is satisfied by expressing the quasi-static part of the ESF through the polynomial expansion:

$$F_{Bx,es}(z)=\bar{F}_x(h)\sum_{k=0}^{n-1}p_k\left(\frac{z}{h}\right)^k \tag{57}$$

where p_k (k=0,1,..n-1) are n non-dimensional coefficients used to impose that Eq. (57) provides the correct values of the quasi-static parts of n specified load effects.
Generalising this method to the global ESF provides the following expression:

$$F_{x,es}(z)=\bar{F}_x(h)\sum_{k=0}^{n-1}q_k\left(\frac{z}{h}\right)^k \tag{58}$$

where q_k (k=0,1,..n-1) are n non-dimensional coefficients used to impose that Eq. (58) provides the correct values of n specified maximum load effects.

3 Dynamic 3-D response

This section provides a general framework of the dynamic alongwind, crosswind and torsional (3-D) response of structures, by analysing the sequence of four classical problems. Section 3.1 deals with the aerodynamic actions on a slender fixed cylinder of infinite length, immersed in an ideal bi-dimensional turbulent wind field. Section 3.2 generalises these actions to the atmospheric boundary layer and determines the 3-D wind-excited response of a slender structure or structural element with a dominant mode of vibration per each direction of motion; accordingly, it defines an appropriate set of equivalent static actions based on a quantity referred to as the 3-D gust response factor. Section 3.3 evaluates the wind-excited load effects and the corresponding equivalent static actions for a slender vertical structure, introducing a quantity referred to as the 3-D gust effect factor. Section 3.4 provides some introductory notes concerning the alongwind, crosswind and torsional response of three-dimensional buildings.

3.1. Aerodynamic wind actions on slender fixed cylinders

Let us consider a slender fixed cylinder of infinite length and cross-section S, immersed in an ideal bi-dimensional turbulent wind field. Let us consider three Cartesian reference systems coplanar with S and with the same origin O in a suitable point of S (Figure 3). The first reference system, x_o, y_o, has unit vectors $\mathbf{i}_o$, $\mathbf{j}_o$; it is fixed and independent of the wind field. The second reference system, x, y, has unit vectors $\mathbf{i}$, $\mathbf{j}$; $\mathbf{i}$ is aligned with the mean wind velocity $\bar{u}$ rotated β

with respect to $\mathbf{i}_o$. The third reference system, x', y', has unit vectors, $\mathbf{i}'$, $\mathbf{j}'$; $\mathbf{i}'$ is aligned with the instantaneous vectorial wind velocity $\mathbf{U}$ rotated δ with respect to $\mathbf{i}$. Thus:

$$\mathbf{U}(t) = u(t)\mathbf{i} + \upsilon(t)\mathbf{j} \tag{59}$$

$$u(t) = \bar{u} + u'(t) \qquad \upsilon(t) = \upsilon'(t) \tag{60}$$

$$\delta(t) = \mathrm{arctg}\left\{\frac{\upsilon'(t)}{\bar{u} + u'(t)}\right\} \qquad \gamma(t) = \beta + \delta(t) \tag{61}$$

$$U(t) = \|\mathbf{U}(t)\| = \sqrt{[\bar{u} + u'(t)]^2 + [\upsilon'(t)]^2} \tag{62}$$

where u' and υ' are referred to as the longitudinal (parallel to $\bar{u}$) and transversal (orthogonal to $\bar{u}$) turbulence components. It is worth noting that $\bar{u}$, u' and υ' are dealt with as orthogonal to the axis z of the cylinder. A generalisation of the formulation reported below to yawed circular cylinders has been proposed by Carassale et al (2005).

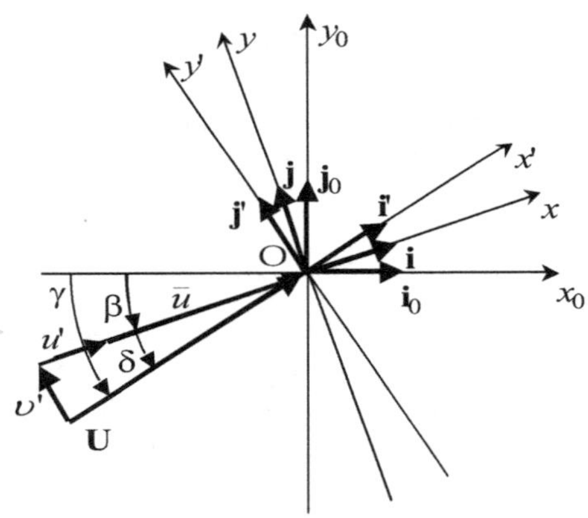

Figure 3. Bi-dimensional turbulent wind field.

Due to the wind velocity $\mathbf{U}$, the cylinder is subjected to a force $\mathbf{F}$ applied in O and to a torque M_z per unit length. The Cartesian components of $\mathbf{F}$ with respect to x, y are denoted by F_x and F_y; the Cartesian components of $\mathbf{F}$ with respect to x', y' are denoted by F_d and F_l (Figure 4). Thus:

$$\mathbf{F}(t) = F_x(t)\mathbf{i} + F_y(t)\mathbf{j} = F_d(t)\mathbf{i}' + F_l(t)\mathbf{j}' \tag{63}$$

$$F_x(t) = \bar{F}_x + F_x'(t) \qquad F_y(t) = \bar{F}_y + F_y'(t) \tag{64}$$

$$F_d(t) = \bar{F}_d + F_d'(t) \qquad F_l(t) = \bar{F}_l + F_l'(t) \tag{65}$$

$$M_z(t) = \bar{M}_z + M_z'(t) \tag{66}$$

where $\overline{F}_x$, $\overline{F}_y$, $\overline{F}_d$, $\overline{F}_l$, $\overline{M}_z$ and F'_x, F'_y, F'_d, F'_l, M'_z are, respectively, the mean values and the zero-mean fluctuations of F_x, F_y, F_d, F_l, M_z.
The drag coefficiecient c_d, the lift coefficient c_l and the torsional moment coefficient c_m of the bluff cylinder depend on the direction β of the mean wind velocity $\overline{u}$. They are provided by the relationships:

$$c_d(\beta) = \frac{2\overline{F}_x}{\rho \overline{u}^2 b} \qquad c_l(\beta) = \frac{2\overline{F}_y}{\rho \overline{u}^2 b} \qquad c_m(\beta) = \frac{2\overline{M}_z}{\rho \overline{u}^2 b^2} \tag{67}$$

where ρ is the density of air and b is a reference size of S. Based on Eq. (67), the mean values of F_x, F_y, M_z result:

$$\overline{F}_x = \frac{1}{2}\rho \overline{u}^2 b c_d(\beta) \qquad \overline{F}_y = \frac{1}{2}\rho \overline{u}^2 b c_l(\beta) \qquad \overline{M}_z = \frac{1}{2}\rho \overline{u}^2 b^2 c_m(\beta) \tag{68}$$

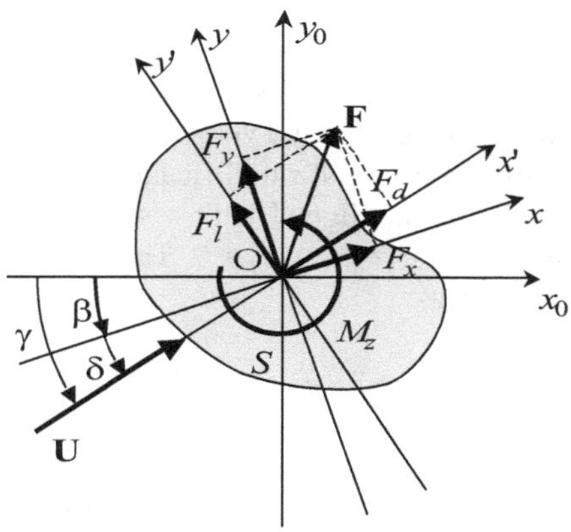

Figure 4. Aerodynamic wind actions on a slender cylinder.

Using the quasi-steady theory, the instantaneous values of the aerodynamic wind actions F_d, F_l, M_z are expressed by generalising Eq. (68) as:

$$F_d(t) = \frac{1}{2}\rho U^2(t) b c_d(\gamma(t)) \quad F_l(t) = \frac{1}{2}\rho U^2(t) b c_l(\gamma(t)) \quad M_z(t) = \frac{1}{2}\rho U^2(t) b^2 c_m(\gamma(t)) \tag{69}$$

where $U(t)$ and $\gamma(t)$ are, respectively, the instantaneous wind velocity (Eq. (62)) and its direction (Eq. (61)). The instantaneous values of F_x and F_y are linked with F_d and F_l (Eq. (69)) through the relationships:

$$F_x(t) = F_d(t)\cos\delta(t) - F_l(t)\sin\delta(t) \qquad F_y(t) = F_d(t)\sin\delta(t) + F_l(t)\cos\delta(t) \tag{70}$$

where $\delta(t)$ denotes the instantaneous direction of $\mathbf{U}(t)$ with respect to $\overline{u}$ (Eq. (61), Figure 3).

Replacing Eq. (69) into Eq. (70), using Eqs. (61) and (62), and assuming that turbulence is small with respect to the mean wind velocity, i.e. $u'/\bar{u} \ll 1$ and $\upsilon'/\bar{u} \ll 1$, the zero-mean fluctuations of F_x, F_y, M_z result (Solari 1994):

$$F_x'(t) = \rho\bar{u}bc_d u'(t) + \frac{1}{2}\rho\bar{u}b(c_d' - c_l)\upsilon'(t) \tag{71}$$

$$F_y'(t) = \rho\bar{u}bc_l u'(t) + \frac{1}{2}\rho\bar{u}b(c_d + c_l')\upsilon'(t) \tag{72}$$

$$M_z'(t) = \rho\bar{u}b^2 c_m u'(t) + \frac{1}{2}\rho\bar{u}b^2 c_m'\upsilon'(t) \tag{73}$$

where:

$$c_k = c_k(\beta) \qquad c_k' = \left.\frac{\partial c_k(\gamma)}{\partial\gamma}\right|_{\gamma=\beta} \qquad (k = d,l,m) \tag{74}$$

It is worth noting that F_x', F_y', M_z' are linear combinations of u', υ' through proportionality factors depending on the aerodynamic coefficients c_d, c_l, c_m and their prime derivatives c_d', c_l', c_m' with respect to the direction of the mean wind velocity.

Eq. (68) and Eqs. (71)-(73) involve substantial simplifications in noteworthy particular cases. Firstly, if the mean wind velocity $\bar{u}$ is aligned with a symmetry axis of the cylinder cross-section, i.e. x is a symmetry axis of S, $c_l = c_m = c_d' = 0$. Therefore $\bar{F}_y = \bar{M}_z = 0$ (Eq. (68)); moreover, Eqs. (71)-(73) can be rewritten as:

$$F_x'(t) = \rho\bar{u}bc_d u'(t) \qquad F_y'(t) = \frac{1}{2}\rho\bar{u}b(c_d + c_l')\upsilon'(t) \qquad M_z'(t) = \frac{1}{2}\rho\bar{u}b^2 c_m'\upsilon'(t) \tag{75}$$

It is apparent that the classic analysis of the dynamic alongwind response based on considering only the longitudinal turbulence (Section 2) is rigorously correct provided that the mean wind velocity occurs along a symmetry axis. In all the other cases, the fluctuating alongwind force depends on both the longitudinal turbulence and its transversal component (Eq. (71)).

Secondly, if the cylinder is endowed with polar symmetry, the previous simplifications apply and, moreover, $c_l' = c_m' = 0$. Thus, Eq. (75) can be rewritten as:

$$F_x'(t) = \rho\bar{u}bc_d u'(t) \qquad F_y'(t) = \frac{1}{2}\rho\bar{u}bc_d\upsilon'(t) \qquad M_z'(t) = 0 \tag{76}$$

It is apparent that, in this case, the drag coefficient is the sole aerodynamic parameter relevant to quantify the wind actions. Clearly, this coefficient does not depend on the wind direction.

The application of the quasi-steady theory involves a fundamental limit: it takes into account the actions induced by the oncoming turbulence, but it does not consider the effects of the signature turbulence, i.e. the turbulence caused by the cylinder in its wake due to the flow separation. This problem is not relevant for aerodynamic shapes as the airfoils, which do not give rise to flow separation. Instead, it is determinant for all the bluff shapes, as the slender structures and structural elements, which produce a vortex wake.

The schematisation of the wind actions produced by the vortex wake, i.e. the correction of the wind loading model obtained by the quasi-steady theory, can be performed in two different ways depending on the frequency content of the wake.
If the cylinder sheds vortices whose harmonic content is well above the frequency range of the oncoming turbulence, it is reasonable to admit that the fluctuating part of the wind loading is increased by an additive term associated with the vortex shedding. In such case, Eqs. (71)-(73) may be rewritten as (Solari 1994):

$$F_x'(t) = \rho\bar{u}bc_d u'(t) + \frac{1}{2}\rho\bar{u}b(c_d' - c_l)\upsilon'(t) + \frac{1}{2}\rho\bar{u}^2 b\tilde{c}_{ds} s_x^*(t) \tag{77}$$

$$F_y'(t) = \rho\bar{u}bc_l u'(t) + \frac{1}{2}\rho\bar{u}b(c_d + c_l')\upsilon'(t) + \frac{1}{2}\rho\bar{u}^2 b\tilde{c}_{ls} s_y^*(t) \tag{78}$$

$$M_z'(t) = \rho\bar{u}b^2 c_m u'(t) + \frac{1}{2}\rho\bar{u}b^2 c_m'\upsilon'(t) + \frac{1}{2}\rho\bar{u}^2 b\tilde{c}_{ms} s_z^*(t) \tag{79}$$

where $\tilde{c}_{ds}$, $\tilde{c}_{ls}$, $\tilde{c}_{ms}$ are the rms drag, lift and torsional moment wake coefficients, respectively; s_x^*, s_y^*, s_z^* are stationary Gaussian random processes with zero mean and unit rms values, whose psdfs describe the harmonic content of the alongwind (x), crosswind (y) and torsional (z) actions induced by the vortex wake on the bluff cylinder (Vickery and Clark 1972).
On the other hand, if the cylinder sheds vortices whose harmonic content is embedded within the frequency range of the oncoming turbulence, the vortex wake modifies the harmonic content of oncoming turbulence and the above formulation does not represent the physical phenomenon. In this case, Eqs. (71)-(73) can be rewritten in the frequency domain and the role of the vortex wake may be schematised by a suitable set of aerodynamic admittance functions that, multiplied by the power spectral density functions of u' and υ', take into account the modifications corresponding to each component of the wind actions.
The situation depicted above completely changes when the bluff cylinder is free to vibrate in the wind field. This condition may lead to wind-structure interaction or aeroelastic phenomena that can be schematised by a suitable set of motion-induced actions; such actions are added to the mean wind actions defined by Eq. (68) and to the fluctuating wind actions defined by Eqs. (77)-(79). A basic framework of this topic may be found, for instance, in (Solari 1994). A specific treatment for the aeroelastic phenomena induced by the resonant shedding of vortices is given by Vickery and Basu (1983).

3.2. Wind-excited response of slender structures or structural elements

Dealing with the simple but fundamental case in which the harmonic content of the signature turbulence is well above the harmonic content of the oncoming turbulence, the model in Eqs. (68), (77)-(79) is easily generalized to any slender cylinder in the atmospheric boundary layer.
With this aim, let us consider a structure or a structural element whose length l is much greater than the reference size b of its cross-section S. Let x, y, z be a local Cartesian reference system with origin at o (Figure 5); z coincides with the structural axis, x is aligned with the mean wind velocity $\bar{u}$, o lies on the face of the structure with $z = 0$, at height h_1 above the ground. Let X, Y, Z be a global Cartesian reference system with origin at O; X and Y are coplanar with the ground; Y and Z are coplanar with y and z; X is parallel to x; Z is directed upwards and passes

through o; z is rotated ϕ with respect to Z. Let u', v', w' be the longitudinal, lateral and vertical turbulence components parallel to X, Y, Z, respectively; creating a link with the ideal turbulence field depicted in Figure 3, $\upsilon' = v'$ for $\phi=0$, $\upsilon' = w'$ for $\phi=\pi/2$.
The wind loading is schematized by a 3-variate 2-dimensional stationary Gaussian random process, whose α-th component ($\alpha = x, y, \theta$) is defined as:

$$F_\alpha(z;t) = \bar{F}_\alpha(z) + F'_\alpha(z;t) \tag{80}$$

where $0 \le z \le l$; F_x, F_y, $F_\theta = M_z$ are the alongwind force, the crosswind force and the torsional moment around z; $\bar{F}_\alpha$ and F'_α are the mean value and the zero-mean fluctuation of F_α. These quantities may be expressed as:

$$\bar{F}_\alpha(z) = \frac{1}{2}\rho\bar{u}^2(z)\, b\lambda_\alpha c_{\alpha u}\gamma_{\alpha u}(z) \tag{81}$$

$$F'_\alpha(z;t) = \sum_\varepsilon \frac{1}{2}\rho\bar{u}^2(z) J_\varepsilon(z)\, b\lambda_\alpha c_{\alpha\varepsilon} f^*_{\alpha\varepsilon}(z;t)\gamma_{\alpha\varepsilon}(z) \tag{82}$$

where Σ_ε is the sum of four loading terms with indices $\varepsilon = u, v, w, s$, associated with the three turbulence components (u, v, w) and with the wake excitation (s); $c_{\alpha\varepsilon}$ and $f^*_{\alpha\varepsilon}$ are the α,ε-th elements of the matrices:

$$[c] = \begin{bmatrix} c_d & (c'_d - c_\ell)\cos\phi & (c'_d - c_\ell)\sin\phi & \tilde{c}_{ds} \\ c_\ell & (c_d + c'_\ell)\cos\phi & (c_d + c'_\ell)\sin\phi & \tilde{c}_{\ell s} \\ c_m & c'_m \cos\phi & c'_m \sin\phi & \tilde{c}_{ms} \end{bmatrix} \tag{83}$$

$$[f^*(z;t)] = \begin{bmatrix} u^*(z;t) & v^*(z;t) & w^*(z;t) & s^*_x(z;t) \\ u^*(z;t) & v^*(z;t) & w^*(z;t) & s^*_y(z;t) \\ u^*(z;t) & v^*(z;t) & w^*(z;t) & s^*_\theta(z;t) \end{bmatrix} \tag{84}$$

$I_u=\sigma_u/\bar{u}$, $I_v=\sigma_v/\bar{u}$, $I_w=\sigma_w/\bar{u}$ are the intensities of the three turbulence components, $\sigma_u, \sigma_v, \sigma_w$ being the rms values of u', v', w'; $u^*=u'/\sigma_u$, $v^*=v'/\sigma_v$, $w^*=w'/\sigma_w$ are the reduced turbulence components; s^*_α is the α-th reduced component of the wake excitation; $J_u = 2I_u$, $J_v = I_v$, $J_w = I_w$, $J_s = 1$; $\gamma_{\alpha\varepsilon}$ is a non-dimensional function of z that makes Eqs. (81) and (82) suitable for variable aerodynamic properties and non-prismatic elements.

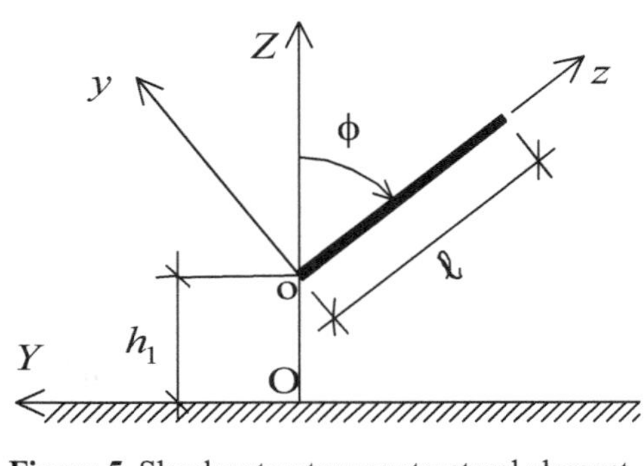

Figure 5. Slender structure or structural element.

Let us assume that structure has a linear behavior and three un-coupled components of motion, the alongwind and crosswind displacements, towards x and y, and the θ torsional rotation around z. The generalized displacement $\alpha = x, y, \theta$ is a stationary Gaussian random process provided by the relationship:

$$\alpha(z;t) = \bar{\alpha}(z) + \alpha'(z;t) \tag{85}$$

where $\bar{\alpha}$ and α' are, respectively, the mean value and the zero-mean fluctuation of α. Generalizing Eq. (32) to the dynamic 3-D response, the mean value of the maximum generalised displacement α is given by (Piccardo and Solari 1996, 1998a, 2000):

$$\bar{\alpha}_{max}(z) = \bar{\alpha}(z) + g_\alpha(z)\sigma_\alpha(z) = \bar{\alpha}(z) + g_\alpha(z)\sqrt{\left[\sigma^2_{B\alpha}(z)\right] + \left[\sigma^2_{R\alpha}(z)\right]} = G_\alpha(z)\bar{\alpha}^x(z) \tag{86}$$

where σ_α and g_α are, respectively, the rms value and the peak factor of α'; $\sigma_{B\alpha}$ and $\sigma_{R\alpha}$ are the rms values of the quasi-static part and resonant parts of the fluctuating response, respectively; $\bar{\alpha}^x$ is the static generalized displacement due to the application of the generalized force $\lambda_\alpha \bar{F}_x$ in the α direction, being $\lambda_x = \lambda_y = 1$, $\lambda_\theta = b$; G_α is referred to as the 3-D GRF:

$$G_\alpha(z) = \frac{\bar{\alpha}(z)}{\bar{\alpha}^x(z)} + g_\alpha(z)\frac{\sigma_\alpha(z)}{\bar{\alpha}^x(z)} = \mu_\alpha(z) + 2g_\alpha(z)I_u\sqrt{B^2_\alpha(z) + R^2_\alpha(z)} \tag{87}$$

where:

$$\mu_\alpha(z) = \frac{\bar{\alpha}(z)}{\bar{\alpha}^x(z)} \qquad B_\alpha(z) = \frac{1}{2I_u}\frac{\sigma_{B\alpha}(z)}{\bar{\alpha}^x(z)} \qquad R_\alpha(z) = \frac{1}{2I_u}\frac{\sigma_{R\alpha}(z)}{\bar{\alpha}^x(z)} \tag{88}$$

Coherently with the formulation developed in Section 2.4, it is admitted here that the generalised α displacement depends on only the fundamental mode shape $\psi_{\alpha 1}$ in the α direction. Thus:

$$\bar{\alpha}(z) = \frac{\psi_{\alpha 1}(z)}{m_{\alpha 1}(2\pi n_{\alpha 1})^2}\int_0^h \bar{F}_\alpha(\zeta)\psi_{\alpha 1}(\zeta)d\zeta \quad \bar{\alpha}^x(z) = \frac{\psi_{\alpha 1}(z)}{m_{\alpha 1}(2\pi n_{\alpha 1})^2}\lambda_\alpha\int_0^h \bar{F}_x(\zeta)\psi_{\alpha 1}(\zeta)d\zeta \tag{89}$$

$$\sigma_{B\alpha}(z) = \frac{\psi_{\alpha 1}(z)}{m_{\alpha 1}(2\pi n_{\alpha 1})^2}\sqrt{\int_0^\infty\left[\int_0^h\int_0^h S_{F_\alpha F_\alpha}(\zeta,\zeta';n)\psi_{\alpha 1}(\zeta)\psi_{\alpha 1}(\zeta')d\zeta d\zeta'\right]dn} \tag{90}$$

$$\sigma_{R\alpha}(z) = \frac{\psi_{\alpha 1}(z)}{m_{\alpha 1}(2\pi n_{\alpha 1})^2}\sqrt{\int_0^h\int_0^h S_{F_\alpha F_\alpha}(\zeta,\zeta';n_{\alpha 1})\psi_{\alpha 1}(\zeta)\psi_{\alpha 1}(\zeta')d\zeta d\zeta'}\sqrt{\frac{\pi n_{\alpha 1}}{4\xi_{\alpha 1}}} \tag{91}$$

$n_{\alpha 1}$, $\xi_{\alpha 1}$, $m_{\alpha 1}$ are the natural frequency, the damping ratio and the modal mass of the first mode of vibration in the direction α; $S_{F_\alpha F_\alpha}(z,z';n)$ is the cross-psdf of $F'_\alpha(z;t)$ and $F'_\alpha(z';t)$ (Eq.(82)):

$$S_{F_\alpha F_\alpha}(z,z';n) = \sum_\varepsilon\sum_\eta \frac{1}{4}\rho^2\bar{u}^2(z)\bar{u}^2(z')J_\varepsilon(z)J_\eta(z')b^2\lambda^2_\alpha c_{\alpha\varepsilon}c_{\alpha\eta}S^*_{\alpha\varepsilon\eta}(z,z';n)\gamma_{\alpha\varepsilon}(z)\gamma_{\alpha\eta}(z') \tag{92}$$

where $S^*_{\alpha\varepsilon\eta}(z,z';n)$ is the cross-psdf of $f^*_{\alpha\varepsilon}(z;t)$ and $f^*_{\alpha\eta}(z';t)$.

Based upon this formulation, likewise for the analysis of the alongwind response (Section 2.4), also in this case B_α, R_α, g_α and G_α become constant quantities. A closed form solution for G_α is reported by (Piccardo and Solari 1998a, 2000)

The α-th ESF is the generalized force that, statically applied to the structure in the α direction, determines the mean value of the maximum generalised displacement $\bar{\alpha}_{max}(z)$. Generalising Eq. (42), it is given by:

$$F_{\alpha,es}(z) = \lambda_\alpha G_\alpha \bar{F}_x(z) \tag{93}$$

For α=*x*, i.e. dealing with the alongwind response, the 3-D GRF (Eq. (87)) and the related ESF (Eq. (93)) coincide, respectively, with the classical GRF (Eq. (22)) and the corresponding ESF (Eq. (42)); for α=*y* and α=θ, G_y and G_θ denote, respectively, the crosswind gust factor and torsional gust factor. Thus, the 3-D GRF technique is a generalisation of the GRF technique; in its turn, the GRF technique may be regarded as a particular case of the 3-D GRF technique.

3.3. Wind-excited load effects for slender vertical structures

Let us consider a vertical structure schematised as a continuous cantilever beam whose height *h* is much greater than the reference size *b* of its cross-section *S*. The scheme depicted in Figure 6 coincides with that shown in Figure 2(a) provided that $l = h$, $h_1 = 0$ and $\phi = 0$.

Wind actions are schematized by Eq. (80), where $\bar{F}_\alpha$ and F'_α are expressed by Eqs. (81) and (82). In this case, the third column of the matrix [*c*] (Eq. (83)) is null and the structure is not sensitive to the vertical turbulence component. Thus, only three loading mechanisms excite the structure: the longitudinal turbulence component (*u*), the lateral turbulence component (*v*) and the vortex shedding (*s*).

It is admitted here that the structure has a linear behavior and three un-coupled components of motion, the alongwind and crosswind displacements, directed towards *x* and *y*, and the θ torsional rotation around *z*. The load effect e_α at the height *r* associated with the generalized direction α of the motion is a stationary Gaussian random process defined as:

$$e_\alpha(r;t) = \bar{e}_\alpha(r) + e'_\alpha(r;t) \tag{94}$$

where $0 \le r \le h$; $\bar{e}_\alpha$ and e'_α are, respectively, the mean value and the zero-mean fluctuation of e_α. Generalising Eqs. (44) and (86), the mean value of the maximum load effect e_α is given by (Piccardo and Solari 2002):

$$\bar{e}_{\alpha,max}(r) = \bar{e}_\alpha(r) + g^e_\alpha(r)\sigma^e_\alpha(r) = \bar{e}_\alpha(r) + g^e_\alpha(r)\sqrt{\left[\sigma^e_{B\alpha}(r)\right]^2 + \left[\sigma^e_{R\alpha}(r)\right]^2} = G^e_\alpha(r)\bar{e}^x_\alpha(r) \tag{95}$$

where σ^e_α and g^e_α are, respectively, the rms value and the peak factor of e'_α; $\sigma^e_{B\alpha}$ and $\sigma^e_{R\alpha}$ are the rms values of the quasi-static part and of the resonant part of e_α, respectively; $\bar{e}^x_\alpha$ is the static load effect due to the application of the generalized force $\lambda_\alpha \bar{F}_x$ in the α direction; G^e_α is referred to as the 3-D GEF:

$$G^e_\alpha(r) = \frac{\bar{e}_\alpha(r)}{\bar{e}^x_\alpha(r)} + g^e_\alpha(r)\frac{\sqrt{\left[\sigma^e_{Q\alpha}(r)\right]^2 + \left[\sigma^e_{D\alpha}(r)\right]^2}}{\bar{e}^x_\alpha(r)} = \mu^e_\alpha(r) + 2g^e_\alpha(r)I_u\sqrt{\left[B^e_\alpha(r)\right]^2 + \left[R^e_\alpha(r)\right]^2} \tag{96}$$

where:

$$\mu_\alpha^e(r) = \frac{\bar{e}_\alpha(r)}{\bar{e}_\alpha^x(r)} \qquad B_\alpha^e(r) = \frac{1}{2I_u}\frac{\sigma_{B\alpha}^e(r)}{\bar{e}_\alpha^x(r)} \qquad R_\alpha^e(r) = \frac{1}{2I_u}\frac{\sigma_{R\alpha}^e(r)}{\bar{e}_\alpha^x(r)} \tag{97}$$

Coherently with the formulation developed in Section 2.5, it is admitted here that the static and the quasi-static parts of the load effect depend on all the modes of vibration; instead, the resonant parts of the load effect are related to only the corresponding fundamental mode. Thus, adopting the influence function technique, these quantities can be expressed as:

$$\bar{e}_\alpha(r) = \int_0^\ell \bar{F}_\alpha(z)\eta_\alpha^e(r,z)\mathrm{d}z \qquad \bar{e}_\alpha^x(r) = \lambda_\alpha \int_0^\ell \bar{F}_x(z)\eta_\alpha^e(r,z)\mathrm{d}z \tag{98}$$

$$\sigma_{B\alpha}^e(r) = \sqrt{\int_0^\infty \left[\int_0^h\int_0^h S_{F_\alpha F_\alpha}(z,z';n)\eta_\alpha^e(r;z)\eta_\alpha^e(r;z')\mathrm{d}z\mathrm{d}z'\right]dn} \tag{99}$$

$$\sigma_{R\alpha}^e(r) = \frac{m_\alpha^e(r)}{m_{\alpha 1}}\sqrt{\int_0^h\int_0^h S_{F_\alpha F_\alpha}(z,z';n_{\alpha 1})\psi_{\alpha 1}(z)\psi_{\alpha 1}(z')\mathrm{d}z\mathrm{d}z'}\sqrt{\frac{\pi n_{\alpha 1}}{4\xi_{\alpha 1}}} \tag{100}$$

$$m_\alpha^e(r) = \int_0^h \mu_\alpha(z)\psi_{\alpha 1}(z)\eta_\alpha^e(r;z)\mathrm{d}z \tag{101}$$

where $\bar{F}_\alpha(z)$ and $S_{F_\alpha F_\alpha}(z,z';n)$ are given by Eqs. (81) and (92), respectively; $\mu_x=\mu_y$ and μ_θ are, respectively, the mass and the torsional mass moment of inertia per unit height; $\eta_\alpha^e(r,z)$ is the influence function of $e_\alpha(r)$, i.e. the value of e_α at the height r due to a unit static action applied at height z in direction α. Focusing attention on the generalised displacements, bending moments and shear forces, they are provided by Eq. (51), with x replaced by α. A closed form solution for G_α^e is reported by Piccardo and Solari (2002).

Based on this formulation, the ESF is defined as the force that, statically applied to the structure in the direction α, produces a specified load effect e_α at the height r. Likewise in Section 2.5, this definition is not unique and three criteria deserve a special concern: the GEF technique, the LC technique and the GL technique (Repetto and Solari 2004).

Using the GEF technique, the ESF is defined as:

$$F_{\alpha,es}^e(r;z) = \lambda_\alpha \bar{F}_x(z) G_\alpha^e(r) \tag{102}$$

where the ESF implies a unique load pattern $\bar{F}_x(z)$. In the alongwind and crosswind directions, $\bar{F}_x(z)$ is scaled by the alongwind and crosswind GEF, $G_x^e(r)$ and $G_y^e(r)$, respectively; in the torsional direction, $\bar{F}_x(z)$ is multiplied by the eccentricity $bG_\theta^e(r)$, $G_\theta^e(r)$ being the torsional GEF. This definition is very useful for engineering applications: firstly, one has to apply only three static loading conditions, i.e. the force $\bar{F}_x(z)$ along x, the force $\bar{F}_x(z)$ along y, and the torsional moment $b\bar{F}_x(z)$ around z; the reference load effects $\bar{e}_x^x$, $\bar{e}_y^x$ and $\bar{e}_\theta^x$ are then calculated by equilibrium relationships; the maximum load effects are finally obtained by scaling these patterns by the appropriate 3-D GEF (Eq. (96)).

It is worth noting that, for $\alpha=x$, the 3-D GEF technique coincides with the GEF technique presented in Section 2.5. For $e=d$, the 3-D GEF technique coincides with the 3-D GRF technique presented in Section 3.2; in particular, since $B_\alpha^d(r)=B_\alpha$, $R_\alpha^d(r)=R_\alpha$, $g_\alpha^d(r)=g_\alpha$, the 3-D

GEF for the displacement is independent of r and coincides with the 3-D GRF, i.e. $G_\alpha^d(r) = G_\alpha$. For $\alpha=x$ and $e=d$, the 3-D GEF technique coincides with the GRF technique illustrated in Section 2.4. Thus, the 3-D GEF technique is the full generalisation of all the previous techniques; on the other hand, the GRF, the GEF and the 3-D GRF techniques may be regarded as particular cases of the 3-D GEF technique.

Analogously to Section 2.5, the LC technique assignes the ESF as a combination of three distinct load patterns associated, respectively, with the static, the quasi-static and the resonant parts of the response. Also in this case it is assumed, conventionally, that the quasi-static and the resonant parts of the ESF give rise to $\sigma_{B\alpha}^e$ and $\sigma_{R\alpha}^e$, respectively. Thus, the mean value of the maximum load effect $\overline{e}_{max}$ can be evaluated through Eq. (95).

The static and the resonant parts of the ESF are conveniently defined, respectively, by the mean static action $\overline{F}_\alpha(z)$ and by the inertial load that produces the rms value of the resonant response related to the vibration mode $\psi_{\alpha 1}$, i.e.:

$$F_{R\alpha,es}(z) = \mu_\alpha(z)(2\pi n_{\alpha 1})^2 \sigma_{R\alpha}(z) = 2\mu_\alpha(z)(2\pi n_{\alpha 1})^2 \overline{d}_\alpha^x(z) I_u R_\alpha \tag{103}$$

where $\overline{d}_\alpha^x$ is the reference displacement due to the application of $\lambda_\alpha \overline{F}_x(z)$ in the α direction.

The quasi-static part of the ESF may be evaluated by the LRC method (Kasperski 1992), the GEF technique and the GL technique (Repetto and Solari 2004). Generalising Eqs. (54)-(57):

$$F_{B\alpha,es}^e(r;z) = 2\lambda_\alpha \overline{F}_x(z) I_u B_\alpha^e(r) \tag{104}$$

$$F_{B\alpha,es}^e(r;z) = 2\lambda_\alpha \overline{F}_x(z) I_u B_\alpha^e(r) \Delta_\alpha^e(r,z) \tag{105}$$

$$\Delta_\alpha^e(r;z) = \frac{\overline{e}_\alpha^x(r)}{\lambda_\alpha \overline{F}_x(z)\left[\sigma_{B\alpha}^e(r)\right]^2} \int_0^\infty \left[\int_0^h S_{F_\alpha F_\alpha}(z,z';n)\eta_\alpha^e(r;z)\mathrm{d}z'\right]\mathrm{d}n \tag{106}$$

$$F_{B\alpha,es}(z) = \lambda_\alpha \overline{F}_x(h) \sum_{k=0}^{n-1} p_k \left(\frac{z}{h}\right)^k \tag{107}$$

where p_k (k=0,1,..n-1) are n non-dimensional coefficients used to impose that Eq. (107) provides the correct values of the quasi-static parts of n specified load effects.

Finally, using the GL technique:

$$F_{\alpha,es}(z) = \lambda_\alpha \overline{F}_x(h) \sum_{k=0}^{n-1} q_k \left(\frac{z}{h}\right)^k \tag{108}$$

where q_k (k=0,1,..n-1) are n non-dimensional coefficients used to impose that Eq. (108) provides the correct values of n specified maximum load effects.

3.4. 3-D response of vertical three-dimensional structures

Alongwind, crosswind and torsional actions on three-dimensional structures are deeply affected by the complexity of the flow circulation and separation from the bluff-body. Especially the signature turbulence in the rear of the structure is so complex to prevent the formulation of simple and general models. Thus, while alongwind actions on three-dimensional structures (Sections 2.4 and 2.5) and 3-D actions on slender structures (Section 3.2 and 3.3) may be dealt with analytical and numerical approaches, the determination of crosswind and torsional actions and response for three-dimensional structures need the support of wind tunnel tests. Making recourse to such tests, they are obviously extended to evaluating the full 3-D wind-excited behaviour of structures.

Wind tunnel tests aimed at determining the 3-D response of vertical three-dimensional structures may be classified into two techniques involving rigid and flexible or aeroelastic models.

Wind-tunnel tests involving rigid models may be further distinguished into two classes.

The first class involves the use of pressure taps distributed on the structural surface to detect the local pressure induced by wind on a grid of discrete points (Rosati 1968, Akins and Paterka 1977, Blessmann and Riera 1979, Reinhold et al 1979, Kareem 1982, Isyumov and Poole 1983). The results of measurements, reported to the full-scale, are then applied to finite element models to evaluate the dynamic response in the time domain (Patrickson and Friedmann 1979, Torkomani and Pramono 1985) or in the frequency domain (Sidarous and Vanderbilt 1979, Kareem 1981, 1985, Solari 1985, 1986).

The second class involves the use of high-frequency force-balances put at the base of the model to detect the base reactions. Base reactions are later processed to determine modal actions or pressure distributions. Using the criteria introduced by Tschanz and Davenport (1983), base overturning moments coincide with the quasi-static part of the modal forces related to linear modes. Literature is rich of analytical correction methods to take into account nonlinear mode shapes (Xu and Kwok 1993). Techniques were also proposed to determine the first modal torque based on measured base torsional moments (Tschanz and Davenport 1983, Marukawa et al. 1985, Xu and Kwok 1993). Indirect methods to derive equivalent pressure fields on the structural surface starting from the base reactions are reported by Ohkuma et al (1995) and Solari et al (1997).

The joint use of pressure taps and force balances is becoming more and more frequent.

Wind tunnel tests involving flexible or aeroelastic models are necessary when relevant effects due to flow-structure intercation occur. They may be distinguished into two classes.

The first class includes dynamically scaled multi-degree-of-freedom aeroelastic models aimed at simulating the relevant structural modes of vibration (Isyumov 1982). Measures carried out on such models, reported to the full-scale, directly provide the structural response, especially the accelerations to consider with regard to human comfort.

The second class includes rigid stick models supported by flexible balances that simulate the fundamental frequencies and damping ratios of the structure (Whitebread and Scruton 1965, Saunders and Melbourne 1975, Isyumov 1982). Measurements usually provide top generalised displacements and acceleration. Results are later processed analytically or numerically to take into account the actual structural properties (Zhou and Kareem 2003).

Using this large amount of experimental knowledge (Solari 1999), some novel techniques were proposed to evaluate the 3-D wind-excited response of buildings, without making recourse to wind tunnel tests. Choi and Kanda (1993) and Tamura et al (1996) derived analytical formulae of the maximum alongwind, crosswind and torsional responses of buildings with rectangular plan and un-coupled modes of vibration, and the corresponding ESFs, by fitting the results of a wide campaign of wind tunnel tests. An alternative method implemented by Zhou et al (2003) provides the 3-D response through the interactive use of a database that collects the results of high-frequency force-balance wind tunnel tests carried out on a wide set of building shapes.

4 Applications

The example reported below concerns the alongwind and crosswind response of a reinforced concrete chimney submitted to full-scale tests by Müller and Nieser (1976).
The structure is h = 180 m high and has a reference width b = 5.6 m. The equivalent mass per unit height is $\mu_x = \mu_y$ =10.686 kg/m , the fundamental frequencies are $n_{x1} = n_{y1} = 0.26$ Hz , the normal modes may be schematised as $\psi_{x1}(z)=\psi_{y1}(z)=(z/h)^{2.15}$, the corresponding damping ratios are $\xi_{x1} = \xi_{y1} = 0.005$. The drag and rms lift wake coefficients are $c_d = 0.8$ and $\tilde{c}_{ls} = 0.28$, respectively. The structure lies on a flat homogeneous terrain with roughness length $z_0 = 0.1$ m ; the intensities of the longitudinal and of the lateral turbulence components are, respectively, $I_u(z)=1/\ln(z/z_0)$ and $I_v(z)=0.75 I_u(z)$. Further details on the input parameters may be found, for instance, in (Piccardo and Solari 2002).
Due to the polar symmetry with reference to the vertical axis, the structure is excited by the longitudinal turbulence in the alongwind direction and by the lateral turbulence and the vortex shedding in the crosswind direction; no loading mechanism excites the torsional rotation. The analyses of the dinamic response have been carried out both numerically and using the Closed Form Solution (CFS) developed by Piccardo and Solari (2002).
Figure 6 shows the rms values of the alongwind and crosswind displacements at the top of the structure; solid lines represent the CFS, dotted lines correspond to numerical integrations; the figure is enriched by the results of the measurements carried out by Müller and Nieser (1976). At low mean wind velocities, due to the vortex shedding resonance with the fundamental mode of vibration, the crosswind response is higher than the alongwind response. At high mean wind velocities, the alongwind response due to the longitudinal turbulence exceeds the crosswind response caused by the lateral turbulence.

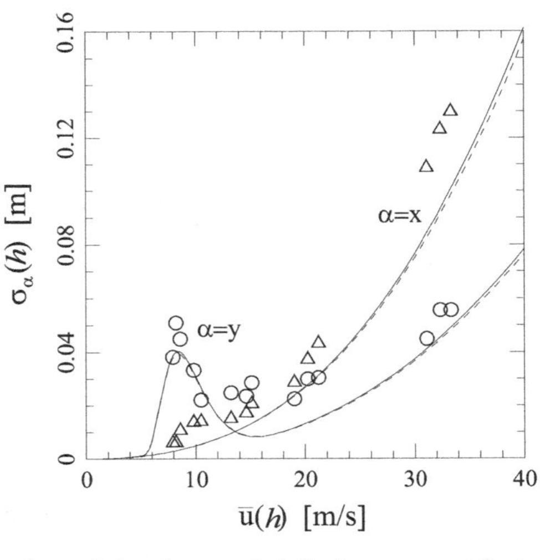

Figure 6. Rms values of the alongwind and crosswind displacements at the top of the structure (Δ and ○ denote measures along x and y, respectively).

Figure 7 shows the 3-D GEFs for the displacements, the base bending moments and the base shear forces (Eq. (96)), G_α^d, $G_\alpha^b(0)$ and $G_\alpha^s(0)$, versus the mean wind velocity at the chimney top; α=x corresponds to the alongwind vibrations (a), α=y corresponds to the crosswind vibrations (b). The agreement between the analytical and the numerical resuts is very good.

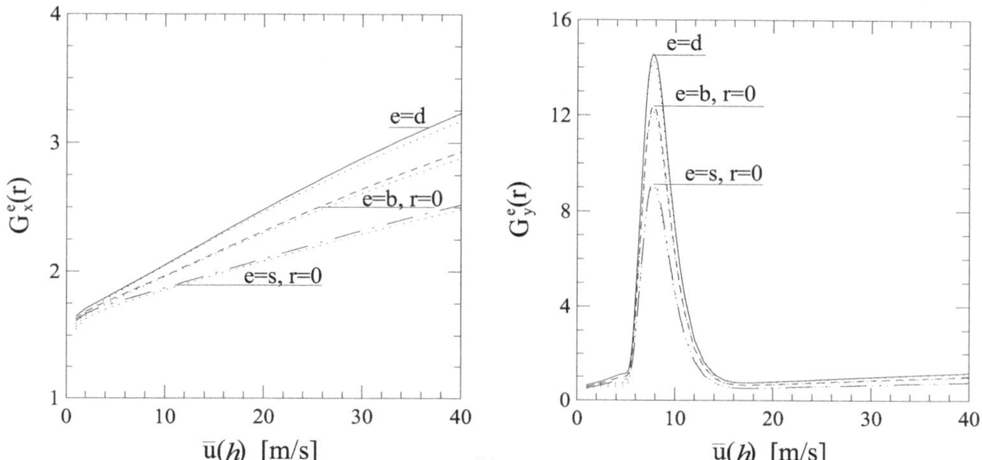

(a) (b)

Figure 7. Gust effect factors: (a) $\alpha = x$; (b) $\alpha = y$

Figure 8 shows the dependence of the 3-D GEF on the height r, for $\overline{u}(h)$= 8 m/s (a) and $\overline{u}(h)$= 20 m/s (b). Solid, dashed and dash-dotted lines correspond, respectively, to $e = d, b, s$ using the CFS; dotted lines denote numerical integrations. It is worth noting that, while G_d is independent of height, G_b and G_s vary with height; in particular, at least in this case, $G_b(h) = G_s(h) > G_d$; $G_s(0) < G_b(0) < G_d$. Thus, the GRF technique underestimates the top internal forces, while it overestimates the base internal forces and the base reactions (Solari and Repetto 2002). Also in this case the agreement between the analytical and the numerical results is very good.

Figure 9 shows some alongwind ESF evaluated through the GF technique (Eqs. (52) and (102)) (solid lines) and the GL technique (Eqs. (58) and (108)) (dotted line), for $\overline{u}(h)$= 20 m/s. The load patterns provided by the GF technique are proportional to the mean static force $\overline{F}_x(z)$ and depend on the load effect considered. The ESF evaluated by the GL technique is calibrated with the aim of providing correct estimates of the base bending moment, of the base shear force, and of the limit values of the top bending moment and shear force.

Figure 10 shows some alongwind ESF evaluated through the LC technique, for $\overline{u}(h)$ = 20 m/s. Figure 10(a) shows the diagrams of the mean static force $\overline{F}_x(z)$ (Eqs. (38) and (81)) and of the resonant part of the ESF $F_{Rx,es}(z)$ (Eqs. (53) and (103)). Figure 10(b) shows four load patterns of the quasi-static parts of the ESF: $F_{Bx,es}^d(z)$ (Eqs. (54) and (104)) (thick solid lines); $F_{Bx,es}^b(0,z)$ based on the GEF technique (Eqs. (54) and (104)) (thick dashed lines) and on the LRC method (Eqs. (55) and (105)) (thin dashed lines); and $F_{Bx,es}(z)$, based on the GL technique (Eqs. (57) and (107)), with the aim of providing correct estimates of the base bending moment, of the base shear force, and of the limit values of the top bending moment and shear force (thin dotted lines). The GL technique provides a parabolic load pattern different from all the other shapes. The GEF technique furnishes similar results when the ESF is calibrated to the displacement or the base bending moment. Using the LRC method focused on the base bending moment provides lower actions than the GEF technique in the upper structural part; on the contrary, it provides higher actions in the lower structural part due to larger turbulence effects. This result depends on the profile of the LRC factor $\Delta_x^b(0,h)$ defined by Eqs. (56) and (106).

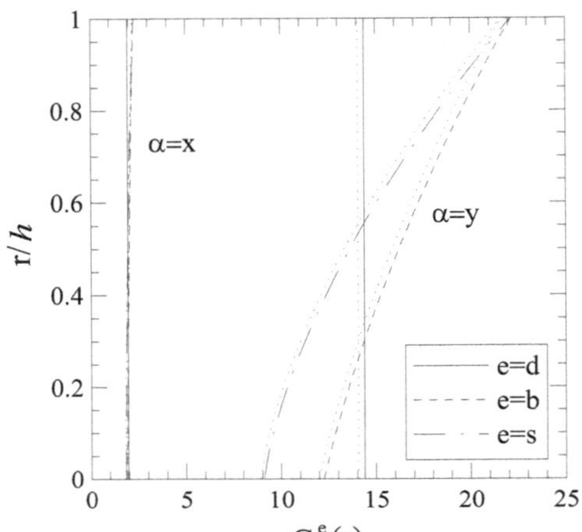

(a)

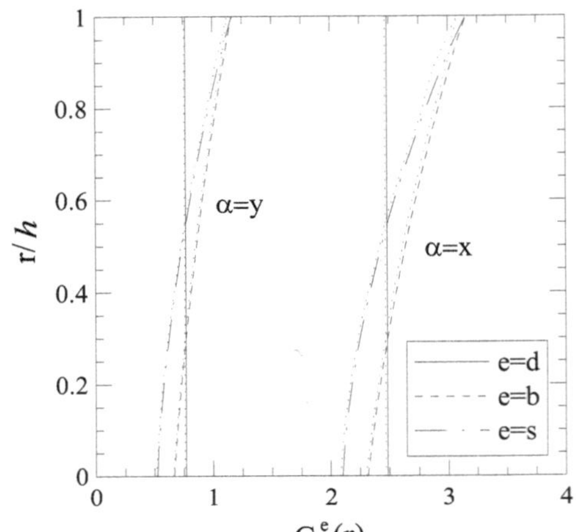

(b)

Figure 8. Gust effect factors: (a) $\bar{u}(h) = 8$ m/s; (b) $\bar{u}(h) = 20$ m/s.

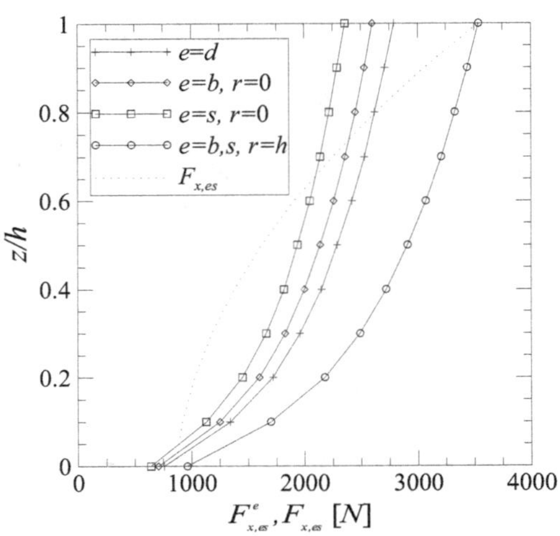

Figure 9. Alongwind ESFs evaluated by means of the GF and GL techniques.

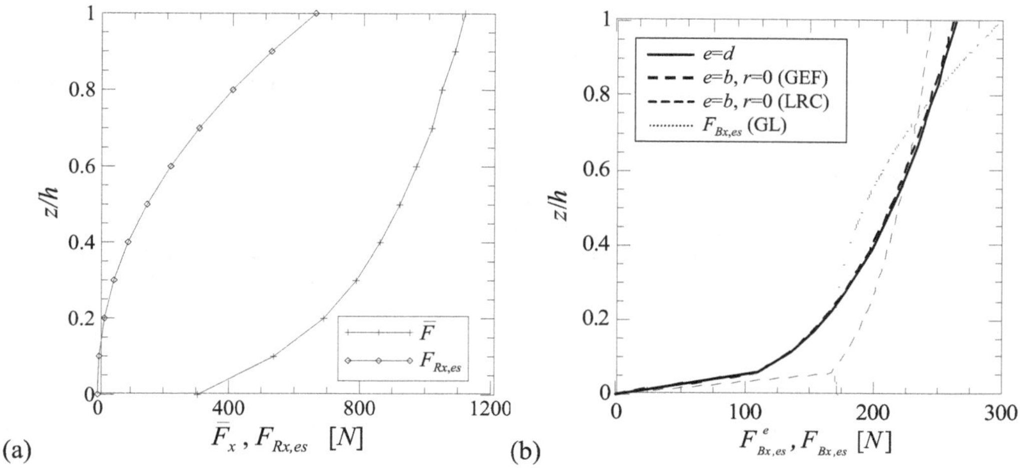

Figure 10. Alongwind ESFs evaluated by means of the LC technique: (a) mean static part and resonant part of the ESF; (b) quasi-static parts of the ESF.

5 Conclusions

The determination of the wind-excited response is a focal step in the design of structures characterised by relevant height, slenderness, flexibility, lightness and low damping properties. Wind engineering has developed a broad band of analytical, numerical and experimental methods aimed at evaluating the maximum values of the wind-induced load effects and suitable equivalent static actions. A noteworthy aspect of this discipline is the joint use of different methods to obtain robust and precise results.

The choice of the most appropriate methods to determine the wind-excited behavior of structures firstly depends on the aims of the analysis: in the preliminary stage of the project, analytical methods represent basic tools for the designer; in the stage of the final controls for the structural safety and serviceability, making recourse to more detailed procedures is often necessary. The decision among different approaches, or their joint application, usually depends on several factors including the shape, the size, the mechanical properties and the importance of the examined structure. Likewise for every design process, this decision involves to know, for each alternative method, merits and defects, limits and implications, costs and benefits.

This paper illustrates a general formulation of the basic methods to carry out the dynamic analysis of the wind-excited response of structures, focusing attention on methods developed at the University of Genoa in the prospect of deriving analytical solutions suitable for engineering applications and code provisions. The simplicity of these solutions with respect to wind tunnel tests cannot dim the necessity of using such methods within their validity range, taking always into account the aims of the analysis.

6 References

Akins, R.E. and Paterka, J.A. (1977). Mean force and moment coefficients for buildings in turbulent boundary layers. *Journal of Industrial Aerodynamics* 2: 195-209.

Blessmann, J. and Riera, J.D. (1979). Interaction effects in neighbouring tall buildings", *Proceedings of the 5th International Conference on Wind Engineering*, Fort Collins: 381-395.

Carassale, L., Freda, A. and Piccardo, G. (2005). Aeroelastic forces on yawed circular cylinders: quasi-steady modeling and aerodynamic instability. *Wind and Structures* 8: 373-388.

Chen, X. and Kareem, A. (2005). Dynamic wind effects on buildings with 3D coupled modes: Application of high frequency force balance measurements. *Journal of Enginnering Mechanics* ASCE 131: 1115-1125.

Choi, H. and Kanda, J. (1993). Proposed formulae for the power spectral densities of fluctuating lift and torque on rectangular 3-D cylinders. *Journal of Wind Engineering and Industrial Aerodynamics* 46&47: 507-516.

Dalgliesh, W.A. (1971). Statistical treatment of peak gust on cladding. *Journal of the Structural Division* ASCE 97: 2173-2187.

Davenport, A.G. (1961). The application of statistical concepts to the wind loading of structures. *Proceedings Institution of Civil Engineers*, London, UK; 19: 449-472.

Davenport, A.G. (1964). Note on the distribution of the largest value of a random function with application to gust loading. *Proceedings Institution of Civil Engineers*, London, UK; 24: 187-196.

Davenport, A.G. (1967). Gust loading factors. *Journal of the Structural Division* ASCE 93: 11-34.

Davenport, A.G. (1995). How can we simplify and generalize wind loads. *Journal of Wind Engineering and Industrial Aerodynamics* 54-55: 657-669.

Engineering Sciences Data Unit (1976). The response of flexible structures to atmospheric turbulence. *ESDU 76001*. London.

European Convention for Constructional Steelworks (1978). *Recommendations for the calculation of wind effects on buildings and structures*. Brussels: ECCS.

Holmes, J.D. (1975). Pressure fluctuations on a large building and alongwind structural loading. *Journal of Industrial Aerodynamics* 1: 249-278.

Holmes, J.D. (1994). Along-wind response of lattice towers: part I - derivation of expressions for gust response factors. *Engineering Structures* 16: 287-292.

Holmes, J.D. (1996). Along-wind response of lattice towers - III. Effective load distributions. *Engineering Structures* 18: 489-494.

Holmes, J.D. (2002). Effective static load distributions in wind engineering. *Journal of Wind Engineering and Industrial Aerodynamics* 90: 91-109.

Isyumov, N. (1982). The aeroelastic modelling of tall buildings. *Proceedings of the International Workshop on Wind Tunnel Modelling for Civil Engineering Applications*, Gaithersburg: 373-407.

Isyumov, N. and Poole, M. (1983). Wind induced torque on square and rectangular building shapes, *Journal of Wind Engineering and Industrial Aerodynamics* 13: 183-196.

Kareem, A. (1981). Wind-excited response of buildings in higher modes. *Journal of the Structural Division* ASCE 107: 701-706.

Kareem, A. (1982). Fluctuating wind loads on buildings. *Journal of the Engineering Mechanics Division* ASCE 108: 1086-1102.

Kareem, A. (1985). Lateral-torsional motion of tall buildings to wind loads. *Journal of Structural Engineering* ASCE 111: 2479-2496.

Kasperski, M. (1992). Extreme wind load distributions for linear and nonlinear design. *Engineering Structures* 14: 27-34.

Katsumura, A., Tamura, Y. and Nakamura, O. (2004). Universal wind load distribution reproducing maximum load effects on structural members. *Proceedings of the 5th International Colloquium on Bluff Body Aerodynamics & Applications*, Ottawa, Canada: 351-354.

Marukawa, H., Kato, N., Fujii, K., Tamura, Y. (1996). Experimental evaluation of aerodynamic damping of tall buildings. *Journal of Wind Engineering and Industrial Aerodynamics* 59: 177-190.

Muller, F.P. and Nieser H. (1976). Measurements of wind-induced vibrations on a concrete chimney. *Journal of Industrial Aerodynamics* 1: 239-248.

Ohkuma, T., Marukawa, H., Yoshi,e K., Niwa, H., Teramoto, T. and Kitamura, H. (1995). Simulation method of simultaneous time-series of multi-local wind forces on tall buildings by using balance data. *Journal of Wind Engineering and Industrial Aerodynamics* 54/55: 115-123.

Patrickson, C.P. and Friedmann, P.P. (1979). Deterministic torsional building response to winds. *Journal of Structural Division* ASCE 105: 2621-2637.

Piccardo, G. and Solari, G. (1996). A refined model for calculating 3-D equivalent static wind forces on structures. *Journal of Wind Engineering and Industrial Aerodynamics* 65: 21-30.

Piccardo, G. and Solari, G. (1998a). Closed form prediction of 3-D wind-excited response of slender structures. *Journal of Wind Engineering and Industrial Aerodynamics* 74-76: 697-708.

Piccardo, G. and Solari, G. (1998b). Generalized equivalent spectrum technique. *Wind and Structures* 1: 161-174.

Piccardo, G. and Solari, G. (2000). 3-D wind-excited response of slender structures: Closed form solution. *Journal of Structural Engineering* ASCE 126: 936-943.

Piccardo, G. and Solari, G. (2002). 3-D gust effect factor for slender vertical structures. *Probabilistic Engineering Mechanics* 17: 143-155.

Reinhold, T.A., Sparks, P.R., Tieleman, H.W. and Maher, F.J. (1979). Mean and fluctuating forces and torques on a tall building model of square cross-section. *Report VPI-E-79-11, Department of Engineering Science and Mechanics*, Virginia Polytechnic Institute and State University, Blacksburg, Virginia.

Repetto, M.P. and Solari, G. (2004). Equivalent static wind actions on vertical structures. *Journal of Wind Engineering and Industrial Aerodynamics* 92: 335-357.

Rosati, P. (1968). An experimental study of the response of a square prism to wind load. *Report BLWT-II-68, Faculty of Engineering Science*, The University of Western Ontario, London, Canada.

Saunders, J.W. and Melbourne, W.H. (1975). Tall rectangular building response to cross-wind excitation. *Proceedings of the 4th International Conference on Wind effects on buildings and structures*, Cambridge University Press, Cambridge, U.K.: 369-379.

Sidarous, J.F.Y. and Vanderbilt, M.D. (1979). An analytical methodology for predicting dynamic building response to wind. *Proceedings of the 5th International Conference on Wind Engineering*, Fort Collins, Colorado: 709-724.

Simiu, E. (1976). Equivalent static wind loads for tall buildings design. *Journal of the Structural Division* ASCE; 102: 719-737.

Simiu, E. (1980). Revised procedure for estimating alongwind response. *Journal of the Structural Division* ASCE; 106: 1-10.

Solari, G. and Piccardo, G. (2001). Probabilistic 3-D turbulence modeling for gust buffeting of structures. *Probabilistic Engineering Mechanics* 16: 73-86.

Solari, G. (1982). Alongwind response estimation: closed form solution. *Journal of the Structural Division* ASCE; 108: 225-244.

Solari, G. (1983). Analytical estimation of the alongwind response of structures. *Journal of Wind Engineering and Industrial Aerodynamics* 14: 467-477.

Solari, G. (1988). Equivalent wind spectrum technique: theory and applications. *Journal of Structural Enginnering* ASCE 114: 1303-1323.

Solari, G. (1993). Gust buffeting. I: Peak wind velocity and equivalent pressure. II: Dynamic alongwind response. *Journal of Structural Enginnering* ASCE 119: 365-382, 383-398.

Solari, G. (1994). Gust-excited vibrations, in *Wind-excited vibrations of structures*, Sockel H., Editor, Springer Verlag, Wien, New York: 195-291.

Solari, G. (1999). Progress and prospects in gust-excited vibrations of structures. *Engineering Mechanics* 6: 301-322.

Solari, G. and Repetto, M.P. (2002). General tendencies and classification of vertical structures under wind loads. *Journal of Wind Engineering and Industrial Aerodynamics* 90: 1299-1319.

Solari, G. (1985). Mathematical model to predict 3-D wind loading on buildings, *Journal of Engineering Mechanics* ASCE 111: 254-276.

Solari, G. (1986). 3-D response of buildings to wind action. *Journal of Wind Engineering and Industrial Aerodynamics* 23: 379-393.

Solari, G., Pagnini, L.C. and Piccardo, G. (1997). A numerical algorith for the aerodynamic identification of structures. *Journal of Wind Engineering and Industrial Aerodynamics* 69: 719-730.

Tamura, Y., Kawai, H., Uematsu, Y., Marukawa, H., Fujii, K. and Taniike, Y. (1996). Wind loads and wind-induced response estimations in the Recommendations for loads on buildings, AIJ. Engineering Structures; 18: 399-411.

Torkomani, M.A.M. and Pramono, E. (1985). Dynamic response of tall building to wind excitation. *Journal of Structural Engineering* ASCE 111: 805-825.

Tschanz, T. and Davenport, A.G. (1983). The base balance technique for the determination of dynamic wind loads. *Journal of Wind Engineering and Industrial Aerodynamics* 13: 429-439.

Vellozzi, J. and Cohen, E. (1968). Gust response factors. *Journal of the Structural Division* ASCE 94: 1295-1313.

Vickery, B.J. (1970). On the reliability of gust loading factors. Proc Techn Meet concerning Wind Loads on Buildings and Structures. *National Bureau of Standards*, Washington, DC: 93-104.

Vickery, B.J. and Basu R.I. (1983). Across-wind vibrations of structures of circular cross-section. Part I: Development of a mathematical model for two-dimensional conditions. *Journal of Wind Engineering and Industrial Aerodynamics* 12: 49-74.

Vickery, B.J. and Clark, W. (1972). Lift or across-wind response of tapered stacks. *Journal of the Structural Division* ASCE: 98: 1-20.

Vickery, B.J. and Davenport, A.G. (1967). A comparison of theoretical and experimental determination of the response of elastic structures to turbulent flow. *Proceedings of the International Research Seminar on Wind effects on buildings and structures*, Ottawa, Canada, 1: 705-738.

Whitebread, R.E., Scruton, C. (1965). An investigation of the aerodynamic stability of a model of the proposed tower blocks for the World Trade Center, New York. *National Physical Laboratory Aerodynamics* Rep. No. 1165, Teddington, U.K.

Xu, Y.L. and Kwok, K.C.S. (1993). Mode shape corrections for wind tunnel tests of tall buildings. *Engineering Structures* 15: 387-392.

Zhou Y., Gu M. and Xiang H. (1999). Alongwind static equivalent wind loads and response of tall buildings. Part I: Unfovarable distributions of static equivalent wind loads. Part II: Effects of mode shape. *Journal of Wind Engineering and Industrial Aerodynamics* 79: 135-150, 151-158.

Zhou, Y. and Kareem, A. (2001). Gust loading factor: new model. *Journal of Structural Engineering* ASCE 127: 168-175.

Zhou, Y. and Kareem, A. (2003). Aeroelastic balance. *Journal of Engineering Mechanics* ASCE 129: 283-292.

Zhou, Y., Kijewski, T. and Kareem, A. (2003). Aerodynamic loads on tall buildings: Interactive data-base. *Journal of Structural Engineering* ASCE 129: 394-404.

Bridge Aerodynamics and Aeroelastic Phenomena

Claudio Borri[1,2] and Carlotta Costa[1,2]

[1] Dipartimento di Ingegneria Civile, Università di Firenze, Firenze, Italy.
[2] CRIACIV, Centro Interuniversitario di Aerodinamica delle Costruzioni ed Ingegneria del Vento, Prato, Italy.

Abstract. This memory deals with wind-structure interaction phenomena involving atmospheric flow investing bridges and roofs. Main load models describing the different components of the wind action are introduced and described, and the conditions under which instability phenomena can occur are defined. In particular, for bridges, it is possible to express the wind load as a steady, a quasi-steady or an unsteady action. The solution of the equations of motion in the different cases allows one to identify the critical condition for galloping, torsional divergence and flutter. The effect and the representation of the turbulence action on a bridge deck are also discussed, through the introduction of indicial functions representative of an assigned admittance. Concerning roofs, the representation of the wind action via pressure coefficients is discussed and the relevant investigations in wind tunnel are presented.

1 Introduction

The description of the mechanical behavior of structures under wind action is a stimulating challenge for applied and theoretical reasons.

First of all, the turbulent and fluctuating nature of wind flow in the atmospheric boundary layer produces complex flow fields around bodies. Flow separation and eventual re-attachments give rise to highly fluctuating pressure fields, and therefore to loads essentially of dynamic nature. This kind of ambient load is usually referred to as "aerodynamic" load. Moreover, the structures oscillate in natural flows, originating interaction or "aeroelastic" phenomena. Resonance-type as well as instability phenomena can occur, depending on the geometry of the structure, on their mechanical properties, on the characteristics of turbulent flow and on its mean speed.

The complex pressure field generated by the interaction of atmospheric flow and structural response is commonly simplified by adopting load schemes taking into account different aspects: (i) a steady component, dependent on the mean wind velocity and representing the very low fluctuating action of the wind; (ii) a buffeting component, dependent on the turbulent wind fraction or gustiness in the free-stream flow, (iii) a vortex-shedding component, due to the synchronized flow separation and (iv) a self-excited component, dependent on the structural motion. It is assumed that such different aspects, which could arise contemporarily during the life of the structure, are related to different mean speeds and frequency ranges and, therefore, their interaction is negligible. This is, in principle, not true, but such a scheme has proven to be satisfying for the description of several phenomena.

Attention is focused here on the description of main load models for each of these contributions, with reference to special structures like bridges and roofs, after the introduction of basic concepts and variables necessary to characterize the wind action. Moreover, the uncertainties related to local wind characteristics, load models and structural details should be included in the description of the wind load, due to the fact that they could modify the expected response. For this reason, a model is discussed, in which the turbulence effect can be included in a time-domain framework for the structural analysis. Relevant resonance-type or instability-type events are classified and, eventually, identified through solution of the equations of motion. In particular, galloping, torsional divergence, flutter, buffeting and vortex shedding will be addressed.

2 Mechanical schemes and aerodynamic loads

In order to sketch the dynamic phenomena involved in fluid-structure interaction, a simple two-dimensional rigid body immersed in a uniform flow of velocity U is considered. The body has mass m, moment of inertia I and static moment of inertia S_x evaluated with respect to the local x axis. It is fixed in horizontal direction and spring-supported with respect to vertical motion z and rotation α (Figure 1). It can represent the section of a body, like a bridge or a building.

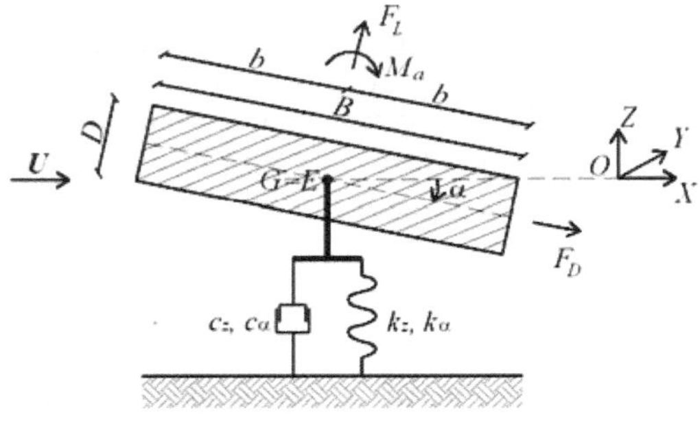

Figure 1. Structural reference system.

Torsion and heaving elastic properties are symbolized by a spring, characterized by stiffness k_z and k_α, and by a viscous damping device, characterized by damping coefficients c_z and c_α. Dampers and springs are applied in the elastic centre E (i.e. the shear center) coincident, in this case, with the centre of mass G. The equations of motion describing the dynamic system per unit length are given by

$$\begin{aligned} m\ddot{z} + c_z\dot{z} + k_z z + S_x\ddot{\alpha} &= F_z \\ I\ddot{\alpha} + c_\alpha\dot{\alpha} + S_x\ddot{z} + k_\alpha\alpha &= M_y \end{aligned} \tag{1}$$

with dot denoting the substantial derivative with respect to time t.

The external loads acting along the two active Degrees of Freedom (DoFs) are represented by a vertical force F_z and an aerodynamic moment M_y. With reference to a local coordinate system attached to the body in motion, the vertical force can be expressed as the combination of a drag force F_D directed along the main characteristic dimension on the body (in this case it is assumed that it corresponds to B) and of a lift force F_L acting in the orthogonal direction (in this case, along the direction of D):

$$F_z = F_L \cos\alpha - F_D \sin\alpha \tag{2}$$

The angle α is also referred to as the *angle of attack* and identifies the relative position of the section with respect to the incoming flow. For the sake of simplicity, the resultant wind flow U represented in Figure 1 is assumed directed along the global direction X. If a turbulent flow needs to be accounted for, the angle of attack should account for the contribution of horizontal and vertical turbulent components. If the section is in motion, the effective angle of attack between the flow and the section should take into account not only the static angle between the direction of the wind and the section, but also the relative velocities of the section, both vertical and torsional, as in the following equation,

$$\tilde{a} = \alpha - \arctan\left(\frac{R\dot{\alpha}}{U}\right) - \arctan\left(\frac{\dot{z}}{U}\right) \tag{3}$$

and, eventually, the presence of turbulence. The distance between the elastic center and the investigated point is indicated by R.

The representation of the steady wind forces is achieved by means of dimensionless coefficients, called *aerodynamic coefficients*, allowing the easy switch from wind tunnel and full-scale measurements. If the representation in terms of drag and lift components rather than in terms of vertical and horizontal forces is chosen, the coefficients C_L, C_D and C_M are adopted, allowing the following formulation of aerodynamic forces per unit length:

$$F_D = \frac{1}{2}\rho U^2 D C_D(\alpha)$$

$$F_L = \frac{1}{2}\rho U^2 B C_L(\alpha) \tag{4}$$

$$M_a = \frac{1}{2}\rho U^2 B C_M(\alpha)$$

The aerodynamic coefficients are evaluated as a function of the static angle of incidence between the mean flow and the section. As example, the aerodynamic coefficients extracted for a footbridge deck section model tested in *CRIACIV* wind tunnel are shown in Figure 2 (Bartoli et al., 2005).

In order to solve in time domain the system of differential equations (1), the initial conditions for displacements and velocities at starting time t_0 are required. In the most general case, the two equations of the system (1) are coupled by a structural point of view (via the static moment of inertia) and by an aerodynamic point of view, because of the presence of terms on the right-hand side of the equation depending on the motion of the body.

The dynamic nature of the phenomenon can be, eventually, neglected, if a steady approach is employed and only wind load of Eqs. (4) is applied. With the expression "steady aerodynamics", one refers to the study of systems in which wind forces do not vary in time, but are only a function of the position. More properly, a quasi-steady (or quasi-stationary) approach can be employed. The evolution of the wind forces in time is accounted for via the time history of the angle of attack (which includes also the effect of the actual structural motions). By this way, the system is moving but it is considered at each instant as steady, with forces depending only on the actual configuration. The effects on phase and force magnitude related to the effective interaction between structural motion and fluid are included only if an "unsteady" approach is adopted.

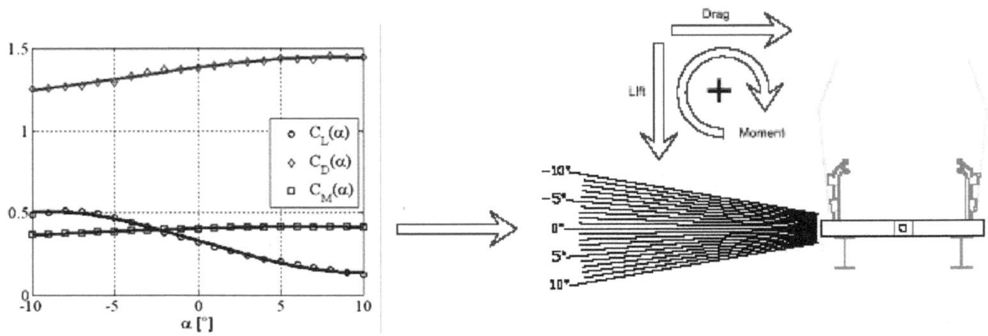

Figure 2. Aerodynamic coefficients for a footbridge deck section (left) with relevant section model and wind tunnel test conditions (right) (Bartoli et al., 2005).

3 Galloping

Galloping is an aeroelastic instability phenomenon concerning non-circular sections and manifesting itself with large amplitude oscillations, mainly in the crosswind direction. Tall buildings and high-rise structures, isolated structural elements, lighting pole and cables subjected to icing conditions are typical examples of structures susceptible to gallop when the wind velocity exceeds certain critical values.

Galloping can manifest itself as a one-DoF instability or as a coupled phenomenon. First observations on this type of phenomenon have been made mainly for one-DoF systems, while first studies on coupling in galloping are due to Desai et al. (1990) for coupled torsional-vertical and to Sarkar et al. (1992) for coupled translational galloping. Nowadays, the study of galloping involves more complicated patters, typically with coupling phenomena [see, e. g., Luongo and Piccardo (2005)].

In the following, the critical galloping condition is derived for one-DoF systems, both translational and torsional. The Glauert-Den Hartog criterion is defined, based on the quasi-steady load expressions. Some insights are given for galloping within systems with more than one DoF.

3.1 Equations of motion for galloping bodies. The Glauert-Den Hartog criterion for galloping instability

The first stability criterion to estimate the possibility for a section to gallop for wind exceeding certain critical values has been established by Den Hartog in 1932, for a system with only one DoF, in particular in order to specify the aerodynamic loading conditions under which a horizontal conductor can gallop. In this case, only one equation of motion is considered, for example the one describing the vertical motion:

$$m\ddot{z} + c_z\dot{z} + k_z z = F_z \tag{5}$$

By expressing the vertical force in terms of drag and lift force as in Eq. (2), and by expressing lift and drag forces as in Eqs. (4), the following equation is obtained

$$m\ddot{z} + c_z\dot{z} + k_z z = \frac{1}{2}\rho U^2 B C_L(\alpha)\cos\alpha - \frac{1}{2}\rho U^2 B C_D(\alpha)\sin\alpha \tag{6}$$

An analogous approach has been proposed, among the other authors, by Parkinson and Brooks (1961) and by Novak (1969), based on wind-tunnel experiments of the simplest kind, with no distinction made between the inclination of relative wind velocity U_r (accounting for the velocity of the body) and the angle of attack, specifying the steady wind direction with respect to the body at rest.

Taking, for example, as steady reference position the zero angle of attack, the aerodynamic coefficients can be linearized as follows:

$$C_L(\alpha) \cong C_L(0) + \left.\frac{dC_L}{d\alpha}\right|_0 \alpha + o(\alpha^2) \cong C_L(0) + C_{L,\alpha}(0)\alpha$$

$$C_D(\alpha) \cong C_D(0) + \left.\frac{dC_D}{d\alpha}\right|_0 \alpha + o(\alpha^2) \cong C_D(0) + C_{D,\alpha}(0)\alpha \tag{7}$$

$$C_M(\alpha) \cong C_M(0) + \left.\frac{dC_M}{d\alpha}\right|_0 \alpha + o(\alpha^2) \cong C_M(0) + C_{M,\alpha}(0)\alpha$$

If position (7) is introduced into Eq. (6) and the slope of drag coefficients is considered negligible around the zero angle of attack, equation of motion (6) reduces to

$$m\ddot{z} + c_z\dot{z} + k_z z = \frac{1}{2}\rho U^2 B\{C_L(0) + [C_{L,\alpha}(0) - C_D(0)]\alpha\} \tag{8}$$

This equation has been derived in steady conditions but, if the section is in motion, a quasi-steady representation is required and the angle of attack should be the effective one of Eq.(3) accounting for the velocity of the section:

$$\tilde{\alpha} = \arctan\left(\frac{\dot{z}}{U}\right) \tag{9}$$

or, if the angle of attack remains small,

$$\tilde{\alpha} \cong \frac{\dot{z}}{U} \tag{10}$$

The following equation of motion is achieved

$$\ddot{z} + 2\nu_z\omega_z\dot{z} + \omega_z^2 z = \frac{1}{2m}\rho U^2 B\left\{C_L(0) + \left[C_D(0) + C_{L,\alpha}(0)\right]\frac{\dot{z}}{U}\right\} \tag{11}$$

If all terms related to vertical velocity of the section are grouped, such final equation is obtained:

$$\ddot{z} + 2\omega_z\left[\nu_z - \frac{\rho UB}{4m\omega_z}\left[C_D(0) + C_{L,\alpha}(0)\right]\right]\dot{z} + \omega_z^2 z = \frac{1}{2m}\rho U^2 B C_L(0) \tag{12}$$

Therefore, the wind force induces an additional aerodynamic damping on the body which adds itself to the structural damping. The system is stable if

$$\left[C_D(0) + C_{L,\alpha}(0)\right] \leq 0 \tag{13}$$

Square sections are, for example, prone to gallop and a sufficient structural damping should be provided to counteract such behavior. For a quite large table of drag and lift slope coefficients, see Blevins (1990).

The equation (12) describes a nonlinear system in which the nonlinearity is owing to the damping term. By assuming that the amplitude is nearly constant in the period of one vibration cycle, Parkinson and Brooks (1961) and Novak (1969) were able to treat the amplitude dependence of the stability and the hysteresis (or bifurcation) found in experiments by analytical means. The term hysteresis is used to describe that the maximum vibration amplitude of the wind tunnel model can differ if the flow speed is decreased to or increased to the critical value.

The critical wind velocity corresponding to the gallop phenomenon corresponds to

$$U_{cr} = \frac{4\nu_z m\omega_z\left[C_D(0) + C_{L,\alpha}(0)\right]}{\rho D} \tag{14}$$

3.2 Torsional one DoF galloping

The problem of torsional uncoupled oscillations can be treated by an analogous approach. In particular, the uncoupled torsional one DoF system can be represented by the following equation of motion

$$I\ddot{\alpha} + c_\alpha \dot{\alpha} + k_\alpha \alpha = M_a \tag{15}$$

with aerodynamic load expressed following Eqs. (4). By linearizing up to the first order following the approximation proposed in (7), by considering the expression of the angle of attack as in Eq. (3) and by neglecting the vertical velocity of the section, the equation of motion can be written as

$$\ddot{\alpha} + 2\nu_\alpha \omega_\alpha \dot{\alpha} + \omega_\alpha^2 \alpha = \frac{1}{2I} \rho U^2 B \left\{ C_M(0) + C_{M,\alpha}(0) \left[\alpha - \left(\frac{R\dot{\alpha}}{U} \right) \right] \right\} \tag{16}$$

If only torsional velocity is different from zero, Eq. (16) becomes

$$\ddot{\alpha} + \left(2\nu_\alpha \omega_\alpha + C_{M,\alpha}(0) \frac{R}{U} \right) \dot{\alpha} + \omega_\alpha^2 \alpha = \frac{1}{2I} \rho U^2 B C_M(0) \tag{17}$$

The vanishing of the total torsional damping corresponds to the occurring of torsional gallop, at the critical speed

$$U_{cr} = -\frac{4I\nu_\alpha \omega_\alpha}{\rho B^2 R} \frac{1}{C_{M,\alpha}(0)} \tag{18}$$

3.3 Coupled galloping

For coupled oscillations, which can be of vertical-torsional type (Desai et al., 1990) or coupled translational (Sarkar et al., 1992), a two-DoFs system needs to be solved. In the field of vertical-torsional galloping, a recent attempt of investigating the instability trends of the dynamic behavior of a single iced conductor is due to Yu and Ren (2005), while, concerning the mechanisms governing coupled translational galloping, a generalized Den Hartog criterion has been proposed by Sarkar et al. (1992) and a formula defining the onset wind velocity for the bi-dimensional coupled galloping oscillations of tower buildings has been presented by Shuguo et al (1993).

Galloping phenomenon can occur also for sectional models having a higher number (in general, three) of coupled DoFs, as evidenced by Solari (1994) and Piccardo (1993).

The effects of the upstream turbulence on galloping have been studied, for example, by Novak et al. (1969), who suggested that significant modifications in the behavior of scaled models in wind tunnel can be modeled through changes in steady aerodynamic coefficients.

4 Torsional divergence

Torsional divergence is a static instability phenomenon which occurs when the total torsional stiffness, obtained as the sum of structural and aerodynamic stiffness, vanishes. In particular, with reference to the equation of motion (15), it can be obtained that

$$k_\alpha - \frac{1}{2} \rho U^2 D^2 C_{M,\alpha}(0) = 0 \tag{19}$$

which corresponds to the following critical wind velocity:

$$U_{cr} = \sqrt{\frac{2k_\alpha}{\rho D^2 C_{M,\alpha}(0)}} \tag{20}$$

For bridges, wind velocity corresponding to torsional divergence is normally higher with respect to the design wind speed.

5 Flutter instability

Generally speaking, flutter is an aeroelastic instability phenomenon involving coupling of vibration modes. Different types of flutter instabilities can be identified, as described in detail by Matsumoto et al. (1996), depending on flow distribution and structural geometry. In every case, structural damping and stiffness, and therefore natural frequencies, are modified by the aeroelastic terms due to the wind action. Different incoming wind speeds can play a different role on aerodynamic damping and stiffness by modifying structural behavior and driving, eventually, structure to failure.

The identification of the flutter mechanism and of the conditions under which it can occur represents a fundamental step in bridge and footbridge design. In fact, a critical wind speed should be provided during the design of a structure sensitive to the wind action. It should be noted that flow separation is not necessary for the occurrence of flutter and that the phenomenon occurs at any flow velocity above the critical one (critical velocity is also referred to as flutter boundary). This fact clearly distinguishes the flutter from resonance problem, occurring only at specific ranges of wind velocity.

In the following, a qualitative description of flutter instability is presented, distinguishing both "classical" and "torsional" flutter. Flutter can be captured via an adequate representation of self-excited loads on the body. The classical formulation in frequency domain is introduced and discussed as well as the indicial approach in time domain. Frequency-domain formulation adopts non-dimensional parameters called *flutter* or *aeroelastic derivatives*, while *indicial functions* are used in time-domain approach. The correspondence between flutter derivatives and indicial function is established and several results for different sections are shown. Spectral approach is useful under the hypotheses of linear behavior and small displacements. On the other hand, if the structure presents nonlinear and/or nonstationary features, or an evolution of its dynamic properties with varying ambient loadings, only time-domain analyses can estimate appropriately the response. A tool for calculation of critical condition through flutter derivatives is also given.

5.1 Qualitative description of flutter instability

Flutter is characterized by a harmonic and usually coupled motion. "Quality" of flutter is mainly dependent on the geometry of the section. In particular, for a quite streamlined section, coupling occurs between vertical and torsional modes (*classical flutter*, see Figure 3). Coupling frequency strongly depends on the ratio between the two natural ones. In a full bridge, mode shapes influence also the possibility of coupling. For a bluff section, one-DoF *torsional flutter* is possible (see Figure 4), in which torsional motion is undamped, while flexural motion maintains its frequency and its damping.

Flutter boundary can be shifted, clearly, by acting on the geometry of the section, but also on the damping system and on its natural frequencies. In particular, danger of coupled flutter can be avoided by increasing the ratio between natural frequencies supposed to couple.

As already pointed out, flutter instability can occur also in uniform flow, without the presence of an external disturbance (like turbulence). The effect of turbulence is difficult to be quantified on flutter threshold, but it seems to result in an increase of critical speed, in the sense that, even if coupling of frequencies occurs, the motion still remains undamped, because of the presence of the energy at many others levels different from the coupling one.

The wind forces responsible for flutter are called self-excited, being due only to the motion of the system. In particular, self-excited forces can be expressed by means of appropriate coefficients applied to the body motion components. These coefficients are commonly extracted through wind tunnel tests, as a function of the incoming wind speed and frequency of the system.

In this case, the load model is clearly unsteady, because it includes loads varying with structural response, accounting therefore for phase and magnitude effects.

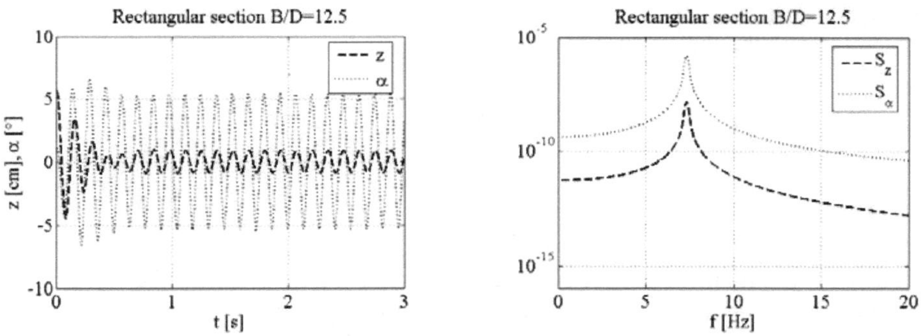

Figure 3. Classical flutter for a streamlined section (numerical simulation).

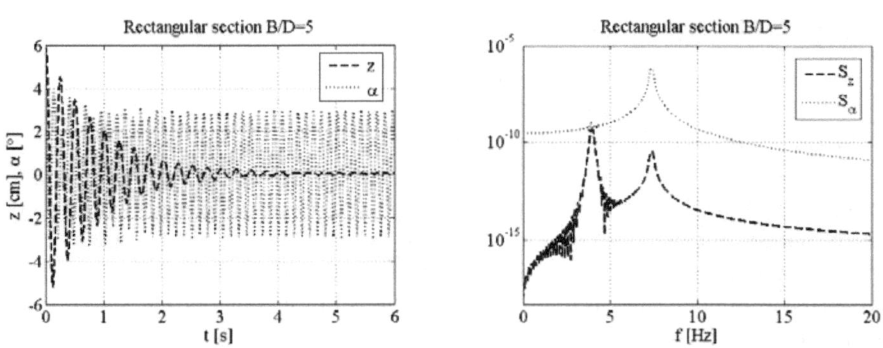

Figure 4. Torsional flutter for a bluff section (numerical simulation).

In the following, self-excited forces and their characteristic parameters are discussed for a two-dimensional scheme.

5.2 Self-excited forces on a 2 DoFs section: classical approach in frequency domain

Load models classically used in frequency domain to pattern the aeroelastic loads on bridges follow the work of Scanlan and Tomko (1971), who discussed the modeling of buffeting and self-excited forces. In particular, self-excited load expressions are built by taking in mind a strong parallelism between the behavior of a thin airfoil and a streamlined bridge deck. Nevertheless, the main hypotheses on which thin airfoil theory is based, as existence of potential flow, no separation of shear layers and full gust coherence, do not apply strictly to bodies with different geometrical shapes. Therefore, the formulation of the external loads follows the guidelines of aerodynamic theory, but adapts the theory itself to the case of each bridge with the introduction of proper experimental data. Self-excited forces are included in a time-domain framework, common also to thin airfoil aerodynamics, and characterized by frequency-dependent functions. In particular, the self-excited forces for a thin airfoil, in the simplified case of an airfoil with the center of mass coincident with the shear center and neglecting added mass and inertia, are expressed as

$$\begin{aligned} F_L(t) &= \rho\pi b^2 U\dot{\alpha} + 2\pi\rho U^2 b C(k)\left(\frac{\dot{z}}{U} + \alpha + \frac{1}{2}\frac{b\dot{\alpha}}{U}\right) \\ M_a(t) &= -\frac{1}{2}\rho\pi b^3 U\dot{\alpha} + \pi\rho U^2 b^2 C(k)\left(\frac{\dot{z}}{U} + \alpha + \frac{1}{2}\frac{b\dot{\alpha}}{U}\right) \end{aligned} \tag{21}$$

The quantity within brackets is the so-called *downwash* w and represents the vertical velocity of the fluid particle attached to the airfoil at the rear point. It generates the circulation (unsteady) lift, if multiplied by the Theodorsen's circulation function $C(k) = F(k) + iG(k)$ (defined as a function of the reduced frequency k). The real and the imaginary parts of the Theodorsen's function are given in closed form by a combination of Bessel functions of first and second kind, J_i and Y_i, respectively:

$$\begin{aligned} F(k) &= \frac{J_1(k)[J_1(k) + Y_0(k)] + Y_1(k)[Y_1(k) - J_0(k)]}{[J_1(k) + Y_0(k)]^2 + [Y_1(k) - J_0(k)]^2} \\ G(k) &= -\frac{J_1(k)J_0(k) + Y_0(k)Y_1(k)}{[J_1(k) + Y_0(k)]^2 + [Y_1(k) - J_0(k)]^2} \end{aligned} \tag{22}$$

For a bridge deck section, an analogous formulation in terms of frequency-dependent parameters can be provided. But, as no theoretical function is available to describe to unsteady lift acting on a bluff section at a certain frequency, experimental frequency-dependent functions need to be identified for each bridge deck sections. Such functions are commonly measured as discrete-frequency parameters in wind tunnel tests and are the flutter derivatives.

The original formulation due to Scanlan and Tomko (1971) proposes, for a two-dimensional bridge deck section, these expressions for self-excited lift and moment:

$$
\begin{aligned}
F_L(t) &= qB\left[KH_1^*(K)\frac{\dot{z}(t)}{U} + KH_2^*(K)\frac{B\dot{\alpha}(t)}{U} + K^2H_3^*(K)\alpha(t) + K^2H_4^*(K)\frac{z(t)}{U}\right] \\
M_a(t) &= qB^2\left[KA_1^*(K)\frac{\dot{z}(t)}{U} + KA_2^*(K)\frac{B\dot{\alpha}(t)}{U} + K^2A_3^*(K)\alpha(t) + K^2A_4^*(K)\frac{z(t)}{U}\right]
\end{aligned}
\tag{23}
$$

Height flutter derivatives H_p^* and A_p^* (with p=1,…,4) are obtained experimentally as function of the reduced velocity U_{red}, with $U_{red}=U/fB=2\pi/K$ (being K the reduced frequency) and introduced in a time-domain framework. $q = 0.5\rho U^2$ is the dynamic pressure.

It is worth to mention at this point that different conventions can be found in literature, with regard to self-excited loads. By the way, discrepancies can be related to the versus assumed as positive for vertical displacements and lift force. The positive aerodynamic moment is ordinarily clockwise (nose-up), as the positive rotation angle. The following three different conventions are recalled, with respect to the sign of aeroelastic derivatives, to allow one the conversion of the available experimental values:

1. original Scanlan's convention: lift force positive downwards and vertical displacement positive downwards, used, for instance, by Scanlan and Tomko (1971) and by Matsumoto et al. (1996);
2. "experimental" convention: lift force positive upwards and vertical displacement positive upwards, followed, for example, by Dyrbye and Hansen (1996);
3. thin airfoil's convention, i.e. convention common to aeronautical sciences: lift force positive upwards and vertical displacement positive downwards.

With reference to the original Scanlan's convention, the lift-related flutter derivatives H_p^* (p=1,…,4) have to be all changed in sign to obtain a consistent representation in thin airfoil's convention. In order to achieve the same values of the experimental convention, all mixed derivatives must change sign.

For a thin airfoil, it is possible to express the different contribution to the self-excited loads by means of theoretical flutter derivatives, calculated from the Theodorsen's circulation function (Figure 5). In particular, relationships (24) and (25) hold true.

$$
\begin{aligned}
H_1^*(K) &= \pi\left[\frac{2F(k)}{K}\right] \\
H_2^*(K) &= \pi\left[\frac{2G(k)}{K^2} + \frac{1}{2K} + \frac{F(k)}{2K}\right] \\
H_3^*(K) &= \pi\left[\frac{2F(k)}{K^2} - \frac{G(k)}{2K}\right] \\
H_4^*(K) &= \pi\left[-\frac{1}{2} - \frac{2G(k)}{K}\right]
\end{aligned}
\tag{24}
$$

$$
\begin{aligned}
A_1^*(K) &= \pi\left[\frac{F(k)}{2K}\right] \\
A_2^*(K) &= \pi\left[-\frac{1}{8K}+\frac{G(k)}{2K^2}+\frac{F(k)}{8K}\right] \\
A_3^*(K) &= \pi\left[\frac{1}{64}+\frac{F(k)}{2K^2}-\frac{G(k)}{8K}\right] \\
A_4^*(K) &= \pi\left[-\frac{G(k)}{2K}\right]
\end{aligned}
\tag{25}
$$

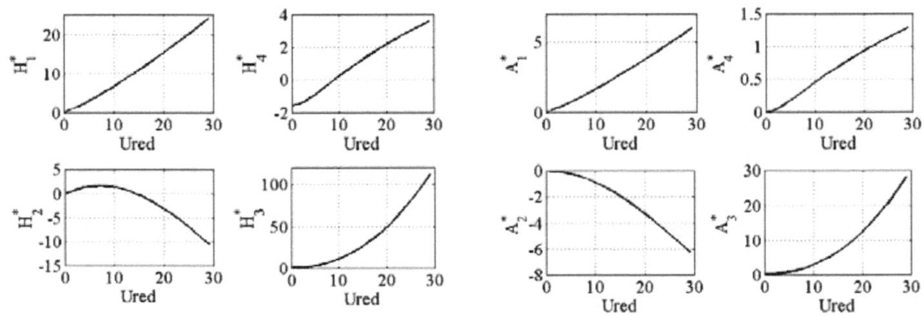

Figure 5. Theoretical flutter derivatives corresponding to Theodorsen's circulation function.

5.3 Self-excited forces on a 2 DoFs section: the approach in time domain

The flutter derivatives offer the great advantage of being extracted through a quite consolidate approach, but they are not well suited for time domain simulations, being expressed as a function of frequency. A full time-domain approach can help to represent eventual nonlinearities in the mechanical system as well as the unsteadiness of the fluid load on the body.

Among time-domain approaches, an interesting one is represented by the indicial theory. Indicial theory is noteworthy in aeronautical field, while his extension to bridge engineering represents a quite recent challenge. Its application, in fact, needs to be properly calibrated and discussed.

The counterpart in time domain of the frequency formulation of Eq. (21) for the unsteady lift on a thin airfoil is given by

$$
F_L(s) = 2\pi\rho U b w \phi(s) \tag{26}
$$

where the function $\phi(s)$ is called Wagner's function and is expressed as a function of the dimensionless time $s=Ut/b$ (that is the inverse of the reduced frequency). In particular Wagner's function describes the evolution of self-excited circulatory lift and moment for an infinitesimal variation of the angle of attack of the section (Figure 6). A common approximation of Wagner's function is due to Jones (1940), as the sum of exponential functions practical for Fourier-transforming, i.e. $\phi(s)=1-0.165e^{-0.0455}-0.335e^{-0.30}$. More general functions are proposed by Bisplinghoff et al. (1955), also for the description of the aerodynamic moment and in order to distinguish the weight of the effects of the different components of the motion on the self-excited loads.

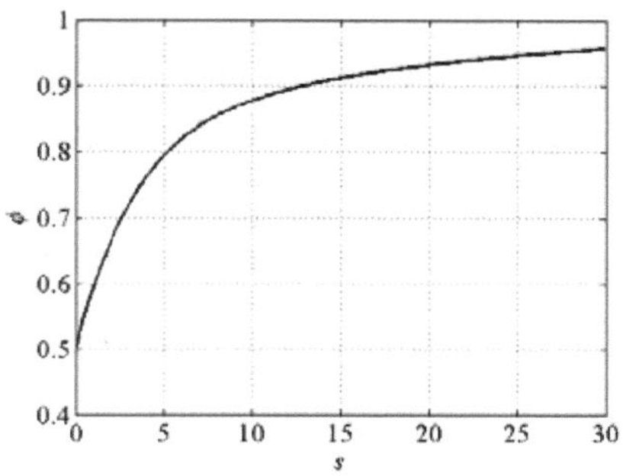

Figure 6. Wagner's function.

When a generalized time history is considered, a superposition of elementary lift forces can be accounted for and the total lift force F_L can be written in terms of a convolution integral (Fung, 1969), namely

$$F_L(s) = 2\pi\rho Ub \int_{-\infty}^{s} \phi(s-\tau)w'(\tau)d\tau \tag{27}$$

The prime denotes differentiation with respect to s. The value 2π indicates the slope of the theoretical aerodynamic coefficient C_L, while the slope of the aerodynamic moment is equal to $\pi/4$.

Analogously, a representation of self-excited actions completely in time domain can be provided for bridge deck cross-sections, without explicit dependence on frequency, if appropriate functions are defined. Such functions describe, according to thin airfoil theory, the time development of sectional aeroelastic forces due to instantaneous infinitesimal structural motions and are usually referred to as indicial functions. If arbitrary motions are considered, total action can be calculated via convolution, as in Eq. (27).

The first relevant work suggesting the possibility of using indicial functions for bridges is due to Scanlan et al. (1974), with an expression of indicial functions based on an exponential approximation like the one of Wagner's function. Other main important historical references are due to Lin and Ariaratnam (1980) and Lin and Li (1993). The indicial formulation for self-excited forces has been further discussed by Borri and Höffer (2000) and by Borri et al. (2002). In the specific formulation proposed in (Costa and Borri, 2006) for a two-dimensional bridge deck section, the aeroelastic forces are expressed as follows:

$$F_{Lse}(s) = qBC'_L\left[\Phi_{L\alpha}(0)\alpha(s) + \Phi_{Lz}(0)z'(s) + \int_0^s \Phi'_{L\alpha}(s-\tau)\alpha(\tau)d\tau + \int_0^s \Phi'_{Lz}(s-\tau)z'(\tau)d\tau\right]$$
$$M_{ase}(s) = qB^2C'_M\left[\Phi_{M\alpha}(0)\alpha(s) + \Phi_{Mz}(0)z'(s) + \int_0^s \Phi'_{M\alpha}(s-\tau)\alpha(\tau)d\tau + \int_0^s \Phi'_{Mz}(s-\tau)z'(\tau)d\tau\right] \quad (28)$$

Self-excited forces are calculated via convolution, accounting for motion histories expressed as series of infinitesimal step-wise increments. Each indicial function describes the non-stationary evolution in time of the load due to a unit step change in the angle of attack. Four indicial functions are included in this case, with subscripts identifying, respectively, the load component (L=lift or M=aerodynamic moment) and the motion component that experiences the step change. Torsional displacement and vertical velocity contribute in this model to the definition of the angle of attack.

Indicial functions for bridges can be defined as Wagner-like functions, i.e. as the superposition of m exponential groups

$$\Phi_{il}(s) = 1 - \sum_{k=1}^{m} a_{ilk}\exp(-b_{ilk}s) \quad (29)$$

Subscripts i and l identify, respectively, the load and the motion components.

Indicial functions can be calculated from aeroelastic derivatives or directly identified in wind tunnel tests. The identification from flutter derivatives is performed by comparison of load model of Eqs. (28) with the Scanlan's relationships of Eqs. (23). Relationships obtained are recalled in Eqs. (30) and (31).

$$\frac{2\pi}{U_{red}}H_1^* = \frac{dC_L}{d\alpha}\left[1 - \sum_i a_{iLy}\frac{\pi^2}{b_{iLy}^2U_{red}^2 + \pi^2}\right]$$
$$\frac{2}{U_{red}^2}H_4^* = \frac{dC_L}{d\alpha}\left[\sum_i a_{iLy}\frac{b_{iLy}}{b_{iLy}^2U_{red}^2 + \pi^2}\right]$$
$$\frac{4\pi}{U_{red}^3}H_2^* = \frac{dC_L}{d\alpha}\left[\sum_i - a_{iL\alpha}\frac{b_{iL\alpha}}{b_{iL\alpha}^2U_{red}^2 + \pi^2}\right] \quad (30)$$
$$\frac{4\pi^2}{U_{red}^2}H_3^* = \frac{dC_L}{d\alpha}\left[1 - \sum_i a_{iL\alpha}\frac{\pi^2}{b_{iL\alpha}^2U_{red}^2 + \pi^2}\right]$$

$$\frac{2\pi}{U_{red}} A_1^* = \frac{dC_M}{d\alpha}\left[1 - \sum_i a_{iMy} \frac{\pi^2}{b_{iMy}^2 U_{red}^2 + \pi^2}\right]$$

$$\frac{2}{U_{red}^2} A_4^* = \frac{dC_M}{d\alpha}\left[\sum_i a_{iMy} \frac{b_{iMy}}{b_{iMy}^2 U_{red}^2 + \pi^2}\right]$$

$$\frac{4\pi}{U_{red}^3} A_2^* = \frac{dC_M}{d\alpha}\left[\sum_i - a_{iM\alpha} \frac{a_{iM\alpha}}{b_{iM\alpha}^2 U_{red}^2 + \pi^2}\right] \tag{31}$$

$$\frac{4\pi^2}{U_{red}^2} A_3^* = \frac{dC_M}{d\alpha}\left[1 - \sum_i a_{iM\alpha} \frac{\pi^2}{b_{iM\alpha}^2 U_{red}^2 + \pi^2}\right]$$

As example, flutter derivatives for two sample rectangular cross-sections and their relevant indicial functions are plotted in Figure 7 and Figure 8 (for the rectangular section *B*/*D*=12.5), and in Figure 9 and Figure 10 (for the rectangular section *B*/*D*=5). In particular, for rectangular section *B*/*D*=5, moment indicial functions take a special shape, related to the bluffness of the section. Two exponential groups are required in this case to achieve a good approximation of flutter derivatives, against the one necessary for streamlined sections [for details, see (Costa and Borri, 2006)].

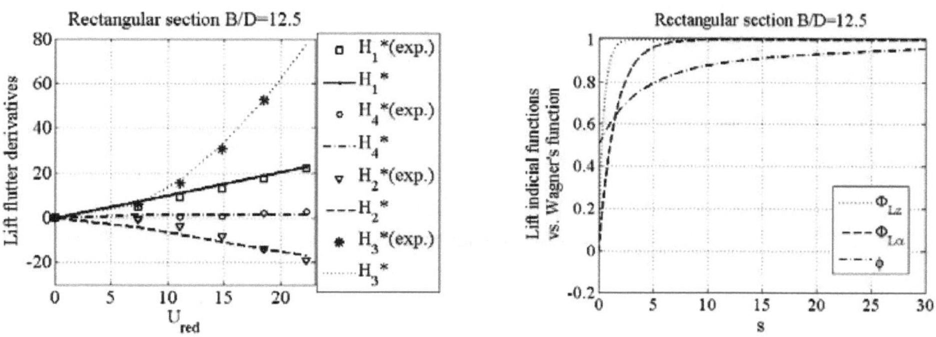

Figure 7. Lift flutter derivatives and relevant indicial functions for a rectangular section *B*/*D*=12.5.

Moreover, by the observation of flutter derivatives, the possibility of arising of classical or torsional flutter can be evinced. In particular, the fact that the coefficient A_2^* tends to become positive as the wind speed increases represents a clear sign of a section prone to torsional flutter (see Figure 10, left).

References to the aeroelastic behavior of rectangular cross-sections are furnished also by Bartoli and Righi (2006).

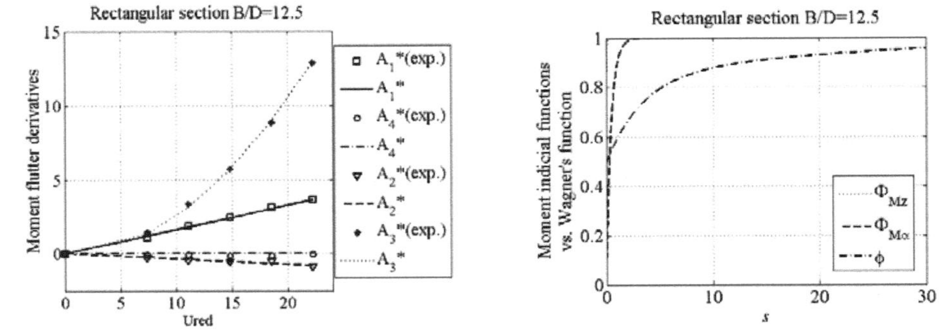

Figre 8 Moment flutter derivatives and relevant indicial functions for a rectangular section *B*/*D*=12.5.

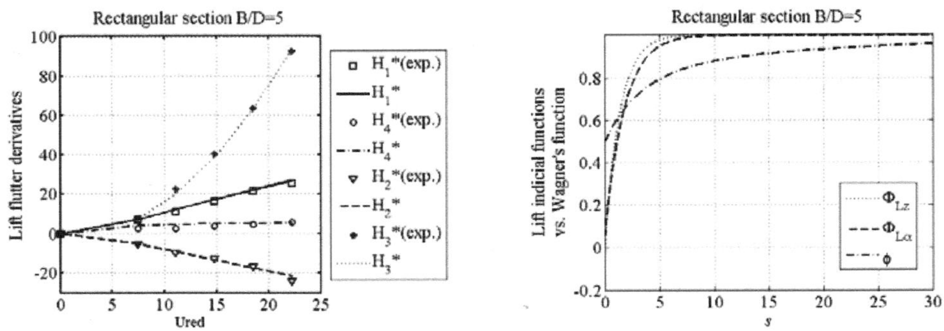

Figre 9 Lift flutter derivatives and relevant indicial functions for a rectangular section *B*/*D*=5.

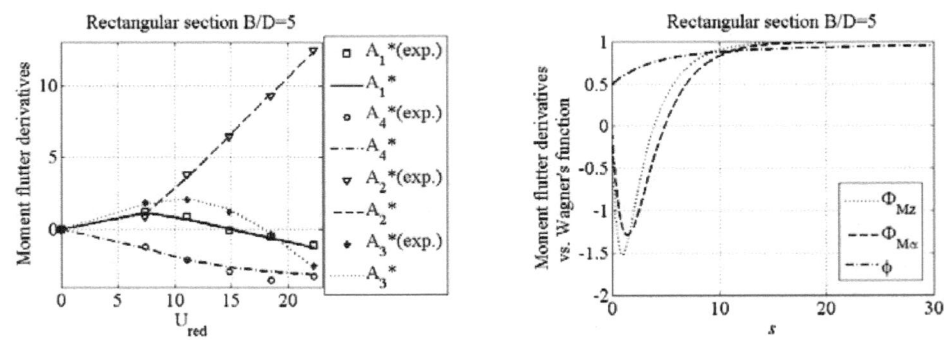

Figre [illegible] Moment flutter derivatives and relevant indicial functions for a rectangular section *B*/*D*=5.

5.4 Calculation of critical flutter condition for a 2 DoFs section in smooth flow (frequency domain)

At classical flutter, it is assumed that coupling between vertical and torsional frequencies is occurred. This fact means that the oscillations of both DoFs have the same frequency, in general different from the natural one and denoted as flutter frequency. The solution of the equations of motion in terms of displacements is, therefore, of this kind:

$$
\begin{aligned}
z &= z_0 e^{i\omega t} \\
\alpha &= \alpha_0 e^{i\omega t}
\end{aligned} \tag{32}
$$

If the solution (32) is introduced in the equations of motion with the wind load assumed completely represented by Eqs. (23), a homogeneous system of the following type is obtained,

$$
\begin{bmatrix} A_1 & A_2 \\ A_3 & A_4 \end{bmatrix} \begin{bmatrix} z_0 \\ \alpha_0 \end{bmatrix} = \begin{bmatrix} 0 \\ 0 \end{bmatrix} \tag{33}
$$

where **A**=[(A_1; A_2): (A_3; A_4)] is a square matrix. This system can be solved to obtain non-trivial solutions, then searching combinations which assure Det(**A**)=0. An unknown variable $X=\omega/\omega_z$ is assumed and a complex equation of the fourth order is obtained. If its real and imaginary parts are separated, two real equations of the fourth order are obtained,

$$
\begin{aligned}
R_4^4 X^4 + R_3^3 X^3 + R_2^2 X^2 + R_1 X + R_0 &= 0 \\
I_4^4 X^4 + I_3^3 X^3 + I_2^2 X^2 + I_1 X + I_0 &= 0
\end{aligned} \tag{34}
$$

whose coefficients have the following expressions:

$$
\begin{aligned}
R_4 &= 1 + \frac{A_3^*}{2\gamma_l} + \frac{H_4^*}{2\gamma_m} - \frac{H_1^* A_2^*}{4\gamma_l \gamma_m} + \frac{A_1^* H_2^*}{4\gamma_l \gamma_m} - \frac{H_3^* A_4^*}{4\gamma_l \gamma_m} + \frac{A_3^* H_4^*}{4\gamma_l \gamma_m} \\
R_3 &= \nu_\alpha \frac{\omega_\alpha}{\omega_z} \frac{H_1^*}{\gamma_m} + \nu_z \frac{A_2^*}{\gamma_l} \\
R_2 &= -1 - \left(\frac{\omega_\alpha}{\omega_z}\right)^2 - 4\nu_z \nu_\alpha \frac{\omega_\alpha}{\omega_z} - \frac{A_3^*}{2\gamma_l} - \frac{\omega_\alpha}{\omega_z} \frac{H_4^*}{2\gamma_m} \\
R_1 &= 0 \\
R_0 &= \left(\frac{\omega_\alpha}{\omega_z}\right)^2
\end{aligned} \tag{35}
$$

and

$$
\begin{aligned}
I_4 &= \frac{A_2^*}{2\gamma_I} + \frac{H_1^*}{2\gamma_m} + \frac{H_1^* A_3^*}{4\gamma_I\gamma_m} - \frac{A_1^* H_3^*}{4\gamma_I\gamma_m} - \frac{H_2^* A_4^*}{4\gamma_I\gamma_m} + \frac{A_2^* H_4^*}{4\gamma_I\gamma_m} \\
I_3 &= -2\nu_z - 2\nu_\alpha \frac{\omega_\alpha}{\omega_z} - \nu_z \frac{A_3^*}{\gamma_I} - \nu_\alpha \frac{\omega_\alpha}{\omega_z}\frac{H_4^*}{\gamma_m} \\
I_2 &= -\frac{A_2^*}{2\gamma_I} - \left(\frac{\omega_\alpha}{\omega_z}\right)^2 \frac{H_1^*}{2\gamma_m} \\
I_1 &= 2\nu_z \left(\frac{\omega_\alpha}{\omega_z}\right)^2 + 2\nu_\alpha \frac{\omega_\alpha}{\omega_z} \\
I_0 &= 0
\end{aligned}
\qquad (36)
$$

Mass and inertial properties of the section are resumed by the dimensionless quantities

$$
\gamma_m = \frac{m}{\rho B^2}, \quad \gamma_I = \frac{I}{\rho B^4} \qquad (37)
$$

The equation corresponding to the imaginary part is of the third order, being the term I_0 equal to zero.

The solutions of Eqs. (34) are indicated, respectively, as X_r and X_i and are evaluated as function of the reduced velocity U_{red}. The first solution common to the two equations corresponds to the frequency at which classical flutter occurs (see, as example, Figure 11).

At the classical flutter, for the two DoFs system, vertical and torsional displacements have the same frequency and show constant amplitude in time, while in the case of torsional flutter only the torsional damping vanishes.

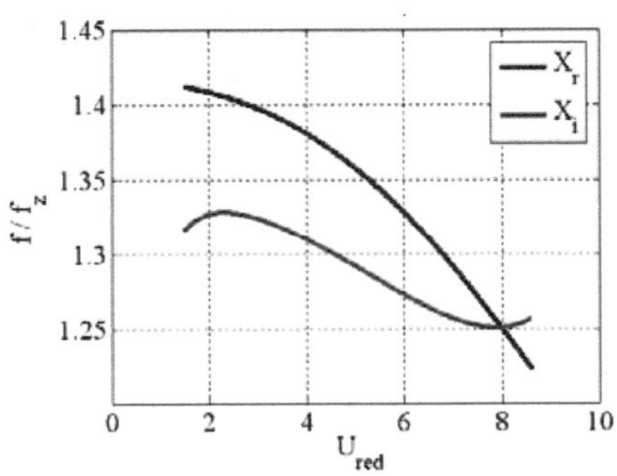

igre 1 Solution of the real and imaginary equations for evaluation of flutter condition.

6 Aeroelastic phenomena in turbulent boundary layer flow

Galloping, torsional divergence and flutter have been discussed through load models defined in laminar flow. Turbulence effects are usually included by adding a buffeting load component dependent on turbulence characteristics.

Wind field is modeled, in fact, as a time-space variable field

$$\mathbf{U}_{wind}(M,t) = \mathbf{U}(M) + \mathbf{U}'(M,t) \tag{38}$$

where t is the time, $\mathbf{U}$ is the mean wind speed and $\mathbf{U}'$ represents the turbulent perturbation, acting at a general point M of coordinates $X(M)$, $Y(M)$ and $Z(M)$.

The mean flow is directed along the X axis, while the significant components of the turbulence field are identified by u, v and w. They are directed, respectively, along X, Y and Z axis. The resultant wind field is then given by

$$\begin{aligned} U_{windX}(M,t) &= U(M) + \mathsf{u}(M,t) \\ U_{windY}(M,t) &= \mathsf{v}(M,t) \\ U_{windZ}(M,t) &= \mathsf{w}(M,t) \end{aligned} \tag{39}$$

The mean wind velocity U may be expressed by a scalar value (usually the value adopted in the definition of the dynamic pressure appearing in steady, quasi-steady and unsteady loads), while turbulence components are defined through an assigned spectrum.

Buffeting forces for bridge cross-sections are commonly treated under two main assumptions: 1) *quasi-steady theory* (i.e. the turbulence load is not depending on frequency) and 2) *strip assumption* (the load on each bridge deck segment is not influenced by the loads on the neighboring segments).

If only vertical turbulence is taken into account, buffeting lift and moment can be expressed as (Simiu and Scanlan, 1996)

$$\begin{aligned} F_{Lb}(t) &= qB\left(\frac{dC_L}{d\alpha} + C_D\right)\frac{\mathsf{w}}{U} \\ M_{ab}(t) &= qB^2\frac{dC_M}{d\alpha}\frac{\mathsf{w}}{U} \end{aligned} \tag{40}$$

Several attempts to remove quasi-steady hypothesis and strip assumption can be found in literature, especially in frequency-domain, in order to attain a more realistic representation of wind-bridge interaction. First of all, Davenport (1961) tried to correct the quasi-steady approach in frequency domain with the definition of buffeting loads depending on frequency through aerodynamic admittance functions. By following that intuition, buffeting lift and moment defined in Eqs. (40) can be corrected by aerodynamic admittance functions $\chi_{ij}(k)$, where subscripts i and j identify, respectively, the aerodynamic load component and the turbulence component.

In this case, due to the fact that only one turbulent component is accounted for ($j \equiv \mathsf{w}$), the subscript j is omitted. Buffeting forces take, therefore, the following expressions

$$F_{Lb}(t) = qB\left(\frac{dC_L}{d\alpha} + C_D\right)\chi_L \frac{\mathrm{w}}{U}$$
$$M_{ab}(t) = qB^2 \frac{dC_M}{d\alpha}\chi_M \frac{\mathrm{w}}{U} \tag{41}$$

with admittance functions $\chi_L(k)$ and $\chi_M(k)$. Admittance functions, similarly to flutter derivatives, are expressed as a function of the reduced frequency $k=K/2$ or K and can be measured in wind tunnel. Otherwise, if wind tunnel measurements are not available, a reference admittance function can be adopted, even if theoretically defined only for a thin airfoil, i.e. the Sears' function $\chi(k)$ [see Fung (1969)].

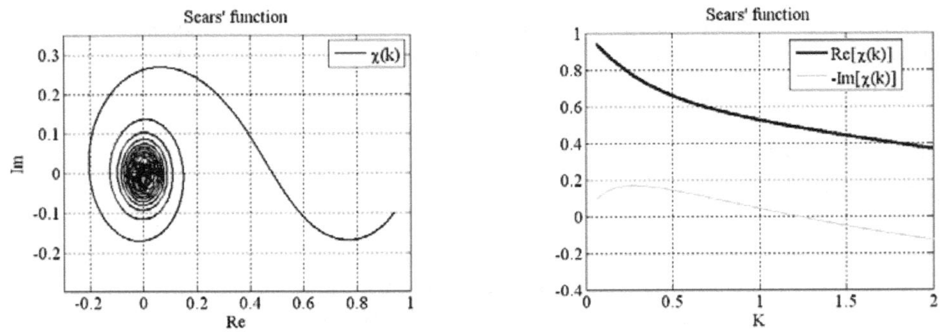

Figure 12. Sears' function.

If a full time-domain formulation is required, Küssner's function $\psi(s)$ should be adopted rather than Sears' function, in order to describe the dimensionless lift developing on an airfoil due to a sharp-edged gust striking the leading edge of the airfoil at $s=0$. An approximation of Küssner's function in incompressible flow, for $s \geq 0$, is given by $\psi(s) \cong 1\text{-}0.50\exp(-0.130s)\text{-}0.50\exp(-s)$.

One can observe the analogy between the approximation proposed for the Küssner's function and the one proposed for the Wagner's function. Therefore, it is possible to think to Küssner-type functions of the following kind to represent the counterpart of aerodynamic admittance function in time domain:

$$\Psi_{iw}(s) = \Psi_i(s) = 1 - \sum_{h=1}^{n} a_{ih}\exp(-b_{ih}s) \tag{42}$$

In analogy with the expressions of Eqs. (28) provided for self-excited forces, buffeting loads could be, therefore, expressed as following

$$F_{Lb}(s) = qB\frac{dC_L}{d\alpha}\left[\Psi_L(0)w(s) + \int_0^s \Psi'_L(s-\tau)w(\tau)d\tau\right]$$
$$M_{ab}(s) = qB^2\frac{dC_M}{d\alpha}\left[\Psi_M(0)w(s) + \int_0^s \Psi'_M(s-\tau)w(\tau)d\tau\right] \quad (43)$$

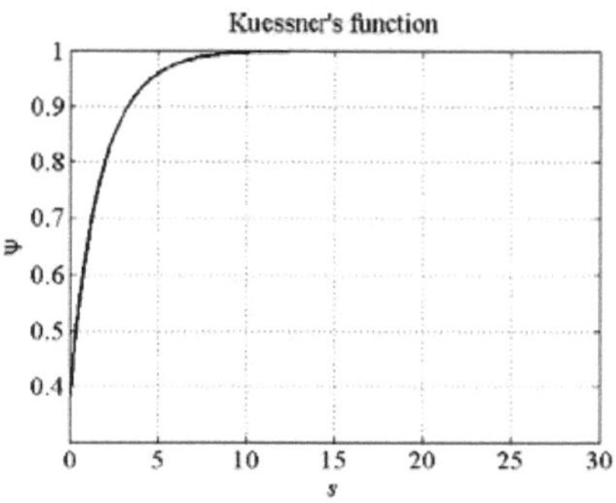

Figure 13. Küssner's function.

Indicial functions Ψ_L and Ψ_M represent the counterpart in time domain of admittance functions χ_L and χ_M used in Eqs. (41). If a *working hypothesis* is made, that the indicial functions calculated from flutter derivatives for self-excited forces *differ not too much* from the indicial functions calculated from admittance functions, the final expression for buffeting forces becomes

$$F_{Lb}(s) = qB\frac{dC_L}{d\alpha}\left[\Phi_{Lz}(0)w(s) + \int_0^s \Phi'_{Lz}(s-\tau)w(\tau)d\tau\right]$$
$$M_{ab}(s) = qB^2\frac{dC_M}{d\alpha}\left[\Phi_{Mz}(0)w(s) + \int_0^s \Phi'_{Mz}(s-\tau)w(\tau)d\tau\right] \quad (44)$$

Physically, this assumption corresponds to the fact of assuming a similar effect in terms of forces on the section if a section is moving sinusoidally in a uniform flow or is steady in a sinusoidal flow. In order to check the validity of such hypothesis, an *a posteriori* check is realized, by calculating admittance functions corresponding to available indicial functions. In order to calculate these admittances, the analogy with thin airfoil is accounted for, considering the expression of Sears' function as a combination of Bessel functions of first and third kind J_0(k) and J_1(k) and Theodorsen's circulation function *C*(*k*) [see Eq. (22)], i.e.

$$\chi(k) = C(k)J_0(k) + iJ_1(k)[1 - C(k)] \quad (45)$$

The definition of equivalent admittance functions is, therefore, modeled with the same analytical structure of Eq. (43). The difference relies on the circumstance that equivalent Theodorsen's functions C_{Leq} and C_{Meq} are defined here for lift and moment:

$$\chi_L(k) = C_{L\text{eq}}(k)J_0(k) + iJ_1(k)[1 - C_{L\text{eq}}(k)]$$
$$\chi_M(k) = C_{M\text{eq}}(k)J_0(k) + iJ_1(k)[1 - C_{M\text{eq}}(k)] \quad (46)$$

Real and imaginary parts of equivalent Theodorsen's functions are indicated, respectively, as F_{eq} and G_{eq}, and can be calculated as

$$F_{\text{eq}}(k) = 1 - \sum_{h=1}^{n} \frac{a_h k^2}{b_h^2 + k^2}$$
$$G_{\text{eq}}(k) = -\sum_{h=1}^{n} \frac{a_h b_h k}{b_h^2 + k^2} \quad (47)$$

with coefficients a_h and b_h taken from lift indicial functions Φ_{Lz} and Φ_{Mz}.

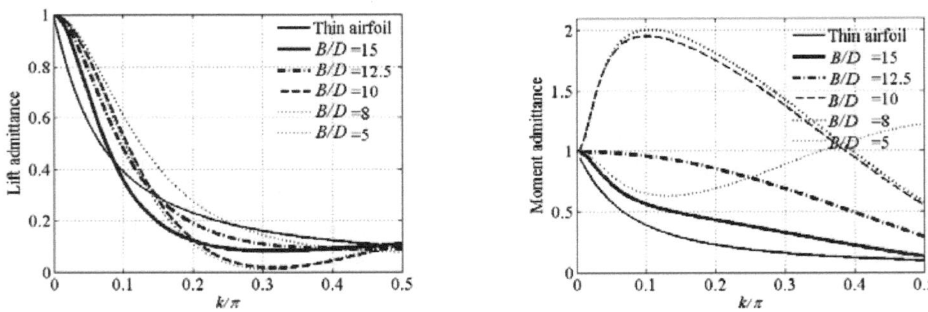

Figure 14. Lift and moment admittance functions for rectangular sections.

Comparison of admittances calculated from equivalent Theodorsen's functions with experimental or literature results can be performed, as shown in Figure 14 and Figure 15. Indicial functions have been calculated, in this case, from available flutter derivatives and admittances have been further estimated. It can be remarked a quite good accord of calculated and experimental admittances, both for streamlined and bluff rectangular sections (Costa, 2006).

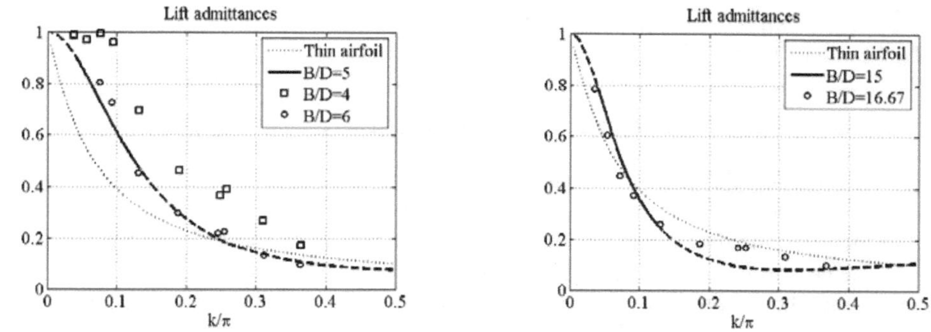

Figure 15. Comparison of lift admittance function calculated by means of indicial functions for a rectangular section with B/D=5 and B/D=15 with Sears' function and experimental results (Jancauskas and Melbourne, 1986).

7 Vortex shedding

Around a bluff body, the flow separates, independently on the turbulence embedded in the flow itself. Pressure fluctuations can be distinguished by those generating by the flow fluctuations in the approaching flow.

In particular, the regular vortex shedding into the wake of a bluff body results from the rolling-up of the separating shear layers alternately one side, then the other, and occurs to bodies of all cross-sections. A regular pattern of decaying vortices, known as the von Kármán vortex "street", appears in the wake. Turbulence in the approaching flow tends to make the shedding less regular, but the strength of the vortices are maintained, or even enhanced. Vibration of the body may enhance the vortex strength, and the vortex shedding frequency may change to the frequency of vibration, in a phenomenon known as *lock-in*.

One of the most important studies concerning lock-in phenomenon is due to Vickery (1983). Its model, concerning bodies with circular cross-sections and their across-wind response, has been extended to full scale problems by Basu and Vickery (1983).

In circular cylinders the shape of the wake is hardly influenced by the non-dimensional Reynolds number Re, defined as the ration between the fluid velocity U, the characteristic dimension of the body D (the diameter in the case of the circular cylinder) and the kinematical viscosity of the fluid itself ν (the air viscosity has the value of $1.5\text{x}10{-}5\ \text{m}^2/\text{s}$) ($=UD/\nu$). In bodies with sharp edges, the dependency on Reynolds number is slighter, because of the strong effect of the body geometry in vortex separation.

The frequency of vortex trail is identified by the Strouhal number fD/U, which corresponds to the reduced frequency k. Strouhal numbers are typically about 0.2 for circular cylinders and 0.1 for bridge decks.

Vibrations due to vortex shedding in bridge aerodynamics are excited at low speeds, normally for winds acting in the direction normal to the bridge span. Flow separation gives rise to harmonic loads in the direction normal to the flow, typically affecting the bending modes of vibration. Stability is not compromised, but relevant comfort and fatigue problems can arise [see the case of the Great Belt Link, (Larsen, 2003)].

8 Aerodynamic and aeroelastic phenomena on large roofs

Aerodynamic and aeroelastic phenomena affect not only bridges but also large roofs, for which wind is often the most influent load in the design. The aerodynamic effects and, sometimes, the interaction between the roof motion and the living loads can be important during erection stages and in service conditions. Often, special geometrical features of the structures require appropriate studies to define the effective wind load and to evidence critical situations. Wind tunnel tests on scaled models represent a powerful tool to detect such situations. Areas of separation and re-attachment of flows can be detected, as well as generic circulation around the structure.

The wind action on the roof is often represented by means of a map of pressures, in particular by a map of peak and average pressure coefficients. Pressure coefficients c_p are defined as the ratio between the effective pressure measured in correspondence of the generic pressure tap and the kinetic pressure at the reference height

$$c_p(t) = \frac{p(t) - p_0}{0.5 \rho U^2} \tag{48}$$

The effective pressure is defined as the difference between $p(t)$, the instantaneous value of the pressure, and p_0, the reference pressure of the undisturbed flow. U is the reference wind speed at the roof level.

The map of pressure coefficients is built up following the subsequent general procedure.

First of all, an appropriate wind field should be reproduced within the wind tunnel, representative of the wind profile characteristic of the site. The length scale of the longitudinal turbulence component is defined as

$$L_u = U(z) \int_0^{\infty} \frac{R_{uu}(\tau)}{\sigma_u^2} d\tau \tag{49}$$

with R_{uu} corresponding to the autocorrelation function and σ_u to the standard deviation. To fit this parameter, artificial roughness and special devices lying on the floor of the "developing fetch" are adopted. As example, in Figure 16, the CRIACIV wind tunnel is shown, with the Couhinan devices and the "spires" at the inlet of the wind tunnel. Moreover, the "carpet" of little cubes is positioned to generate the boundary layer profile. The dimension of the cubes is obtained after several calibration tests and it has to be disposed over the necessary length to develop the target profile.

After a satisfactory reproduction of the wind field within the wind tunnel, the scaled-model is placed and instrumented. The roof is divided into elementary areas and, for each elementary influence area, the time series of the pressure coefficient are recorded. By this way, it is possible to obtain the whole history pressure of the selected area.

Measurements can be done at the outer and at the inner surface. The net pressure coefficient can be defined as

$$c_{pnet}(t) = c_{psup}(t) - c_{pinf}(t) \tag{50}$$

where c_{psup} and c_{pinf} are the results of the measurements for both sides of the roof.

In the following, two case studies are summarized, describing the main results of the wind tunnel campaign: 1) the new Delle Alpi stadium (Turin, Italy) and 2) the Karaiskaki stadium at Piraeus, in Greece. In particular, interference effects motivated the study on Delle Alpi stadium, while structural response was the object of interest in the Karaiskaki case.

Figure 16. Wind tunnel and devices to generate target wind profile stadium.

8.1 The new Delle Alpi stadium (Turin, Italy)

The new design for the Delle Alpi stadium consists in the realization of a new stand within the old structure. The old roof, made by a radial tensile structure, will coexist with the new one, as shown in Figure 17. Interference effects could play an important role, both with the lateral structure and with the pre-existent roofs. For investigating these aspects, a scaled model has been built and studied at the Boundary-Layer Wind Tunnel (BLWT) of *CRIACIV*. The scaled model has been instrumented according to the procedure described in the previous paragraph. The relevant map of the pressure tabs is shown in Figure 18.

Several directions of incoming winds have been considered for construction of pressure maps. In this case, 16 wind directions have been taken into account [for details, see (Borri and Biagini, 2005)]. Main results presented here concerns the comparison between the two configurations of the stadium, the one with the old roof and the one without (Figure 19, left and right). The maps of the pressure coefficients for both conditions are shown in Figure 20.

Once obtained the c_p time histories relevant to all directions measured, the further step is to perform a statistical analysis of the processes. In fact, the samples collected during the wind tunnel tests refer to a short wind storm (30 seconds). The maximum and the minimum values of the single series (for each pressure tap) are not significant enough. According to the procedure proposed by Cook and Maine (1980), suggested also by Eurocode 1, each time history for the pressure coefficients is subdivided in 11 intervals of 2.4-seconds duration. This corresponds to a full scale 10-minutes storm, i.e. the reference duration for the evaluation of the maximum and the minimum values. The 11 extreme values (max or min) are fitted with a Gumbel function and by means of the so-called BLUE correction [Best Linear Unbiased Estimators, see (Box et al., 1994)], then the design value of the pressure coefficients are estimated. These last coefficients are referred to as design pressure coefficients c_{pd} and are considered as the pressure coefficients which, in the extreme values distribution, have the probability to be exceeded of 22%.

Figure 17. New Delle Alpi stadium (Turin, Italy).

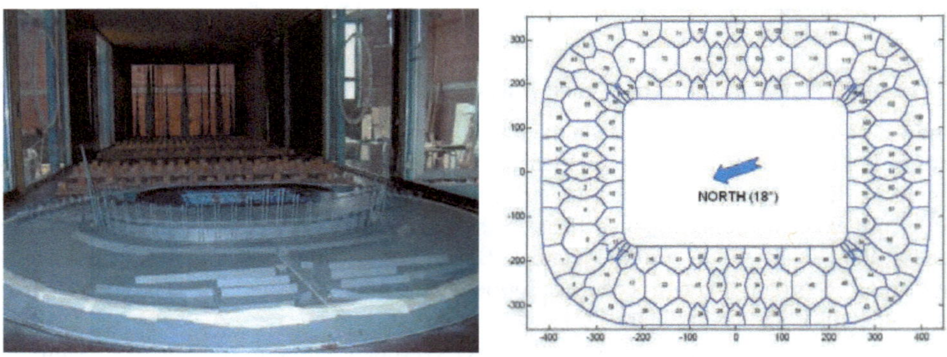

Figure 18. Wind tunnel model and pressure tap scheme for the Delle Alpi stadium (Turin, Italy).

Comfort evaluation. The identification of the circulation around the roof can be relevant by the structural point of view but also to estimate the comfort of the spectators on the stands. As an example of comfort prediction test, a special study on the local "climate/ventilation" has been carried out on the stands of the Turin stadium. For such a test, a section model of the stadium has been obtained from the complete one, in order to analyze the local conditions of the flow motion (Figure 22). The *PIV* (Particle Image Velocimetry) technique has been applied in the investigation, which allows the characterization of vortex structures near the stands by means of the display of the velocity field on a grid over the investigated plane area (red area in Figure 22). A wind velocity-based comfort scale has been defined, for the classification as "comfortable" of a particular site. Common criteria for the definition of the comfort in pedestrian areas are based on the frequency of overcoming a fixed limit values in the wind velocity. Figure 22 shows the mean vectorial maps of the mean wind velocity and the corresponding streamlines for the two analyzed configurations, respectively with and without the old roofing system.

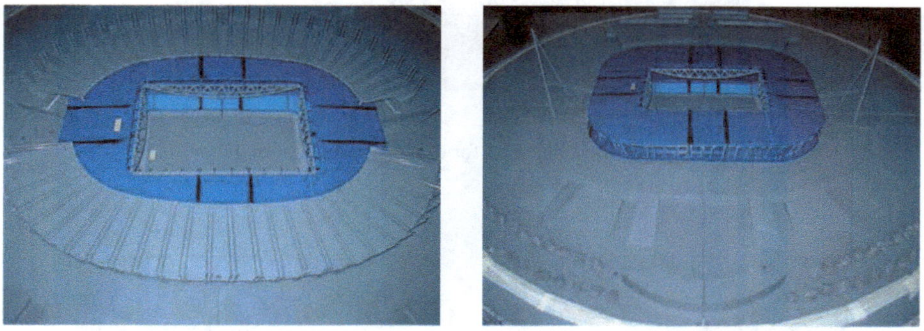

Figure 19. View of two configuration of the Delle Alpi model: with the old roof (left) and without (right).

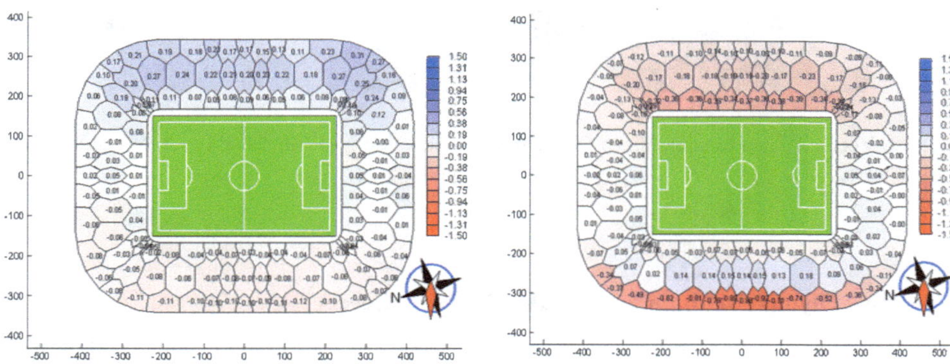

Figure 20. View of the pressure coefficient distribution on the Delle Alpi roof, with the old roof (left) and without (right).

The map of the mean velocity isolines (percentage values) has to be multiplied for the design velocity in order to obtain the resulting map of the actual velocity. The choice of the design velocity is often done in function of the desired performance level. In this case, the reference wind velocity U_{ref} at the roof level has been assumed equal to 3.64 m/s and the motion fields have been recorded obtaining the corresponding isovelocity maps, as a percentage of the reference velocity

(Figure 23). With these maps, once the incoming design velocity value has been chosen, it become possible to detect critical zones of discomfort.

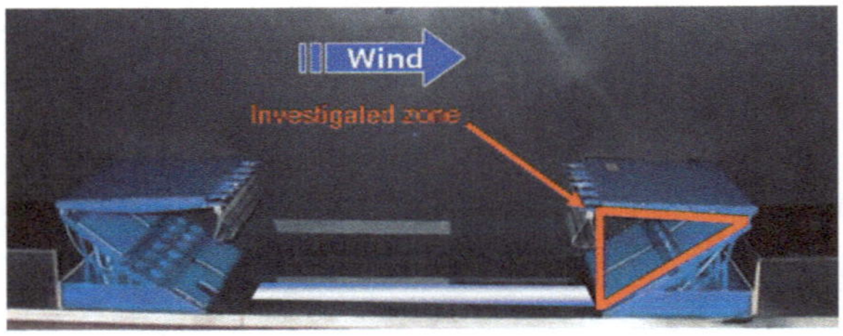

Figure 21. Section model used for the PIV visualization technique.

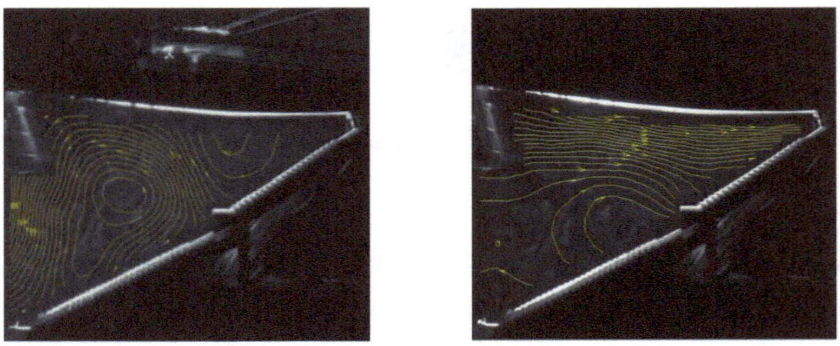

Figure 22. Vectorial map of streamlines for the mean wind velocity (With the old roof, on the left - Without the old roof, on the right).

8.2 The new Olympic stadium Karaiskaki (Piraeus, Greece)

The Karaiskaki roof at Piraeus is 206.5 m long and 167.55 m wide and it is composed by 14 cantilevered structures (35 m span) which sustain the roof panels connected to an internal trussed beam (Figure 24, on the left).

A rigid model of the structure (Figure 24, on the right) has been tested at the *BLWT* of *CRIACIV* (Prato, Italy), in order to define the pressure field acting on the roof in different wind conditions and to evaluate the structural response (Biagini et al., 2006).

As already indicated, the first step of the work is devoted to the characterization of the flow in the *BLWT* through simulation of the oncoming wind profile with appropriate mean wind and turbulence characteristics. The second step consists of the pressure measurements on the model of the roof, which is equipped with an adequate number of pressure taps. The measurements of the pressure field developing on the roof are performed at height different mean wind incident angle. An example of relevant distribution of the pressure coefficients is shown in Figure 25 (the red arrow in the wind rose indicates the incoming wind direction). The time histories of the local pressures are recorded at all the instrumented pressure taps, that is the mean, the standard deviation, the maximum and minimum values are measured, together with other significant parameters of the distribution and spectral densities.

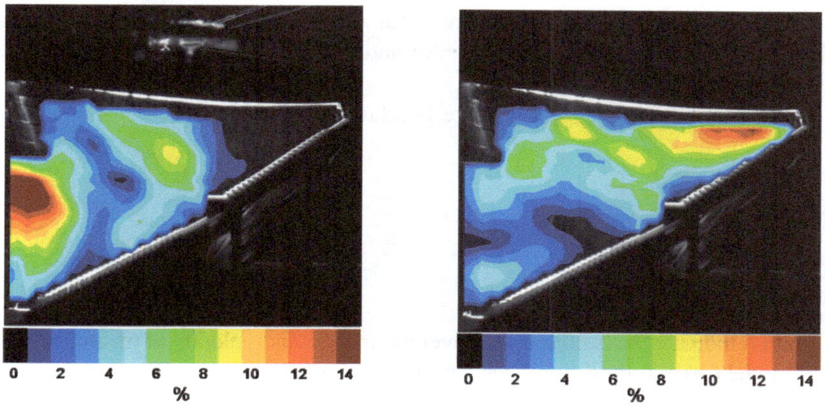

Fgne3 Mean isovelocity maps, as percentage of $U_{ref} = 3.64$ m/s (With the old roof, on the left - Without the old roof, on the right).

Fgne4 View of the Karaiskaki stadium (left) and of its wind tunnel scaled model (right).

Once the measurements of time histories of pressure coefficients are recorded, design calculations have been performed with the effective living loads on a finite-element model. In particular, the structural model of the Karaiskaki stadium has been built, including the roof dead. 1884 nodes, 3741 mono-dimensional elements and 38 link elements, for a total of 9708 DoFs characterize the adopted scheme. A time-domain analysis has been performed, in order to obtain the time histories of the structural response.

Main results of modal analysis performed on the FE model give the first natural frequency around 0.976 Hz (T = 1.085 s), while the second is found at 1.024 Hz (T = 0.976 s). Figure 26 shows the two associated modal shapes. In the first one, the uncoupling of the horizontal displacements between the end of the cantilevered structures and the internal truss is evidenced. That means that, at least for vertical loads on the roof and for wind loads, the most significant natural frequency is the second one.

Velocity scale λ_V and length scale λ_L correspond to

$$\lambda_V = \frac{U_m}{U_r} = \frac{18}{35} = 0.514 \qquad \lambda_L = \frac{1}{250} \tag{51}$$

where U_m and U_r indicate, respectively, the average mean wind velocity at mean roof level for the wind tunnel model (m) and for the full-scale prototype (r).

The frequency scale is given by the ratio between the velocity scale and the geometry scale:

$$\lambda_n = \frac{\lambda_V}{\lambda_L} \cong 129 \tag{52}$$

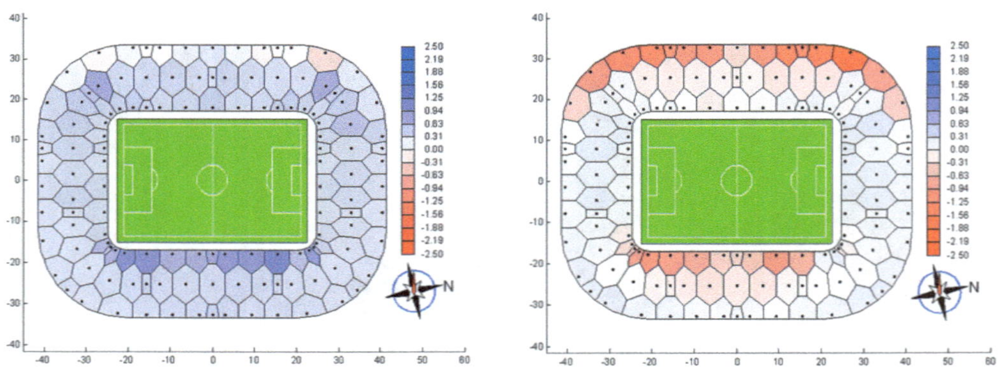

Figure 25. View of the distribution of the cp design value for the Karaiskaki stadium: max (left) and min (right) values.

By using the similarity relationship (52), the sampling frequency of 250 Hz in the wind tunnel corresponds to a frequency of 1.93 Hz in full-scale: 30 seconds-records in the wind tunnel correspond to a 64 minutes-long storm, sampled with an interval step of 0.516 seconds. As the first two natural vibration frequencies of the structure are around 1 Hz, the sampling interval is not small enough to excite dynamically the structure. Therefore, a dynamic analysis performed assuming directly the wind tunnel time histories would assume the meaning of a quasi-static analysis. In this case, one shall consider only the background response of the structure, without considering the resonant response.

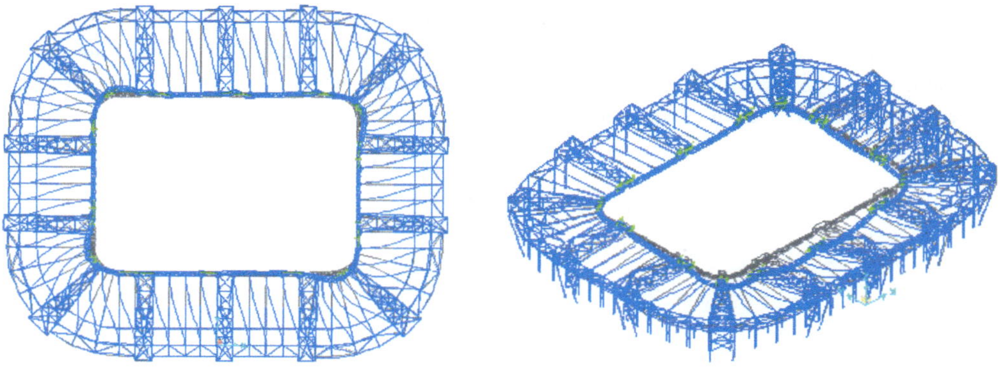

Figure 26. Finite element scheme (top view on the left and assonometric view on the right).

The preliminary study has been carried out by using such an approach, so that a time domain analysis of about 3600 s has been performed. After the analysis, once achieved time histories of the internal forces, the maximum and minimum values have been evaluated by means of the same procedure used for the pressure coefficients. As shown in Figure 27, the maximum increment (+49%) of the total load, with respect to the dead load, is recorded in correspondence of the cantilever 1, for a direction of the wind equal to 259.3°, while the maximum decrement (-58%) is recorded for the cantilever 5, for a 169.3° mean wind direction.

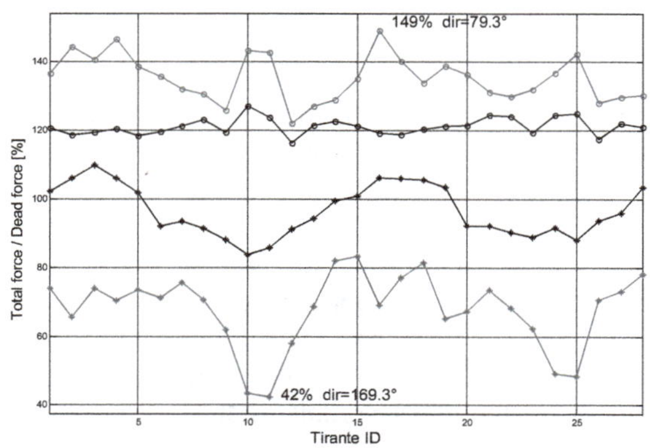

Figure 27. Variation of the normal force on the cantilever tension member due to the wind load (Karaiskaki stadium).

It is worthwhile to note that the resultant of the total load (wind and dead load) on each cantilever is always oriented downwards (no danger for uncontrolled rising of the roof).

Results obtained in terms of maximum, mean and minimum strength in the structural elements for the cantilevered structures have been used to evaluate the design value of the net pressure coefficient to be assumed in the design. In particular, this net coefficient is typically considered as uniformly applied over the whole roof, to obtain (via static analysis) the same maximum, mean and minimum values obtained through the dynamic approach. Results are shown in Figure 28: a uniform pressure coefficient of 0.57 leads about to the same maximum tension strength in the cantilever structure.

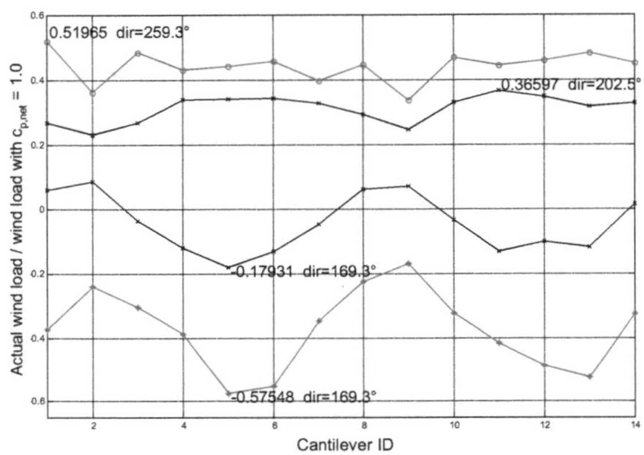

Figure 28. Uniform net pressure coefficient maximizing the tension strength in the cantilever structures (Karaiskaki stadium).

9 Conclusions

In this work, main aeroelastic phenomena involving bridges and large roofs have been summarized. In particular, after the introduction of main concepts necessary to describe the wind load on the structures, main phenomena involving wind loads and bridges have been described. Load models for single DoF galloping, torsional divergence and self-excited forces have been discussed. Vortex-shedding has been roughly described, while buffeting forces for bridges expressed via indicial functions have been presented. A practical tool for the evaluation of critical conditions has been provided. Concerning roofs, the procedure followed in wind tunnel to test a scaled model has been sketched. Two case studies have been analyzed, with reference to two different situations, the first one involving interference problems, the second one oriented on the numerical evaluation of the dynamic response.

In all cases, it has been evidenced as the use of different tools, in particular the possibility of combining measurements obtained in wind tunnel with numerical simulation represents a productive approach for the improvement of design of special structures.

References

Bartoli, G., Borri, C., Costa, C. and Procino, L. (2005). Modelling aeroelastic interactions on lightweight bridges. *Footbridge 2005*, Venice, Italy - 5/8 December 2005.

Bartoli, G. and Righi, M. (2006). Flutter mechanism for rectangular prisms in smooth and turbulent flow. *Journal of Wind Engineering and Industrial Aerodynamics*, 94(5): 275-291.

Basu, R. and Vickery, B. (1983). Across-wind vibrations of structures of circular cross section. Part II. Development of a mathematical model for full scale application. *Journal of Wind Engineering and Industrial Aerodynamics*, 12: 75-97.

Biagini, P., Borri, C., Majowiecki, M., Orlando, M. and Procino, L. (2006). BLWT tests and design loads on the roof of the new Olympic stadium in Piraeus. *Journal of Wind Engineering and Industrial Aerodynamics*, 94 293–307.

Bisplinghoff, R.L., Ahley, H. and Halfman, R.L. (1955). *Aeroelasticity*. Dover Publication, Inc., Mineola, New York.

Blevins, R.D. (1990). *Flow-induced Vibration*. Van Nostrand Reinhold, New York.

Borri, C. and Biagini, P. (2005). Structural response of large stadium roofs due to dynamic wind actions. *Stahlbau*, 74(3): 197 - 206.

Borri, C., Costa, C. and Zahlten, W. (2002). Non-stationary flow forces for numerical simulation of aeroelastic instability of bridge decks. *Computers and Structures*, 80: 1071-1079.

Borri, C. and Höffer, R. (2000). Aeroelastic wind forces on flexible girders. *Meccanica*, 35(10): 1-15.

Box, G., Jenkins, G.M. and Reinsel, G. (1994). *Time Series Analysis: Forecasting & Control*. Prentice Hall.

Cook, N.J. and Mayne, J.R. (1980). A refined working approach to the assessment of wind loads for equivalent static design. *Journal of Wind Engineering and Industrial Aerodynamics*, 6: 125-137.

Costa, C. (2006). Aerodynamic admittance functions and buffeting forces for bridges via indicial functions. *Journal of Fluids and Structures*, In press.

Costa, C. and Borri, C. (2006). Application of indicial functions in bridge deck aeroelasticity. *Journal of Wind Engineering and Industrial Aerodynamics*, 94: 859–881.

Davenport, A.G. (1961). The Dependence of Wind Load upon Meteorological Parameters. *International Research Seminar on Wind Effects on Buildings and Structures*, Toronto: 19-82.

Desai, Y.M., Shah, A.H. and Popplewell, N. (1990). Galloping analysis for two-degree-of-freedom oscillator. *Journal of Engineering Mechanics*, 116 (12): 2583–2602.

Dyrbye, C. and Hansen, S.O. (1996). *Wind Loads on Structures*. John Wiley & Sons, New York.

Fung, Y.C. (1969). *An introduction to the theory of aeroelasticity*. Dover Publication, Inc., Mineola, New York.

Jancauskas, E.D. and Melbourne, W.H. (1986). The aerodynamic admittance of two-dimensional rectangular section cylinders in smooth flow. *Journal of Wind Engineering and Industrial Aerodynamics*, 23: 395-408.

Jones, R.T. (1940). The unsteady lift on a wing of finite aspect ratio. *NACA Technical Report*, 681.

Larsen, A. (2003). Aerodynamic derivatives from simulated indicial functions. *ICWE 2003*, Lubbock, Texas.

Lin, Y.K. and Ariaratnam, S.T. (1980). Stability of bridge motion in turbulent wind. *Journal of Structural Mechanics*, 8(1): 1-15.

Lin, Y.K. and Li, Q.C. (1993). New stochastic theory for bridge stability in turbulent flow. *ASCE - Journal of Engineering Mechanics*, 119: 113-127.

Luongo, A. and Piccardo, G. (2005). Linear instability mechanisms for coupled translational galloping. *Journal of Sound and Vibration*, 288: 1027–1047.

Matsumoto, M., Kobayashi, Y. and Shirato, H. (1996). The influence of aerodynamic derivatives on flutter. *Journal of Wind Engineering and Industrial Aerodynamics*, 60: 227-239.

Novak, M. (1969). Aeroelastic galloping of prismatic bodies. *Journal of the Engineering Mechanics Division ASCE*, 95: 115-142.

Parkinson, G.V. and Brooks, N.P.H. (1961). On the aeroelastic instability of bluff cylinders. *Journal of Applied Mechanics, Transactions of the ASME series E*, 28: 252-258.

Piccardo, G., 1993. *Analysis of Coupled Aeroelastic Phenomena*. Ph. D. Thesis, University of Florence, Italy.

Sarkar, P.P., Jones, N.P. and Scanlan, R.H. (1992). System identification for estimation of flutter derivatives. *Journal of Wind Engineering and Industrial Aerodynamics*, 41-44: 1243-1254.

Scanlan, R.H., Béliveau, J.-G. and Budlong, K. (1974). Indicial aerodynamics functions for bridge decks. *Journal of Engineering Mechanics*, 100: 657-672.

Scanlan, R.H. and Tomko, A. (1971). Airfoil and Bridge Deck Flutter Derivatives. *Journal of Engineering Mechanics*, 97: 1717-1737.

Shuguo, L., Qiusheng, L., Guiqing, L. and Weilian, Q. (1993). An evaluation of onset wind velocity for 2-D coupled galloping oscillations of tower buildings. *Journal of Wind Engineering and Industrial Aerodynamics*, 50: 329–340.

Simiu, E. and Scanlan, R.H. (1996). *Wind effects on Structures*, New York.

Solari, G. (1994). *Gust-excited vibrations*. Wind-Excited Vibrations of Structures. Springer, Wien, 195–291 pp.

Yu, D.-J. and Ren, W.-X. (2005). EMD-based stochastic subspace identification of structures from operational vibration measurements. *Engineering Structures*, 27: 1741-1751.

Design of Wind-Sensitive Structures

Charalampos C. Baniotopoulos

Department of Civil Engineering, Aristotle University of Thessaloniki, Thessaloniki, Greece

Abstract. This chapter deals with certain topics on the design of three characteristic types of wind-sensitive structures and in particular, the steel lattice masts, the steel wind-turbine towers and the aluminium-glass façades. Since nowadays a well-established framework of codes does exist, the previously mentioned wind-sensitive structures can be designed in a proper way with respect to serviceability, strength and safety criteria. The respective design procedures with comments on the codes applied along with certain practical aspects on the application of the wind loading on these structures are herein presented, whereas certain numerical simulations illustrate the respective recommendations.

1 Generalities

The present chapter concerns the design of three typical types of wind-sensitive structures. As a matter of fact, those structures are defined as wind-sensitive structures for which, wind is the principal (i.e. critical) design loading; they are usually light-weight structures, in particular metal or glass-metal structures. For instance, as typical examples of wind-sensitive structures one could mention the wind turbine steel towers of Aeolian parks, the steel guyed or non-guyed lattice masts and the glass-aluminium or steel curtain-walls and roofs in modern buildings (Fig. 1).

Nowadays, a well-established framework of recommendations and codes is available for the design of these structures. This way, such structures can be analyzed and designed in an effective and safe way, whereas their peculiarities can be dealt in a correct way.

In the next paragraphs, certain topics on the design of masts, wind-turbine towers and metal-glass façades are presented along with comments on the respective codes and recommendations.

In the first part of the present chapter, certain notes are presented with the purpose to make up a concise, but complete aid for the computation of wind forces used in the design of telecommunications masts. The subject is confined to telecommunications towers: this is done to discriminate the scope of the present chapter from the power transmission towers where, the presence of the suspended conduits, radically affects the structure of the respective computations as well as the regulatory framework.

Two regulations are presented: the part 7-1 (formerly 3-1) of Eurocode 3 (prENV): Towers, masts and chimneys–Towers and masts and the German DIN 4131.

In the next paragraphs, some basic features of the analysis and the design of the prototype of a steel wind-turbine tower are presented. Such a structure has usually a tubular shape a with varying cross-section and a varying thickness of the wall along its height. It is noteworthy that such a steelwork is almost always manufactured by using high strength steel.

For the simulation of its structural response, a variety of finite element models has been recently developed. Based on the results of the latter analyses, the design of the steel tower for gravity, seismic and wind loadings is performed according to the relevant Eurocodes.

In particular, regarding seismic loading, a dynamic phasmatic analysis of the tower is carried out, whereas the respective Eurocodes methodology on the fatigue checks is applied. In the respective paragraphs, certain points that concern the previous analysis and the relevant design procedure are in details discussed, whereas comments on the design problems that are still open for investigation are presented.

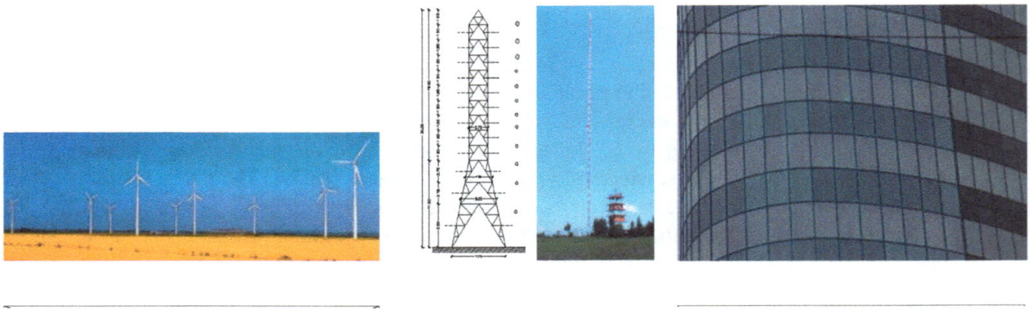

(a) (b) (c)

Figure 1. Three typical wind-sensitive structures: (a) wind mills in Aeolian parks (b) telecommunication masts (c) building façades.

In the last part of the present chapter, certain aspects on the design of glass-aluminium façades are presented; as a matter of fact, the design and the verification of the glass-aluminium façades and their components is an important procedure during the overall design procedure of the building and it must be respectively based on a reliable analysis.

The main question however, is which methodology is the most appropriate and accurate to be applied in order to analyze the performance of such elements and in particular, of the whole glass-metal façades subjected to wind loading.

This last part intends to propose a methodology that will provide to structural engineers the appropriate tools regarding the analysis of façade performance under wind loading. In addition, taking into account what Eurocodes dictate for the design of such structures, the proposed methodology may take also into account the earthquake loading for those cases that the building under investigation is erected in a zone of high seismicity.

To this end, models implemented to simulate the performance of glass-aluminium façades with different grid dimensions including dynamic analyses with regard to wind and earthquakes have been recently developed by Chatzinikos and Baniotopoulos (2005). The proposed methodology is herein illustrated by means of a numerical application where wind effects have been taken into account.

2 Wind Forces on Lattice Telecommunications Masts

E. Koltsakis[1] & C. C. Baniotopoulos

2.1 On the morphology of the lattice masts

First let us comment on the design choice to build lattice towers as opposed to compact cross-section masts: the reason is modularity of construction. As these structures have to be placed in spots of maximum visibility, the choice of hill or mountain peaks for their placement is obvious. Lattice towers can be transported in relatively small modules which makes dealing with difficult terrain, easier. In addition to that, their assembling puts less demands on accuracy, labor and equipment requirements. In addition, the obvious advantages of lattices over compact cross section regarding the effective use of the material apply here as well.

Lattice towers usually come in the form (elevation) of a vertical prism or plano-conical pyramid or combination of the two in an attempt to roughly follow the moments caused by horizontal actions. On a horizontal section, the usual shape is either triangular or a quadrilateral. The triangular shape is associated to mass-produced, prefabricated masts, whereas the quadrilateral cross-section is rather the preferred form for the use of flat-sided sections.

The static system of a tower is that of a vertical cantilever. The vertical actions (self-weight, ancillaries, ice) as well as those causing moments (wind, earthquake) are undertaken by the vertical elements on the corners of the tower henceforth called “columns”.

On the vertical faces of the tower, bracing undertakes to transfer horizontal shear to the foundation. This vertical bracing also controls the buckling lengths of the columns. Horizontal bracings also exist and help control the out-of-plane buckling of the vertical bracing on the faces.

A basic concept for the resistance checks and computation of wind loads is that of a vertical face panel (elementary face lattice). In most cases this lattice has a simple form: X, V, inverted-V being some of the most frequent. A horizontal element usually tops the panel. A special case is that of X-braces where the horizontal element is best attached to the hub of the X-diagonals as, placing it to the top of the X-lattice, it causes additional compressive forces to the diagonals (15%-25% of the N in the columns) due to the redundancy of the resulting structural system.

Usual profiles in lattice towers are angles (equal sides) usually together with UPN's and hollow sections QHS and CHS. The use of hollow sections is closely associated to the problem of accurate machining of the tips before welding, as well as to corrosion problems. A common combination for prefabricated masts is triangular plan form with CHS as columns and vertical braces. The present paragraphs will focus on the case of flat-sided profiles.

2.2 Guidelines for the numerical evaluation of wind forces

In the next paragraphs, the methodologies provided in both the DIN4131 and the EC3-7-1 for the quantitative evaluation of the intensity of the wind loading are presented in a comparative way.

[1] Department of Civil Engineering, Aristotle University of Thessaloniki, Thessaloniki, Greece

The methodology of DIN 4131. Starting the presentation on methodologies for the numerical evaluation of wind forces, we choose to begin with the German regulation being the older one of the two. The central formula for the wind force of an elementary face lattice, the so-called "panel" is given by the formula:

$$W_i = \varphi_B \, c_{fi} \, q_i \, A \qquad (2.1)$$

where

W_i : is the static wind force on tower module corresponding to the panel in question,

φ_B : the dynamic augmentation coefficient,

c_{fi} : the effective aerodynamic coefficient,

q_i : the basic wind pressure at the height of the centre of the panel, and

A_i : the net projected area of the panel elements on a vertical plane (including connection plates, gusset plates etc.)

Before going into the details of the computation of the numerical values of the four factors that yield the wind force W_i, certain issues, having to do with the structure of the DIN regulation have to be commented: although the area A_i of an elementary face lattice (panel) appears as input to the DIN-A.11 formula, the wind force that results thereof, entails the wind drag exercised to all four (or three) faces of the respective module of the tower. Here by "module" is meant the whole of the tower lattices found between the height range covered by the "panel", i.e. a "module" is a space lattice made up of four (or three) face lattices or panels. This is what the wording appearing in the explanation of W_i means.

The authors of the DIN chose to encapsulate the wind forces induced by the four or three face panels in the effective aerodynamic coefficient c_{fi}, so as to simplify the computations. The effect of the wind attack angle is also encapsulated into the value of the effective aerodynamic coefficient as well. This simplifications turns up to be useful under some conditions: uniformity of the face lattices of each module, triangular or quadrilateral plane cross-section of the tower.

The above mentioned relation 2.1 is provided at the part A.2 of DIN4131 where the wind induced dynamic effects are analyzed; the wind force in W_i is termed there a "static equivalent" wind force and, a condition is given that clears the tower for static treatment. Note that for usual communications towers (vertical cantilevers, no stabilising cables) the static treatment can be applied as the ice+wind load case that leads to rather sturdy structures.

The effective aerodynamic coefficient c_{fi}. The effective aerodynamic coefficient is computed as a product of two factors according to the directions of DIN 4131:

$$c_{fi} = \psi \; c_{f0,I} \qquad (2.2)$$

where

ψ : is a reduction coefficient having to do with the height / breadth ratio of the mast, and

$c_{f0,\mathrm{I}}$: the basic value of the aerodynamic coefficient.

For the computation of the value of the basic aerodynamic coefficient, $c_{f0,i}$, the reader is referred to the paragraph A.1.3.2 of which the best part is dedicated to circular cross-sections.

As far as it concerns space lattices made of flat-sided elements, reference is made to the Table A.3 and the nomograph of Fig. A.5 (Figure 2).

Table A.3 gives a clear picture of how to find the net area A_{net} of the panel: if we had to define it for a prismatic part of a tower, we would say that A_{net} is the area of the shadow generated by the panel members on a vertical plane ω parallel to the face when illuminated by an infinitely far

source that lies on the vertical to the face that passes from the panel barycentre. In the case of a non-vertical face, the plane ω must intersect the face along a horizontal line.
Dividing the A_{net} by the area of the convex contour of the shadow, one obtains the coverage ratio φ of the panel, that is the basic parameter whereupon the computation of the wind force is based (for both DIN and EC3).

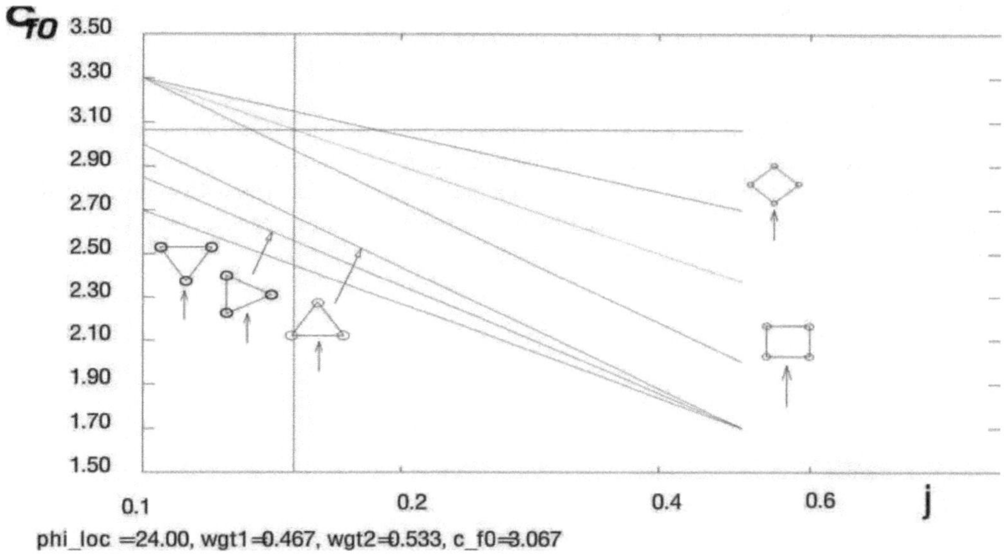

Figure 2. The Nomograph A.5 of DIN4131.

Introducing φ in the abscissa of the diagram of Fig. 2 and choosing the appropriate line for the case at hand (triangular or quadrilateral tower, wind angle of attack), one can find the ordinate that constitutes the basic aerodynamic coefficient of the panel in question (i.e. the panel corresponding to the value of φ that was used). This procedure has to be repeated for every panel configuration in the tower.

This procedure shows one of the drawbacks of this nomograph-based method: the selection of a line in the nomograph for an arbitrary angle of wind attack is difficult. Using the function *sin2θ* for the interpolation is more accurate than the linear interpolation for wind angles other than 0 and 45 degrees.

Here one can see that the nomograph line corresponding to the 45 degrees wind attack angle gives significantly higher values for $c_{f0,i}$: this is due to the fact that the basic aerodynamic coefficient has been chosen to express the larger area of attack that the tower presents to the oblique 45 degrees wind.

Note that during the normal wind attack (zero angle), the totality of the lattice elements on the faces parallel to wind direction are within the aerodynamic shadow of the columns of the windward face.

In Fig. 2 the use of the nomograph A.5 of DIN is presented: a quadrilateral tower is attacked at a 24 degrees from the face normal. A line expressing the given wind attack angle is drawn; using it along with the panel coverage ratio (taken here equal to 0.15) we determine a basic aerodynamic coefficient roughly equal to 3.07.

We now come to the second term of relation 2.2, the reduction coefficient ψ which is a function of the slenderness of the tower and of the coverage ratio φ. The value of the coefficient ψ is found by means of nomograph A.2 of the DIN shown in Fig. 3.

The slenderness of the tower is given by means of the relations:

$\lambda = 0.7\ h/b$ for $h > 50\ m$

$\lambda = h/b$ for $h < 15\ m$

whereas for intermediate h values, linear interpolation may be used. Here h is the height of the tower above ground and b is the breadth of the tower at $h/2$ (b is measured perpendicularly to the wind direction). Here one should not forget that a quadrilateral tower with a 45 degrees wind incidence angle will present an b_{45} increased by a factor of $\sqrt{2}$ with respect to the b of the normal wind incidence.

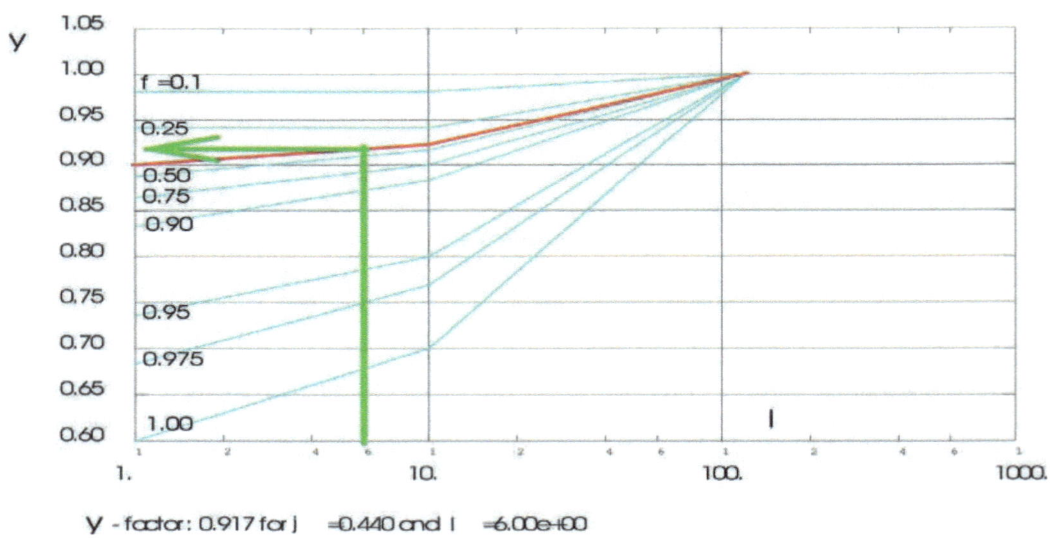

Figure 3. Using nomograph A.2 of DIN4131.

The careful reader may, at this point, wonder which φ may be used in the nomograph for ψ as each panel generates a slightly different φ_i value. Inspecting the lines of the nomograph A.2 one can realise that the range of variation of ψ for the φ usual values of φ_i ($0.2 < \varphi_i < 0.5$) is rather insignificant.

Using the nomograph A.2 depicted in Fig. 3 goes through the following steps:

a. select a line corresponding to the chosen value for φ

b. using the value λ of the slenderness and the φ-line, the factor ψ is calculated.

Variation of the wind basic pressure q with height. Paragraph A.1.2 of DIN 4131 allows to assume a constant value of wind pressure q whenever the height of the tower does not exceed 50m. This pressure should be taken equal to:

$q = 0.75\ (1 + h/100)\ q_0 \quad (q \text{ in kN/m}^2)$ (2.3)

where q_0 is the basic wind pressure resulting from the assumed wind speed via the relation $q(v_w) = v^2 / 1600\ (kN,m,sec)$.

In the case that h exceeds 50m, a linear distribution of the wind pressure exercised on each panel should be taken into account according to the relation:

$q(z) = q_0 + 0.003\ z \quad (q \text{ in kN/m}^2)$. (2.4)

Interesting comments can be found in the relevant paragraph of the DIN where an excess pressure of 0.15 kN/m^2 should be applied whenever the tower is on a rise of the terrain: for communications towers, more often than not, this precisely is the case.

The DIN also prescribes 70% of the calculated wind actions to be taken into account for short, intermediate phases of the construction of the tower.

The dynamic coefficient φ_B. The vibrations induced by the wind flow are taken into account by means of a dynamic coefficient φ_B:

$\varphi_B = \varphi_{B0}\ \eta$ (2.5)

where

$\varphi_{B0} = 1 + (0.042\ T - 0.0019T^2)\ \delta_B^{-0.63}$ (2.6)

and

$\eta = 1.00 \qquad \text{for } h < 50m$ (2.7)

$\eta = 1.05 - h/1000 \quad \text{for } h > 50m$ (2.8)

where T is the fundamental period of vibration and δ_B is the logarithmic damping coefficient of the tower (bolted steel structure). Some computational difficulty may arise for the computation of the fundamental period T of the tower.

For high-seismicity areas, this poses no problem as structural dynamics software is very widely spread. For the case where no such means is available, the DIN provides an approximate formula for the computation of the first mode period based on the displacements of the static solution for horizontal loads.

The methodology of EC3 Part 7-1 (formerly Part 3-1). In this section a very brief introduction into the methodology for wind pressure computation exposed in the Annex A of EC3 Part 7-1 will be attempted. Focus will rather be on the difficulties, than the basic concepts which are the same with those of the DIN (e.g. the coverage ratio φ).

Distribution of the wind pressure over the height of the mast. Cutting right through to the bare essentials, we refer the user to the relation A.13 (Annex A of Part 7-1) where the wind force induced on a face panel is computed as the product between the wind pressure at the respective height and the so called "drag", which has the dimension of L^2 (area) and is symbolised by R_w ; the drag represents the effective area of the panel in question. The numerical computation of the drag constitutes the subject of the bulk of section A.2 of EC3-7-1.

The value of wind force provided by relation A.13 is the *mean* wind force as it is obtained by means of the *mean* wind velocity v_m^2 . For the computation of the design forces of the members of the mast that resist the overturning moment due to the horizontal wind forces (the columns) it is required that the *mean* values be multiplied by the so-called gust response factor $(1 + G_M)$.

The first remark here concerns the fact that the gust factor G_M contains a panel height term $(z_m/h)^2$ (see relation A.14a): this is a clear differentiation from the DIN approach where a constant basic wind pressure applies under 50m. This term also includes the dynamic aspect of the wind action in the sense of the φ_B coefficient of the DIN; however, this correspondence is legitimate only for towers that can be treated via the pseudo-static method, i.e. towers that pass the criterion of relation A.12 of the Eurocode.

It should also be mentioned here that EC3-7-1 provides very limited information about the computation of the gust response term practically making necessary the use of Annex-X of EC1-1-4 (or, in particular, the old Appendix B of EC1-2-4). Nevertheless the gust factor assumes quite significant values ranging from 1.00 to 3.60 for tower heights between 10-20m.

A second very interesting point is the method of application of the gust load in patches as presented on Fig. A.3.1 of Annex A. The patch load approach concerns only towers with a rather intense widening of their cross-section (conical base – prismatic stem). This patch loading method is to be applied only for the resistance checks of the shear braces of the panels following paragraph A.3.3.2.3.

As a final comment one could say that the Eurocode is a more complete set of rules for the design of lattice towers than the (older) DIN 4131. The latter is much closer to the computation by hand (extensive use of nomographs), whereas the Eurocode is suitable for computer implementation. In fact, the complexity of some parts of Annex A makes it rather impractical to use for hand calculations; even creating a code to implement Annex B had better be based on an object-oriented approach as a classic approach would lead to really complex decision trees that would make code maintenance and expansion prohibitive.

Issues relating to the computation of the drag. The computation of the effective aerodynamic area or cross-section is, to the authors' opinion one of the most important steps of the job of calculating the wind forces. In this respect one can not but recognize the obvious qualitative leap of the Eurocode as compared to the DIN. The whole procedure is very clearly described and ready to encode as it does rely on nomographs.

Two methodologies are presented: the simplified (paragraph A.2.1.3, A.2.2) and the analytical (paragraph A.2.7) the difference mainly lying in the phenomenon of aerodynamic shadowing which is omitted in the case of the simplified one. Of course, the analytical method of A.2.7 is meant for use in all the cases where the simple approach can not be applied and that is determined by means of a set of criteria cited in A.2.1.2.

Interesting points where the EC3 methodology is clearly more handy as compared with the DIN approach are:

a. the treatment of the horizontal wind attack angle: the Eurocode makes use of an analytical expression so that we do not, any more, have to choose an intermediate line in a nomograph.

b. ancillaries are treated separately and to rather good detail with respect to their aerodynamic coefficients both for normal and iced conditions.

c. a special section for linear ancillaries (cables) exists.

Last, it must be noted that the main portion of wind induced forces is due to the microwave reflectors that are usually installed on the telecommunication masts. The wind forces of the reflectors (usual diameters between 0.60 – 4.5 m) have to be supplied by the manufacturer in the form of a polar force diagram (actually normal and transverse forces).

To the authors' experience, the presence of even one medium size reflector makes the design values of the members to be determined by the wind force of the reflector rather than the wind force of the lattice itself. This means that a set of closely spaced wind attack angles has to be taken into account.

Whenever no reflectors are present, the design values of the vertical members are usually obtained for the diagonal wind attack.

Combination of ice and wind actions. It should first be noted that this is probably the most common cause of collapse of lattice towers. Note that an ice thickness of 8cm increases the weight of the tower four times over its self-weight.

The first remark is that the coverage coefficient φ must be recomputed due to the presence of the ice coating. The net area A_{net} presented to the action of the wind obviously increases.
The Eurocode states that gaps <75mm between the lattice members must be considered to be compact.

An ice thickness of 8cm is required in South European countries, but this value depends exclusively on local meteorological conditions. Note that regarding the ice specific weight which depends on type of ice, a mean value of 750 kg/m^3 can be assumed.

For the co-existence of the two meteorological actions, a reduction coefficient for the wind actions is provided in both regulations:
the DIN prescribes a 75% reduction in the basic wind pressure, whereas the Eurocode used to provide, in Annex B, the coefficient of reduction k equal to 0.64 "*unless no other, more accurate data exist*". To the authors' knowledge, this proposed value has been dropped from the latest draft of Annex B and reference to ISO 12494 is made instead.

The Annex B also involves two other combination coefficients φ_w and φ_{ice}. The latter quite heavily affect the design values generated by the combined actions of wind and ice (cf. e.g. relations B.1 and B.2).

To the authors' opinion, the values of these three coefficients (k and φ_w, φ_{ice}) would have been more thoroughly defined in the text of the Annex B of the EC3, as it is them that in essence determine the design values of the members of a lattice telecommunications tower.

3 On the Analysis and Design of Steel Wind Turbine Towers

E. Efthymiou[2], I. Lavassas[3], G. Nikolaidis[4], P. Zervas[5] & C. C. Baniotopoulos

3.1 On the wind mills

As long as the traditional methods for energy production are getting both environmentally hazardous and very expensive, the demand for sustainable energy production is becoming more intense. The forecast of the fuels shortage in the near future combined with the negative environmental impacts urged lately all the parts involved in the energy production field to start exploring new directions in energy production, which has led to a plethora of innovative technological solutions. In this framework, the exploitation of so-called renewable energy sources (as are e.g. the wind and the sun) recently became the basic subject of these investigations.

3.2 Types and geometrical characteristics of wind turbine towers

Wind turbine towers are considered as highly wind-sensitive structures often being comparatively light-weight metal structures. Towers for large wind turbines may be either tubular and lattice steel towers or made of concrete. Guyed tubular towers are also used to support wind turbines.

Nowadays in most cases, large wind turbines are delivered with tubular steel towers which are manufactured in sections of 20-30m with flanges at either end and bolted together *in situ* (Fig. 4). The towers are designed as truncated cones with their diameter increasing towards the base in order to increase their strength and at the same time to save material.

Lattice towers are manufactured using welded steel profiles. The basic advantage of lattice towers is cost saving since this type requires only half as much material as a freely standing tubular one of the same stiffness, whereas the basic disadvantage of lattice towers is their appearance. A lot of small wind turbines are built with narrow pole towers supported by guy wires. The advantage is weight savings and thus, cost is diminishing. In this case however, there are disadvantages, such as the difficulty of access, a fact that makes them less suitable for open plain areas. It is also noteworthy that this type of guyed pole towers is more prone to vandalism, thus compromising overall safety. To conclude with it must be noted that there is also the possibility for some towers to be designed with a certain combination of the aforementioned techniques mentioned.

A typical tubular wind turbine tower carries the nacelle and the rotor. In particular, a cantilevered rotor that is characterized by the number of the blades (usually three) and the electricity capacity is mounted on the top of the steel tower. Mostly, the blades are made of reinforced fiberglass polyester material and are fixed on pitched bearings that can be feathered 90° for shutdown purposes, so that both to allow fine-tuning of the maximum power and reduce the

[2,3,4,5] Department of Civil Engineering, Aristotle University of Thessaloniki, Greece

rotor loads under extreme wind conditions. The main shaft is long enough with supports located symmetrically around the tower axis aiming at an optimum transfer of the bending moments to the yaw system and the steel structure.

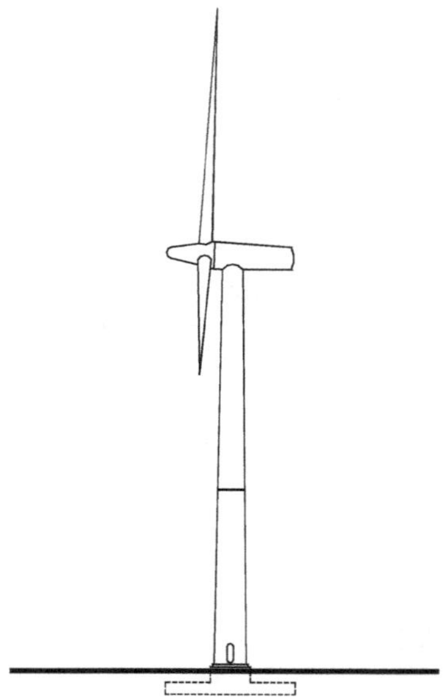

Figure 4. A typical tubular steel wind turbine tower.

For transportation purposes the tower usually is divided in two or three or more parts that can be easily erected on site. The sections are connected each other by means of double flanges with fully preloaded bolts.

The shell thickness of the steel tower varies among the segments usually decreasing from the base to the top. In order to meet the strict requirements of the fatigue design, all welds are designed as full penetration butt welds of high quality. Due to the critical role that local buckling plays in the determination of the shell thickness, the allowable fabrication tolerances of the plates have to be of the excellent quality class. For the same reason, internal stiffening rings should be manufactured and must be located in close distances.

In order to counterbalance the effect of the local concentration of stresses, the opening of the door must be designed with fully rounded corners and be reinforced by both a frame and a number of extra vertical stiffeners. The ascent to the tower is internal usually by means of aluminum ladders, interrupted by platforms made of wooden panels bolted on every second stiffening ring.

3.3 Basic considerations on the loads

Following what Eurocodes dictate, the basic loads under consideration for the design of a steel tower are the gravity loads (i.e. self-weight of the structural elements), the wind pressure and the earthquake loading (cf. e.g. ENV 1991-01-01 (1991) and Dyrbye & Hansen (1997)). With respect to the wind loading (being critical to the dimensioning of the steel structure), the site where the towers are to be constructed, significantly influences some of the wind load characteristics. The effects of the wind action on the blades and the nacelle of the rotor are also taken into account by the existing European Codes and *ad hoc* aeroelastic tests have to be provided by the manufacturer.

The design wind load cases are determined by combining specific external and operating conditions. A series of situations must be investigated as are e.g. the annual or maximum wind gust, the changes of the wind direction, the grid failure, the loss of load and the possibility of fault in the control or the safety system. As a matter of fact, all these factors adequately associated each other within a Eurocodes design framework logic, produce the necessary loading combinations.

In addition to the previously mentioned actions applied to the rotor center, a distributed wind pressure $p_w(z,\theta)$ computed by applying the analytical relations given in ENV 1991-01-01 (1991) must be applied along the height and around the circumference of the tower (Fig. 5).

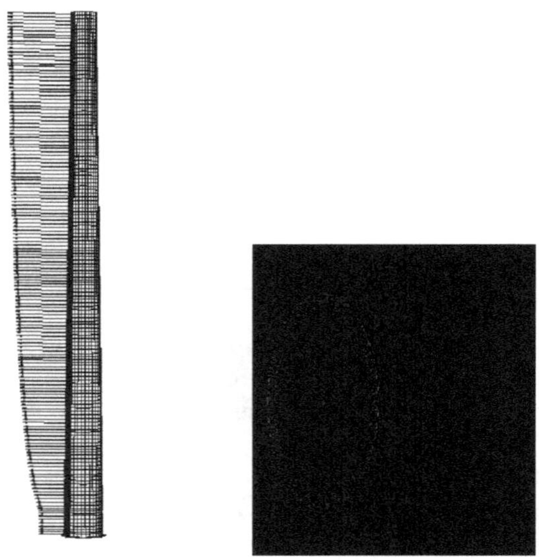

Figure 5. Wind distribution along the height and the circumference of a wind turbine tower.

The tower being a relatively slender structure must be as well checked for vortex excitation, ovalling, galloping and interference effects during cross-wind oscillation. In particular, in order to design this structure against fatigue, the vortex excited vibrations should be taken under consideration. In many cases though, the check with the operational wind loading is considered to establish fully the specified safety level and therefore, the assessment for vortex excitation is unnecessary.

With respect to the safety of the structure against ovalling oscillations that normally occur on unstiffened shells, such a tower has often such design characteristics that meets the safety criteria: the top flange is rigidly bolted on the nacelle and the stiffeners along with the middle flange are welded internally around the shell. The tower can be also considered safe against phenomena such as galloping, as long as the corresponding criterion $v_{CG} > 1, 25\ v_M$ is satisfied. Finally, the geometrical features of the structure and the distance of each tower from any neighboring one must ensure a safe response regarding interference effects during cross-wind oscillations.

Despite the fact that the fatigue check is indifferent for the shell plate thickness selection and the flange design, it often has a significant impact on the detailing and the quality of both the welds and the bolts.

The calculation procedure of the tower fatigue design is based on nominal stress ranges $[\Delta\sigma_i]$ and the structure is analyzed for the load cases of the operational wind-loading spectrum provided by the wind turbine manufacturer.

As a matter of fact, the connections are these parts of the tower being prone to fatigue; therefore, the fatigue checking must cover all the welds, the bolts and the anchors of the tower. In this calculation procedure, only the principal stresses are introduced to the calculations because the contribution of the other components can be considered as negligible in all cases. The welds of the same configuration and type are grouped together and each time the most unfavorable one can be examined.

It is also noteworthy that the shell courses can not be welded directly to the top and to the middle flange rings, but to the free end of the high neck, that is formed after adequate planning of the original steel plate.

The above configuration is an alternative solution taking into account factors such as the importance of the joints to the overall stability of the tower, the poor accessibility of the weld faces, the location of the shell around the outer edge of the ring cross-section and the demand of facilitation of the construction procedure in order to reach the required high quality of the specific welds.

The stress range of the prestressed bolts and anchors must never exceed the cut-off limit, reflecting this way their limited contribution into the stiffness of the overall joint. For each group of connections, the procedure has to be carried out according to ENV1993-01-01 (1993).

The operational loads at the rotor axis depend on the fundamental eigenfrequency of the tower along with both the relevant number of stress cycles and the mean wind velocities. Note that the position of the natural frequency of the structure in relation with the neighboring frequencies of the rotor excitation play a predominant role on the fulfillment of the serviceability criteria for the tower at hand (Fig. 6).

In order to deal with the seismic analysis of the structure, a Spectral Response Analysis can be applied with reference to the Codes (ENV1993-01-01 (1993), ENV1998-01-01 (1998)). For instance, due to the seismic data that correspond to the region where the tower under investigation is to be constructed, the maximum stresses due to earthquake are calculated about 60% smaller compared to those developed due to the wind loading.

In order to design the structure against seismic vibrations, a multimodal Linear Analysis is also applied in terms of the appropriate design response spectrum. The impact of the torsional excitation of the ground usually is negligible and in almost all cases is neglected in the analysis.

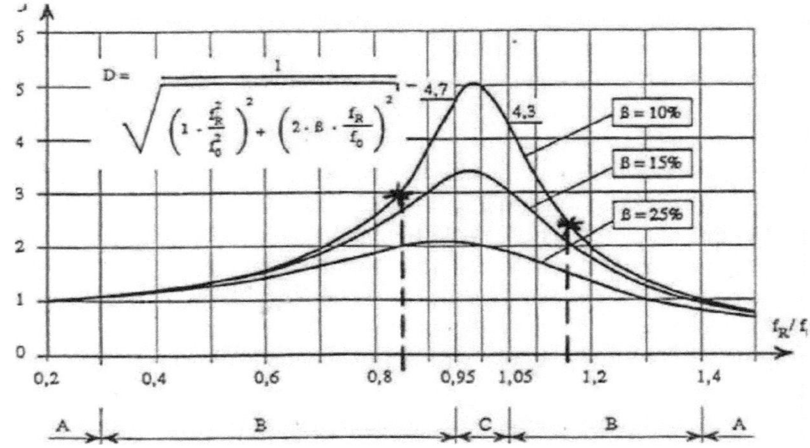

Figure 6. On the natural frequency of the tower in relation to the frequencies of the rotor.

Therefore, following the methodology dictated by the Codes (ENV1993-01-01 (1993), ENV1998-01-01 (1998)), the Spectral Analysis comes as a result of the combination of the translation and rotation spectra. The mathematical expression describing this approach is based on the dynamic similarity of the structure to a single degree of freedom oscillator (i.e. $R(T) = R_d^{(T)} + R^{\theta}_d(T) \cdot h_r$ where $R_d(T)$ is the transitional spectra, $R^{\theta}_d(T)$ the rocking spectra and h_r is the hub height).

3.4 Analysis of the steel tower response

The design of the steel tower is based on a detailed Finite Element analysis performed by applying appropriately chosen linearly or non-linearly elastic material and geometrical laws. In order to assess the reliability and accuracy of the numerical results, a combination of both detailed and simplified FE models is usually needed.

In addition, in order to evaluate the effect of the soil-structure interaction upon the static and dynamic behavior of a steel tower, different FE models should be investigated. In order to take into account any possible local stress concentration, the models should be developed in such a way that they should precisely describe the tower geometry including details as are e.g. the stiffening rings, the flanges and the door stiffeners.

Due to the existence of the latter unilateral contact elements simulating the soil-structure interaction, the numerical treatment of the models requires the application of non-linear algorithms.

Having as scope to evaluate the impact of the 2nd order effects on the structure at hand, the models at hand should be also investigated at a next stage for geometrically nonlinear conditions. In the case of adequate stiffness of the shell and stiff (e.g. rocky) underground, the participation of the aforementioned non-linearities and the soil-structure interaction into the total strain state of the tower often is calculated less than 2%, thus not affecting the overall structural response of the tower. The buckling analysis of the shell structure is based on the stress design method dictated by ENV1993-01-01(1993), Part {§8.5}. An effort to alternatively check the structure according to the direct approach procedures of ENV1993-01-01(1993), Part {§8.5} is practically an impasse. As a matter of fact, the results of such an analysis appear to be very sensitive with respect to both the discretization of the model and the adopted buckling parameters (cf. e.g. Schmidt (2000)). It should be however, mentioned that this type of analysis should be used for preliminary investigation on the shell response and not for its final design (cf. e.g. Lavassas *et al.* (2001), (2002), (2003)).

3.5 Design of the steel tower response

The design of the steel tower prototype is based on a Limit Analysis methodology as proposed by ENV1993-01-01 and in literature (cf. e.g. Karamanos & Karamanos (1998), Papadimitriou *et al.* (2002), Lavassas *et al.* (2003)). The steel wind turbine tower is basically designed against the following four limit states:

1: LS1: Plastic Limit State

2: LS3: Buckling Limit State

3: L S4: Fatigue Limit State

4: Serviceability Limit State

Very often applying a trial-and-error approach, the shell thickness is optimized by several runs so that the best design/resistance ratio to be obtained. For the lower part of the tower, the LS1 state is dominant, whereas in the upper part, the LS3 is the most significant with a remarkable participation of the compressive circumferential stress near the stiffening rings into the overall stress state configuration.

Concerning the LS4 state, the stiffeners (stiffening rings, door stingers, ribs and frame) and the flanges do not need any check against local buckling, since their width to their thickness ratio satisfies the code requirements. The same applies to the parts of the shell around the door bordered by the stringers, the bottom flange and the first stiffening ring. Regarding the rest shell courses, the verification can be done by applying the ENV1993-01-01 (1993) methodology.

The assumptions about the design against buckling worth to be herein presented. In particular, a tower can be divided by the stiffening rings and the flanges into sixteen buckling-reference sections. According to ENV1993-01-01 (1993), the boundary conditions are the bottom, the top flanges (BC1) and the middle flange and stiffening rings (BC2). For simplification reasons, the stepped cylindrical section can be transformed into an equivalent one with uniform thickness.

It must be underlined that representative sample checks have demonstrated the sufficient accuracy of this approach for the specific configuration of the shell approximation (discrepancy less than 2%). The conical section can also be transformed into an equivalent cylinder one. In this case, a specific procedure can be also carried out according to ENV1993-01-01 (1993) for each stress type.

Since the maximum circumferential buckling stresses occur along the stiffeners being the boundary zones of the shell sections, the interaction checks need not to be carried out over the adjacent to the stiffening rings stripes up to a distance equal to the 10% of the height of the section.

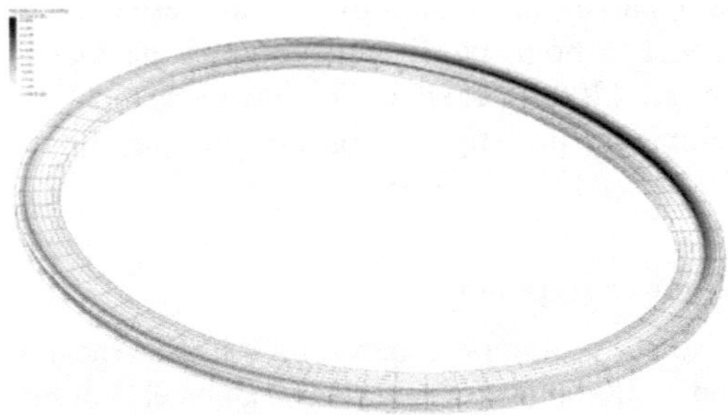

Figure 7. Middle flange model.

For the design of the flanged connections (rings, bolts and anchors), special detailed models should be used, where the interface behaviour between the windmill parts should be described by unilateral contact finite elements interconnected each other (Fig. 7). In these detailed models, bolts/anchors should be modelled as cable elements subjected to the prestressing forces.

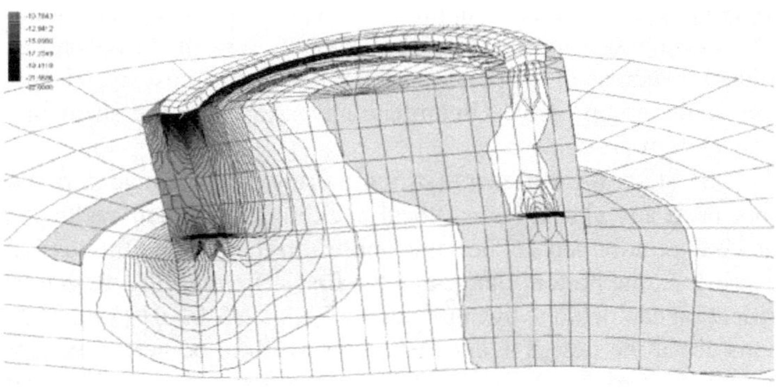

Figure 8. On the foundation model: σ_{zz} stress distribution.

As expected, a complex stress state - resulting from the load combinations transferred to the foundation from the tower and the prestressed anchor bolts - is developed in the foundation slab and the pedestal.

To overcome this difficulty, a detailed brick element model has to be developed, where the anchoring system must be as correctly as possible represented (Fig. 8). The connection between the elements and the support to the ground is realized by means of unilateral contact elements, whereas the prestressed anchors in most cases are modeled as cable (s-unilateral) elements connecting the steel flange with the foundation. The elevation of the designed tower prototype is presented in Fig. 9.

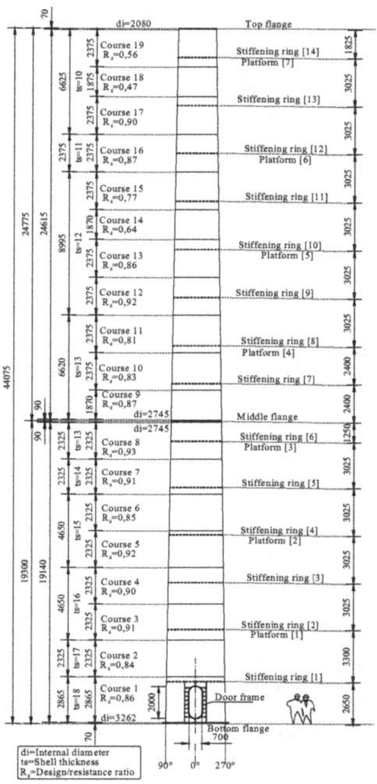

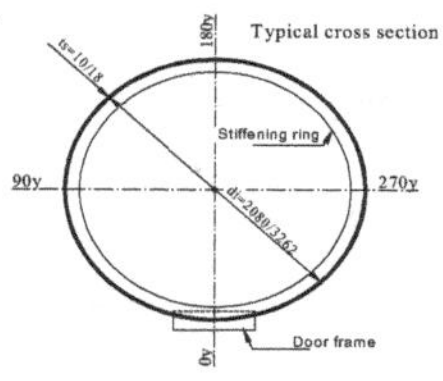

MATERIAL LIST			
No	Description	Quality	Code
1	Steel plates	S 355 J2G3	EN 10025
2	Bolts	10.9	ISO 7411
3	Anchors	8.8	ISO 7411
4	Nuts	---	ISO 4775
5	Washers	---	ISO 7415

PART LIST			
No	Description	Weight (kg)	
1	Shell: t=10mm	3.561	
2	Shell: t=11mm	1.483	
3	Shell: t=12mm	6.538	
4	Shell: t=13mm	7.741	
5	Shell: t=14mm	2.290	
6	Shell: t=15mm	5.070	
7	Shell: t=16mm	5.640	
8	Shell: t=17mm	3.089	
9	Shell: t=18mm	3.921	39.333
7	Stiffening rings	1.891	
8	Door stiffeners & frame	397	2.288
9	Topflange	238	
10	Middleflange	1.281	
11	Bottomflange	2.038	
12	Anchor ring	1.456	5.013
13	Bolts	343	
14	Anchors	1.623	1.966
15	Miscellaneous	1.400	1.400
		Total	50.000

Figure 9. Elevation of a wind turbine tower prototype.

3.6 Remarks on the design of wind turbine towers

The previously presented design procedure for a prototype steel wind turbine tower is based on the rules with respect to serviceability requirements dictated by the Eurocodes. It is noteworthy that the use of a simplified linear static model would be sufficient for the calculation of the basic response and the eigenvalues of the structure, it does not however, suffices for the ULS design due to the fact that local stress concentrations are neglected in this model.

It must be also underlined that, given the present status of the relevant Eurocodes, calculations regarding buckling analysis of the shell of the tower inserts a high level of ambiguity in the results and therefore, buckling related problems have to be handled with utmost care.

With respect to the design loads, the extreme wind leads to the dominant load combination for the design of the structure, whereas the seismic design would become critical only for the case of constructing such steel towers in a seismically hazardous area (zone III or IV), with low intensity wind loads and founded on medium or soft quality soils.

4 On the Analysis of Building Glass-Aluminium Façades

K. Chatzinikos[6] & C. C. Baniotopoulos

4.1 Introductory remarks

During the last decades, the use of glass-steel or glass-aluminium façades being the envelope of buildings is very common in modern construction (cf. e.g. Fig. 10). As a matter of fact, the advantages of their use (natural lighting, better aesthetic results, quality work environment due to transparency etc.) have rendered them an optimal choice for architects. However, the frequent use of glass-aluminium façades nowadays as building envelopes revealed that the latter are rather vulnerable to natural actions as is for instance, the wind. In addition, storms even if not causing major structural damages, may damage secondary structural components (such as the glass-aluminium façades are) and thus, give rise to the necessity of extensive and costly maintenance.

Figure 10. A typical glass-metal façade

4.2 On the analysis model

The glass-aluminium façade consists of a load-carrying grid of aluminium vertical and horizontal elements supporting the glass panels. A parametric model must be implemented in order to formulate the geometry of the model and simulate the performance of glass-aluminium façades with different grid dimensions. The vertical elements (mullions) are attached to the main load-bearing structure on selected points, usually at every second floor. The horizontal elements

[6] Department of Civil Engineering, Aristotle University of Thessaloniki, Greece

(transoms) are usually pin-connected at their edges to the mullions and the connections of the load carrying aluminium grid to the main building can be assumed as being rigid. A typical finite element model of an aluminium load-carrying grid is presented in Fig. 11.

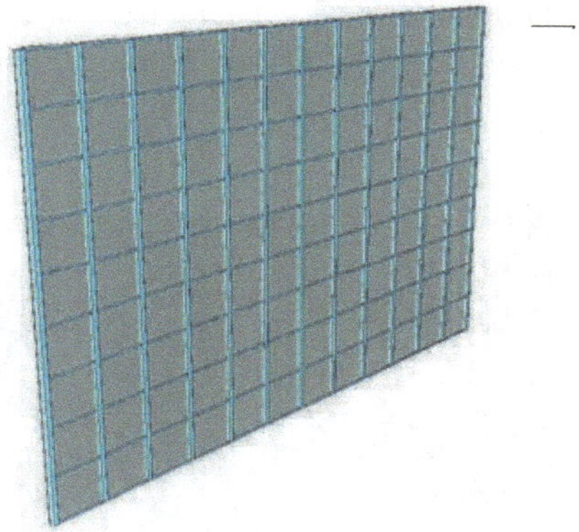

Figure 11. Typical FEM model of a glass-aluminium façade.

In a façade like this at hand, both mullions and transoms could be modeled using two-node beam elements with 6 degrees of freedom at each node. The glass panels, considered as non load-carrying elements, could be modeled as rectangular shell elements defined by the four corner nodes. For consistency purposes, only the four corner nodes of the glass panels are considered to have the same displacements with the load-carrying aluminium grid.

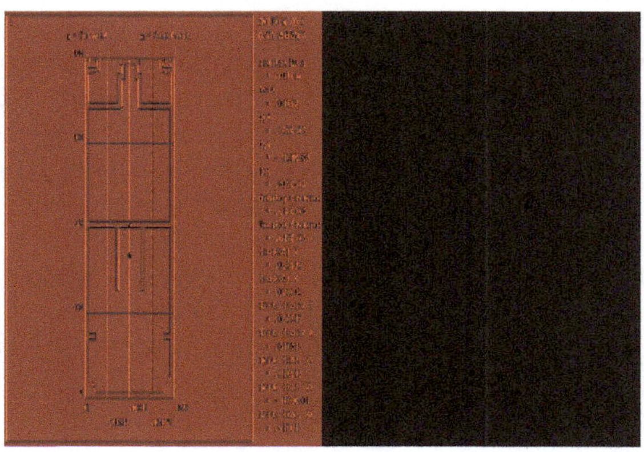

Figure 12. Typical FEM model of the cross-section.

The geometrical properties of the cross-sections of the aluminium elements, such as the area and the moments of inertia used for the analysis, could be determined in two ways. They could be either calculated manually and inserted to the model as parameters of the beam elements or they could be calculated automatically by some specific software in the case that the shape of the section is fully described. A typical cross-section of aluminium elements used in the analysis along with its geometrical properties calculated by the software is illustrated in Fig. 12.

Concerning the glass panels, the thickness of their cross-section is almost always assumed to be uniform (e.g. 10 mm).

Figure 13. Stress-strain curve for the 6082-T4 aluminium alloy.

A finite element model including a dynamic analysis regarding wind loadings is in the sequel formulated. The software employed for the dynamic analysis, automatically formulates the mass matrix of the reference part of the façade using the mass density value of aluminium and glass, whereas the wind load is applied as a force exciting the corresponding masses. Note that the masses are assumed to be concentrated at the joints of the horizontal and vertical aluminium elements.

The structural materials employed in the present analysis were aluminium and glass. The modulus of elasticity of aluminium is 7.0×10^4 MPa and its mass density 2700 kgr/m^3. The stress-strain curve of aluminium used in order to describe aluminium highly depends on the alloy and the treatment employed (cf. e.g. Mazzolani (1985), ENV1999-01-01 (1999)). For structural purposes, alloys 6082 and 6063 are usually employed. Fig. 13 presents the stress-strain curve of aluminium alloy 6082 on T4 condition, as it was input in the finite element model. For this purpose, a multi-linear model comprising of ten points has been used, where the values of y-axis are the stresses and the values of x-axis are the corresponding strains. The curve has been designed based on the following formula proposed by Ramberg-Osgood (Mazzolani (1985), Dwight, (1998)):

$$\varepsilon = \frac{\sigma}{E} + 0.002(\sigma / \sigma_0)^n \tag{4.1}$$

where:

E is the modulus of elasticity, ε is the strain that corresponds to stress level σ, σ is the stress, σ_0 is the actual 0.2% proof stress and n is the index of curvature of the stress-strain curve. The two terms in this expression are the elastic and plastic strain respectively. The role of the index n is to control the curvature of the knee of the curve and depends on the aluminium alloy used. The actual 0.2% proof stress corresponds to a value of plastic strain equal to 0.002. As previously mentioned, for the present analysis alloy 6082-T4 is employed, where the index n is equal to

6.7 and the σ_0 and the ultimate stress σ_u equal to 110 and 205 MPa respectively. The glass behaviour has been obviously considered as brittle and exclusively linear-elastic. The modulus of elasticity of glass is $7.0x10^4$ MPa and the mass density is 2700 kgr/m^3.

4.3 Wind loading

Wind flow varies randomly in time and space and exhibits a turbulent response especially close to the ground level. As is well-known, the characteristic measure of the wind flow is the wind velocity which is described as the sum of a mean velocity and the turbulent components. Assuming a Cartesian coordinate system, the velocity at a given time t is the sum of the mean wind velocity $v_m(z)$ and the turbulence components u(x,y,z,t), v(x,y,z,t) and w(x,y,z,t) in the longitudinal, lateral and vertical direction respectively. The main contributions to the wind actions are given by the mean velocity $v_m(z)$ and the longitudinal turbulent component u(x,y,z,t); this was the reason why in the following analysis only these components have been taken into account. In addition, no consideration was made regarding the fluctuations of the wind flow in space, since only a part of the whole façade is modeled and analyzed, whereas the wind flow was considered dependent only on time.

The mean wind velocity $v_m(z)$ depends only on height z and is given according to ENV1991-01-01 (1995) by the expression:

$$v_m(z) = c_r(z)*c_0(z)*v_b \tag{4.2}$$

where

$c_r(z)$ is the roughness factor calculated as it is indicated in ENV1991-01-01(1991), $c_0(z)$ is the topography factor, which is usually taken as 1.0 and v_b is the basic wind velocity.

In situ measurements showed that the value of the mean wind velocity after a 10-minute period is more or less constant. Therefore, the mean velocity was considered to be constant and the turbulent component to have a mean value equal to zero (Dyrbye & Hansen (1997)). During the 10-minute period, the turbulent component u should be assumed to take random values according to existing anemographic measurements, so that the mean value of the turbulence component to be zero.

For the purposes of the present analysis, the region of Thessaloniki has been considered being one of the most unfavorable regions in Greece regarding the anemographic data. According to the ENV1991-01-01 (1991), the basic wind velocity takes a value of 36 m/sec, while the value of the roughness factor $c_r(z)$ has been calculated equal to 0.7627. Therefore, mean wind velocity for the cases investigated takes a value of 27.45 m/sec.

The value of the static wind pressure applied on the surface of the glass-aluminium façades examined has been determined according to ENV1991-01-01 (1991). The value of the wind pressure applied on a reference surface is the sum of external and internal wind pressure, which are given by the equations (4.3) and (4.4).

$$w_e = q_p(z_e) \text{ x } c_{pe} \tag{4.3}$$

$$w_i = q_p(z_e) \text{ x } c_{pi} \tag{4.4}$$

where

$q_p(z_e)$, $q_p(z_i)$ is the peak velocity pressure (external and internal correspondingly), c_{pe} is the pressure coefficient for the external pressure, c_{pi} is the pressure coefficient for the internal pressure and z_e,z_i is the reference height for the external and the internal pressure correspondingly.

Regarding the dynamic analysis, the wind load should be applied as a force exciting the corresponding masses. Wind force being proportional to the square of wind velocity is generally given by the following equation:

$$F_{tot} = ½*C_D*A_{ref}*\rho*(U_{tot})^2 \tag{4.5}$$

where C_D is the drag coefficient, which is a pressure coefficient of the structure, A_{ref} is the reference area, where the wind pressure is applied and ρ is the air density, which equals to 1.25 kgr/m^3.

Wind velocity in the case of the present study is the sum of the mean value and the turbulence component. As it was aforementioned, the turbulence component was assumed to be dependent only on time. The mean wind velocity is much greater than the turbulence component u and therefore, the quantity $u(t)^2$ is assumed zero and the square of wind velocity is expressed by the following equation:

$$U_{tot}^2 = U(z)^2 + 2U(z)u(t) \tag{4.6}$$

and thus, the wind force that excites the discrete masses of the dynamic analysis model at a given time t is calculated by the following expression:

$$F(t) = C_D A \rho (1/2)\left(U(z)^2 + 2U(z)u(t)\right) \tag{4.7}$$

4.4 Assumptions and primary analysis

A primary static analysis is first performed with regard to wind loading. At first, the surface pressure wind load vertical to the plane of the glass-aluminium façade is calculated. The glass panels are considered as non load-carrying elements and therefore, the surface wind loading is assumed to be transferred directly to the structural aluminium elements as uniform distributed loading. The distribution of the surface wind load depends on the influence area of each corresponding aluminium element. A typical loading distribution scheme is presented in Fig. 14. The load distribution in this model could be produced automatically with the aid of parameters according to the width and height dimensions of the grid. Finally, regarding the glass panels, only their dead load is taken into account.

The present analysis takes also into account possible effects of geometrical and material nonlinearity. In this case, the load is applied step-by-step until it reaches its final value, as it is calculated and introduced to the FE model. The results of such an analysis demonstrate that the aluminium elements of the façade do not show any strong nonlinear effects for typical aluminium grid dimensions (such as 1.0x1.0 m) (cf. e.g. Fig. 15). Moreover, the final values of the maximum deflection are the same as in the analysis, which did not include any non-linearity effects. However, in case of extreme grid dimensions and morphology, the façade may exhibit nonlinear effects. The maximum deflection versus time for a 2.0x2.0 m grid dimension with supports at every two spans is presented in Fig. 16. This figure shows that for such a grid and connection morphology, the façade exhibits a linear response with respect to deformations.

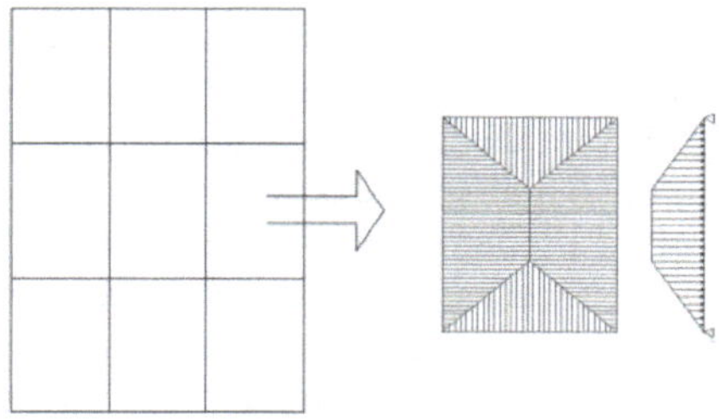

Figure 14. Distribution of the wind load to the grid elements.

Comparing the models that take into account material and geometrical nonlinearities and the linear models, small differences do appear. At this stage, the limits imposed by ENV1999-01-01 (1999) regarding the serviceability limit state - in particular the maximum allowed deflection values of aluminium elements - are investigated. Specifically, the upper limit of 15 mm is put in order to have an insight on whether this value is conservative or not.

For the purposes of the aforementioned investigation, various sections have been employed so as to calculate the maximum deflection of models with nonconservative grid dimensions, such as a 2.0x2.0 m grid dimension supported at every two openings along the height. The comparison of the calculated values for the linear and nonlinear model has proved that for output with maximum deflection close to the limit, the difference between the linear and the nonlinear model is relatively small. For the examined grid dimensions, the difference is 0.2 mm (15 and 15.2 mm) or 1.2%, which is within the scale of safety, although a little bit conservative. Different sections are also regarded for various grid dimensions and regarding outcomes that presented deflections close to the limit value of 15 mm, the difference between the linear and non-linear models is not more than 2.5%.

As previously mentioned, the maximum deflection was measured for different typical grid dimensions varying from 1.0x1.0 m to 1.5x1.5 m. Fig. 15 presents the maximum deformation vertical to the façade plane for typical grid dimensions (1.5x1.5 m). It should be noted that the load-carrying aluminium grid is considered supported to the main building at every three spans along the height.

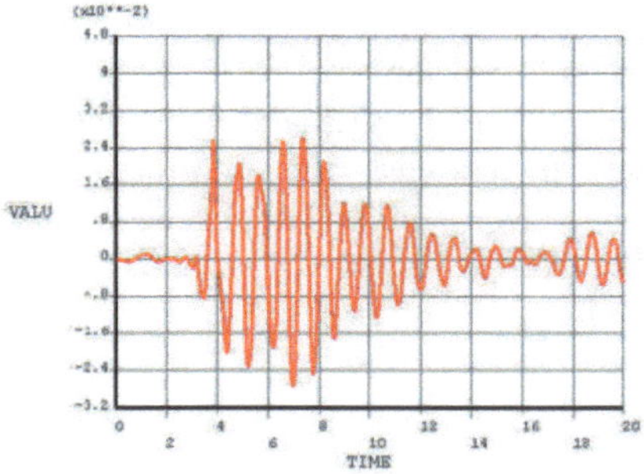

Figure 15. Maximum deformation of the façade versus time for a typical grid (1.5x1.5 m).

4.5 Extreme wind values on the edge areas of façades

Fluctuations of pressure on façade surfaces are caused by flow disturbances generated by the building itself and its surface elements, such as mullions used for cladding requirements or for decorative purposes. In this case, according to earlier research works and wind tunnel tests performed by Stathopoulos and Zhu (1991), a higher local pressure coefficient is proposed for the areas located in a distance of b/5 from the façade edge. In particular, a local pressure coefficient equal to –3.5 is proposed for the tall buildings and equal to –2.8 for the low ones.

The aforementioned values of local pressure coefficients are taken into account in the present analysis for the edge areas of the façades examined. Regarding the analysis of a pilot glass-aluminium façade, only a part of the whole façade is modeled and analyzed and in particular, this one that is loaded by the most critical values of the wind loading. The surface of the glass panels is usually among 1 and 2 m^2 and therefore, the local pressure coefficients are taken into account for the analysis instead of the overall coefficients.

A typical façade is employed for the purposes of the analysis. This façade comprises of a 5x5 grid whose elements have a length equal to 1.0 m. In this case, the model takes into account different values of local pressure coefficient on the edge areas of the façade. First the model is analyzed for a local pressure coefficient equal to -1.4 according to the provisions of ENV1991-01-01 (1993). Then the model is analyzed for a local pressure coefficient equal to -2.8 and -3.5 as indicated at the research work of Stathopoulos & Zhu (1991) for low and tall building respectively. The deformed shapes of the façade for each one of the three investigated cases are depicted in the next Figs.

As previously presented, the increase of the local pressure coefficient at the edge areas of the examined façade leads to an increased deflection at these areas. In addition, a slight increase of the maximum deflection can be observed. In particular, the maximum deflection raises from 9.2 mm (local pressure coefficient equal to –1.4) to 9.3 mm (local pressure coefficient equal to –3.5)

The glass-aluminium façades are in general designed in practice empirically. The present technique intends to contribute to this pilot methodology that provides engineers aiming to design glass-aluminium façades with the appropriate insight regarding their performance under dynamic loadings and in particular, under wind loading. Different models have been formulated in order to achieve the previously mentioned objective.

The computed values are the deflections of the façade elements, since the most critical verification is the serviceability limit state. These values are in good accordance with the restrictions dictated by ENV1999-01-01 (1999) which assigns the limit of L/250 or 15mm (whichever is less) for curtain-wall mullions and transoms.

More specifically, maximum deformation rises to 11.7 mm, for the case that the dimensions of the aluminium load-carrying grid are 1.5x1.5 and the supports to the main building at every two spans at height (4500mm). In this case, the deformation limit value is 15mm (L/250 = 4500/250 = 18mm or 15mm whichever is less) is not exceeded.

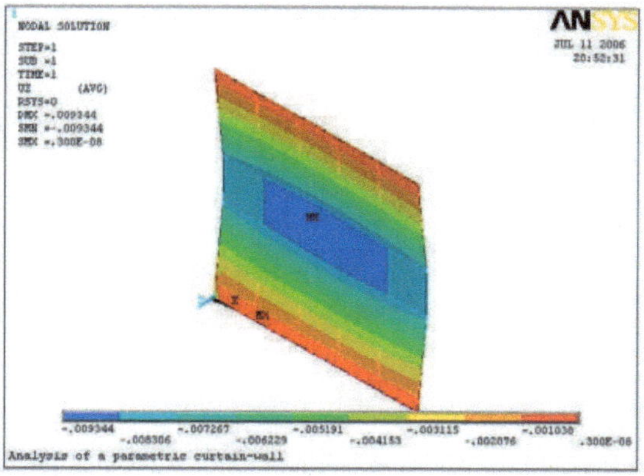

Figure 16. Deformed shape of the façade – local pressure coefficient equal to –3.5.

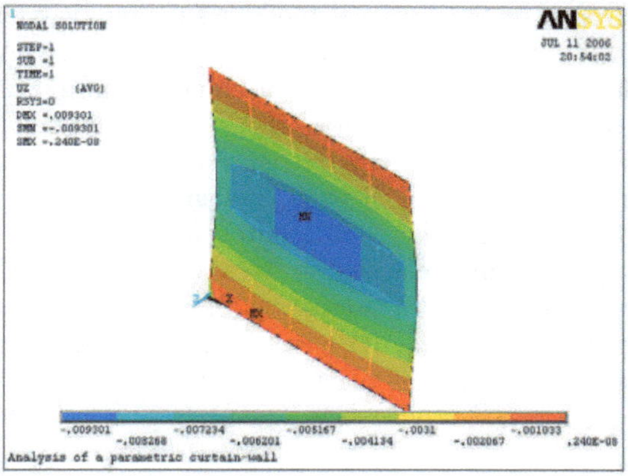

Figure 17. Deformed shape of the façade – local pressure coefficient equal to -2.8.

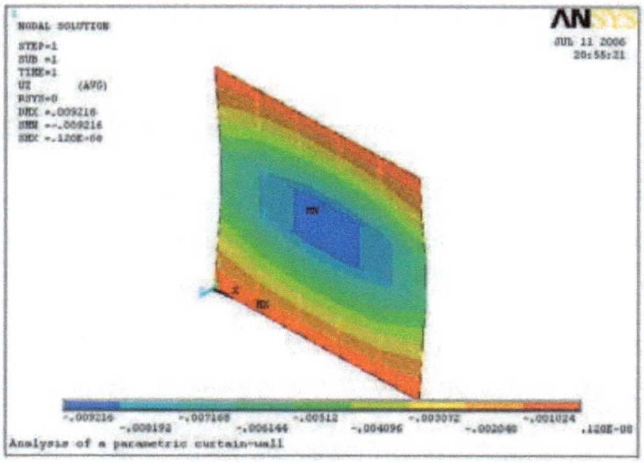

Figure 18. Deformed shape of the façade – local pressure coefficient equal to -1.4.

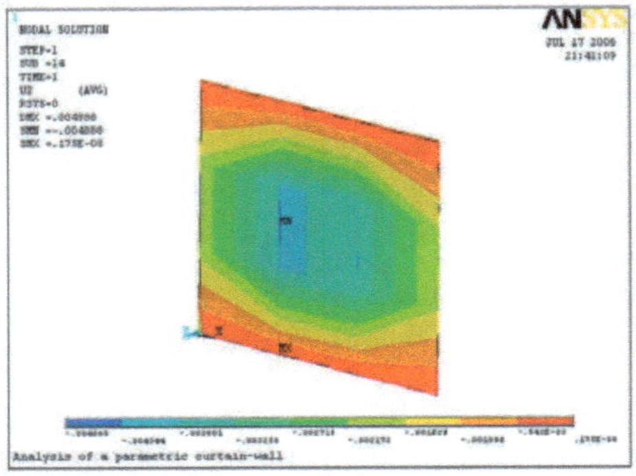

Figure 19. Deformed shape of the façade (2.0x2.0 m grid – support every two spans along height).

5 Conclusive Remarks

In the present chapter, topics on the design methodogies for certain typical wind-sensitive structures have been presented from a structural engineer's point of view. As a matter of fact, these topics concern a plethora of types of important, from any point of view, structures as are e.g. the wind turbine towers, the telecommunication masts and the glass-aluminium façades.

Of special interest are the research efforts to control the structural response of wind-sensitive structures during strong excitations due to wind, so that they behave in a safe way during any strong event (cf. e.g. Stavroulakis *et al.* (2005), Betti *et al.* (2006)).

Another interesting topic tightly connected to the sustainable design of wind sensitive structures is this one of their structural health monitoring (cf. e.g. Nazarko *et al.* (2006)).

Both topics will significantly for sure contribute during the next years towards a more economical and in the same time safe design of such wind-sensitive structures.

6 References

Baniotopoulos, C. C. & Wald, F. (eds.) (2000). *The Paramount Role of Joints into the Reliable Response of Structures*, Kluwer, Dordrecht.

Betti M., Baniotopoulos C.C. & Stavroulakis G.E. (2006). Modelling and Simulation of Piezoelectric-Active Control of Wind-Induced Vibrations on Beams, In: *Proceedings SMART 2006 Conference*, Rome, 26-28.10.2006 (in print).

Chatzinikos K. T. & Baniotopoulos C. C. (2005). Glass - Aluminium Façades in Steel Buildings: Analysis of their Performance under Dynamic Loadings, In: *Proceedings 4TH EUROSTEEL Conference*, Maastricht, 188-196.

Chatzinikos, K.T. & Baniotopoulos, C.C. (2004). Analysis of Building Glass-Aluminium Façades Subjected to Wind Loading, In: *Proceedings 4TH European & African Conference on Wind Engineering*, Prague, 11-15.7.05, 257-265.

Dyrbye, C. & Hansen, S., O. (1997). *Wind Loads on Structures*, John Wiley & Sons, Chichester, N. York.

ENV 1991-01-01: *Basis of Design and Actions on Structures*, CEN, Brussels 1991.

ENV 1992-01-01: *Design of Concrete Structures*, CEN, Brussels 1992.

ENV 1993-01-01: *Design of Steel Structures*, CEN, Brussels 1993.

ENV 1998-01-01: *Design Provisions for Earthquake Resistance of Structures*, CEN, Brussels 1998.

ENV 1999-1-1: *Design of aluminium structures,* CEN, Brussels 1999.

Lavassas I., Nikolaidis G., Zervas P., Baniotopoulos |C. C. & Doudoumis I.N. (2002). Analysis and design of the 1 MW steel wind turbine towers at mount Kalogerovouni Lakonia, In: *Proceedings 4TH National Conference on Steel Structures, Patras*, 272-280.

Lavassas I., Nikolaidis G., Zervas P., Baniotopoulos |C. C., Doudoumis I.N. (2001). Static analysis and seismic design of 1 MW steel wind turbine towers at mount Kalogerovouni Lakonia, In: *Proc. 2ND National Conference on Aseismic Mechanics and Technical Seismology,* Technical Chamber of Greece, Thessaloniki 2001, 183-190.

Lavassas, I., Nikolaidis, G., Zervas, P., Efthimiou, E., Doudoumis, I.N. & Baniotopoulos, C.C. (2003). Analysis and Design of the Prototype of an 1 MW Steel Wind Turbine Tower, *Engineering Structures* 25, 1097-1106.

Mazzolani, F. M. (1985). *Aluminium Alloy Structures*, Pitman Publishing, Boston-London.

Nazarko P., Ziemianski L., Efstathiades C., Baniotopoulos C.C. & Stavroulakis G.E. (2006), Damaged Support Identification in Aluminium Curtain-Walls using Neural Networks, In: *Proceedings SMART 2006 Conference*, Rome, 26-28.10.2006 (in print).

Karamanos A. S., Karamanos S. A.(1998). Structural design of a 40m high steel tubular tower for wind turbine 600 kW, In: *Proceeding 3RD National Conference on Steel Structures, Thessaloniki 1998*, 239-243.

Papadimitriou L. S. , Bazeos N. & Karabalis D. L. (2002). Stability analysis and structural optimization of a prototype wind turbine steel tower,In: *Proceedings 4TH National Conference on Steel Structures, Patras 2002,* 221-229.

Schmidt Herbert (2000). Stability of steel shell structures – General report, *Journal of Constructional Steel Research, 55,* 159 – 181.

Stathopoulos, T. & Zhou, X. (1991). Wind Pressures on Buildings with Mullions. Journal of Structural Engineering 116 (8), 2272-2291.

Stavroulakis, G.E., Foutsitzi, G., Hadjigeorgiou, E., Marinova, D. & Baniotopoulos, C.C. (2005). Design and Robust Optimal Control of Smart Beams with Application on Vibration Suppression, *Advances in Engineering Software* 36, 806-813.

Participants CISM Course

Kolibri 360 GmbH
Ferdinand-Jühlke-Straße 7
99095 Erfurt, Deutschland
produktsicherheit@kolibri360.de